ENVIRONMENTAL SCIENCE AND BIOTECHNOLOGY
THEORY AND TECHNIQUES

ENVIRONMENTAL SCIENCE AND BIOTECHNOLOGY
THEORY AND TECHNIQUES

A. G. Murugesan

Senior Lecturer
Sri Paramakalyani Centre for Environmental Sciences
Division of Industrial Toxicology and Pollution Control
Manonmaniam Sundaranar University
Tamil Nadu

C. Rajakumari

Environmental Researcher
Sri Paramakalyani Centre for Environmental Sciences
Division of Industrial Toxicology and Pollution Control
Manonmaniam Sundaranar University
Tamil Nadu

MJP PUBLISHERS
Chennai 600 005

Cataloguing-in-Publication Data

Murugesan, A.G. (1960–   ).
        Environmental Science and Biotechnology:
Theory and Techniques / by A. G. Murugesan and C. Rajakumari. –
Chennai : MJP Publishers, 2005.
        xiv, 460p. ; 22 cm.
        Includes index.
        ISBN 81-8094-009-8 (pbk.)
        1. Biotechnology. 2. Environment.
I. Rajakumari, C. II. Title
        660.6   MUR   MJP 008

The author and publisher disclaim any liability arising out of improper handling and incorrect procedures while performing the laboratory experiments contained in this book.

ISBN: 978223375753

MJP PUBLISHERS
A unit of Tamilnadu Book House
47, Nallathambi Street
Triplicane, Chennai 600 005

PREFACE

Environmental Science and Biotechnology are the two emerging fields of science today and invariably students of biology are in need of thorough knowledge in these fields. Besides the theoretical aspects, expertise in practical methods demands more skill and concentration. If the students are provided with a clear methodology, principle and objective, they can excel in advanced techniques too. However a complete text that covers the techniques in both environmental science and biotechnology is lacking. Therefore the present book has been proposed to cover a wide range of techniques in both the fields.

A thorough knowledge about the instruments involved in practical methods is mandatory. Chapter 1 deals with the principles and mechanism of various instruments. The pollution status of water, need for water quality standards, basic concepts of pollution studies, and need for water quality analyses are elaborated in chapter 2. The remaining chapters discuss the practical methods in biotechnology, microbiology and waste water anlaysis for the presence of physical, chemical and hazardous pollutants.

We are grateful to Prof. Dr. Cinthia Pandian, Vice chancellor, Manonmaniam Sundaranar University, Tirunelveli, for her constant encouragement, and Prof. Dr. N. Sukumaran, Dean, Faculty of Sciences of Manonamiam Sundaranar University, Tirunelveli, for his support.

A. G. Murugesan
C. Rajakumari

CONTENTS

(1)

INSTRUMENTATION

Analytical chemistry is an art conceived by almost all branches of chemistry (physical, organic, inorganic and biochemical). It evaluates the composition (qualitative) and concentration (quantitative) of substances in concern. The methods by which such analysis is carried out are of two types.

1.   Analysis by chemical methods (volumetric, gravimetric methods)
2.   Analysis using instruments

Physicochemical properties of a compound involve mass, volume, radiation, mass-to-charge ratio, thermal properties and electrical conductance. This can be measured in two ways.

1.   *Direct method*   physicochemical property of the compound is measured as such.
2.   *Indirect method*   fluctuation/change in the physicochemical property is taken as an indicator.

The differences between the two methods are listed in Table 1.1.

Table 1.1   Differences between analysis by chemical methods and by instruments

Chemical methods	Instruments
Bulky sample is required.	Small sample size is sufficient.
Human error is common.	Highly reliable results are acquired.
It takes considerable time period.	Rapid determination is possible.
Relatively cheap.	Expensive with respect to the reagents used and maintenance of instrument.
Can be performed easily.	Skilled personnel are required.
In the case of complex samples, results may be misinterpreting.	Even complex samples can be evaluated.

The principle behind the analytical instruments is that based on the physicochemical property of the compound of interest, it produces a signal which gets transformed and amplified to a presentable form. Transformation of the signal obtained from the compound to a known language is accomplished by a transducer. This transformed signal is amplified electronically and then displayed on a reading scale. Since these steps are automated, it can be performed quickly.

Several instruments are used to measure these properties and they are as follows. Mass and volume of a compound can be determined by gravimetric and volumetric methods respectively, whereas the use of instruments to measure radiation property of a compound depends on its ability to process any one of the following characteristics—absorption, emission, refraction and scattering. Based on these properties, specific instruments are used—absorption (UV spectrophotometer, infrared spectrophotometer, atomic absorption spectrophotometer), emission (emission spectroscopy, flame photometry), scattering (turbidimetry) and refraction (refractrometry). Similarly, to evaluate the compounds with charge-to-mass ratio, mass spectrometry is used. Thermal conductivity methods are preferable for components with thermal properties. In this chapter we have discussed several instruments of commercial significance.

The role of an analyst is not restricted to the instrument operation alone. This is because accuracy of the results can be achieved only when he/she concentrates on three important factors.

1. sample in available/pure form
2. sample preparation
3. interpretation of data

Moreover it is his/her responsibility to choose the method by which the component can be analysed. Again, the selection of specific methods depends on several factors such as

i. nature of material to be analysed
ii. reliability of method
iii. form in which the sample is present
iv. knowledge about the instrument to be handled
v. time limitation

Compounds or elements in solution possess any one of the following characteristics:

1. They may absorb light.
2. They may allow current to pass through the solution.
3. In gaseous form they may conduct heat.

It is these physical properties that form the basis for measurement using any analytical instrument. Accordingly, analysis using an instrument can be classified under any one of the following three methods:

1. optical methods
2. electric methods
3. chromatographic methods

OPTICAL METHODS

Here, the response of a compound/element to radiant energy is measured. The response differs between compounds, i.e. on exposure to radiant energy they may absorb, emit or scatter radiation. However all these interactions bring about changes in the electronic structure of the compound and the change can be subsequently evaluated. Initially instruments were designed to measure the changes in the visible region and hence the analytical method is called optical method. But in recent years with advanced development, interaction of compounds with radiant energy can be measured from X-rays or radiowaves. Based on the reaction of the compound to the radiant energy, several instruments are designed and they are as follows.

1. Absorption methods—UV spectrophotometer, IR spectrophotometer
2. Emission methods—Flame photometry
3. Dispersion and scattering methods

ABSORPTION METHODS

Let us consider a light ray "a" entering the sample and emerging out as "b". Now when you measure the intensity of both "a" and "b", you will observe that the intensity of "b" is lower than that of "a". This clearly shows that the compounds of the sample have absorbed some amount of radiation. Since the nature of the compound determines the absorption at a particular wavelength, a wide range of compounds can be identified by this method. However a single instrument cannot be used to measure absorbed radiation of all wavelengths. Therefore different types of instruments are used and a few are discussed below.

1.1 ULTRAVIOLET SPECTROPHOTOMETER

This instrument (Figure 1.1) is suitable for the measurement of selected organic compounds at low concentrations. This includes compounds containing benzene ring, unsaturated straight chains and those possessing a series of double bonds.

1.1.1 Principle

Whenever a compound with multiple bonds is exposed to the ultraviolet and visible region of electromagnetic radiation, a part of the radiation is absorbed. Absorbed radiation excites the core electrons of atoms of the compound and raises them to high energy orbits to show fairly broad absorption bands in that region. This at times, results in minor molecular changes. Such changes include:

1. rotation of a molecule (rotational energy)
2. vibration of atoms within the molecule (vibrational energy) and
3. excitation of atoms from one orbit to another, i.e. electronic transition (electronic energy).

These electronic transitions depend on the electronic structure of the compound to be analysed. Also, the rate and amount of absorption depends on the wavelength at which the radiation is absorbed and the nature of the compound. Therefore in inorganic substances, selective absorption occurs when incomplete energy level of an atom exhibits a covalent bond with another atom with complete energy level. In the case of organic substances, selective absorption is favoured due to the deployment of electrons, and compounds absorb radiation in the far ultraviolet region.

Figure 1.1   UV - Visible spectrophotometer

1.1.2 Instrument

The main components of the instrument are

1.   Energy source
2.   Monochromator
3.   Sample holder
4.   Photometer

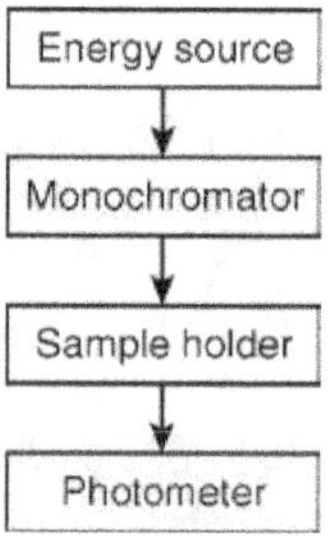

Energy source   The energy source used in ultraviolet spectroscopy should meet the following criteria.

a.   Beam produced should be in the detectable and measurable range.
b.   It should serve as a continuous source of energy.
c.   It should be stable.

Since incandescent tungsten filament lamp is found to satisfy these needs, it is widely used. However, its emitting capacity is directly proportional to the filament temperature and the wavelength of the emitted energy. To attain a stable energy output, the energy source should be equipped with constant voltage transformers.

Tungsten filament incandescent lamp   It works well in the visible and adjacent parts of ultraviolet and infrared regions. Introduction of pressurized iodine vapour makes it a suitable light source even at high temperatures. Quartz iodine lamp with vitreous silica envelope is another option for light source at elevated temperature.

Hydrogen/deuterium discharge lamp   Its optimum wavelength range varies from 160 to 360 nm. Deuterium lamps efficiently provide maximum intensity. Even though they are durable and prove better as light sources than hydrogen lamps, their cost restricts their applicability.

Hydrogen gas lamps   These lamps guarantee the continuous spectrum in the ultraviolet region (180–370 nm).

Monochromators   Filters and monochromators filter the energy source in such a way that a limited portion is allowed to be incident on the sample. Filters allow a wider bound of energy to pass through and they are used in filter photometers whereas, monochromators find their application in spectrophotometers. From the economic and handling point of view, filters are considered to be the best. Generally filters are of three types: a) absorption filters, b) cut-off filters and c) interference filters. In absorption filters the colourless glass with high thermal stability produce narrow bandwidths (30–250 μm) by allowing limited portion of energy source. In the case of cut-off filters, wavelength of maximum absorption occurs for a small portion of visible spectrum whereas towards the end of the spectrum, the transmittance decreases to give wavelength of minimum absorption. Interference filters have vertically placed glass plates whose inside is coated with semitransparent metallic films. The filter is filled with calcium or magnesium theoride.

Sample holder   The selection of material from which the cuvette is constructed is based on the selected range of measurement while its thickness depends on the read intensity of absorption. Cuvettes with varied shapes are used (rectangular, cylindrical or cylindrical with flat ends). However, the main factor is that the windows of the cuvette should be normal to the beam direction. Requirement of cuvette in terms of its make and thickness are as follows.

> UV region—Quartz
>
> Visible region—Glass absorption cells, silica cells and plastic containers
>
> Cell thickness—1, 2 and 5 cm

Photometer/Detector   The mechanism behind the photoelectric devices is the conversion of radiant energy to electric signal. Basically 3 types of photometers are used: a) Photovoltaic cells in which we detect the radiant energy by the current generation between the semiconductor and metal, b) Phototubes in which the energy absorption induces the solid surface to emit electrons and c) Photoconductive cells in which the absorbed energy changes the electrical resistance.

1.1.3 Sample Preparation

Compounds to be separated are converted to either vapour or diluted form. Cyclohexanes, 95% ethanol and, 4-dioxine are solvents commonly used. It should be noted that the solvent used should not absorb radiation in the ultraviolet region. The sample solution should not be concentrated and has to be diluted for reliable results. Dissolution can be done with suitable solvents. The sample holder should be kept clean and it is always better to use a tube as blank. The sides of the sample holder through which the absorption takes place should be wiped clean using blotting paper. This is because oil, perspiration and dust may misinterpret the results. The solvent should not interact with the light-absorbing molecules in the sample.

1.1.4 Procedure

The source emitting light with a wavelength specific for the element of interest is passed through the monochromator to reach the vaporized or diluted sample. The sides of the sample holder through which the absorption takes place should be wiped clean using blotting paper. This is because oil, perspiration and dust may misinterpret the results. The number and arrangement of electrons in the sample determine the intensity of absorption. Absorbed radiation is thus converted into electric signal by the photometer and subsequently recorded. Repeat the procedure without sample. The difference in radiation absorption between the sample and blank is used for the determination of element concentration.

1.1.5 Applications

UV spectrophotometer is widely used in

1. the identification of organic (hydrocarbons, vitamins, steroids, heterocycles and conjugated aliphatics) and inorganic substances (magnesium, molybdenum, nickel, nitrate, aluminium, beryllium, bromide, calcium chloride, copper, cadmium, iron, mercury, potassium and nitrite).
2. the analysis of oils and soaps.
3. the field of biological and pharmaceutical research to evaluate the product quality.
4. the field of biochemistry to identify steroids and enzymes.

1.2 INFRARED SPECTROPHOTOMETRY

The infrared region of the electromagnetic radiation constitutes the wavelength from 0.75 to 300 μm and lies between the visible and microwave region. Infrared spectroscopy is widely used in the quantitative analysis of organic compounds such as pesticides and complex organic chemicals. It is considered to be more advantageous than other spectrophotometry as the instrument shows characteristic spectra for each compound, i.e., different curves are exhibited for different compounds thus making their identification easier.

1.2.1 Principle

The application of this instrument is based on the fact that almost all organic chemicals absorb radiant energy in the infrared region. Being a low-energy radiation, its absorption causes a molecule to show minor changes in its vibration energy. This can be understood by the fact that the absorption of infrared radiation depends on the increasing vibrational or rotational energy of the covalent bond in a molecule and therefore to accomplish more radiation absorption, the molecule has to undergo a change in its dipole moment. Therefore, except diatomic homonuclear species (O_2, N_2, H_2, Cl_2), all the organic molecules with covalent bonds absorb infrared radiation, thus favouring their quantitative analysis. These compounds, in response to absorption of radiation in the infrared region, exhibit an absorption curve which is the characteristic of that compound. This is because infrared radiation absorption at different wavelengths induces different constituents of the same molecule. For example, absorption of short wavelength, long wavelength and still longer wavelength causes the vibration of hydrogen atoms, triple bonds and double bonds in a molecule respectively. IR spectroscopy exhibits the absorption bands as dips called as bands and each represents the radiation absorbed by the sample component at that frequency.

1.2.2 Instrument

The components of infrared spectrophotometers are as follows.

1. Light source
2. Monochromator
3. Sample holder
4. Detector
5. Recorder

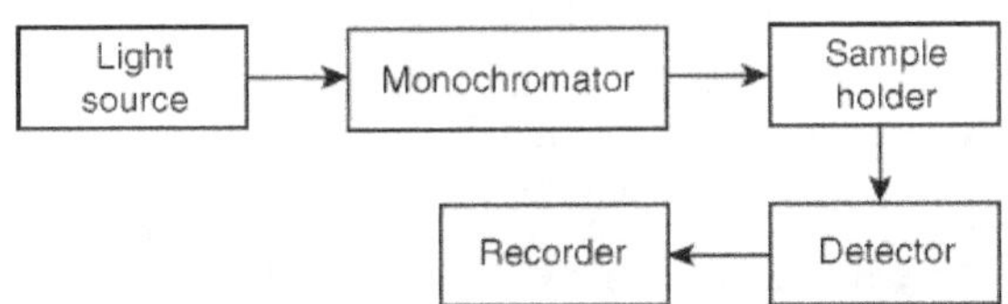

Light source   An incandescent solid with its emission properties resembling the black body radiator is used as a light source for infrared radiation. Nernst glower and Globar are the two sources commonly used. Both of these sources can be maintained at high temperatures by electrical heating and they are the best sources to give wavelength bands which are commonly used in infrared analysis. However, several other sources like carbon arc, nichrome wire, rhodium wire and tungsten filament lamp are also used.

Monochromator and optical material   Of the two types of monochromators—prism and grating—prism instruments are preferred due to ease of handling and greater range. Since materials like quartz and glass do not transmit infrared light, alternative materials like crystals of certain halogen salts should be used for optical parts. In general crystalline sodium chloride, potassium bromide and cerium bromide are used. Most of the infrared-transmitting monochromators are subjected to stretching due to moisture. To prevent this, desiccants are used. For greater convenience, the monochromator is sealed in a desiccated or vacuum box. Here concave mirrors replace the lenses, which are more preferred for the following reasons.

a. They can be easily mounted.
b. They can be made from a variety of materials like metal, aluminized glass, etc.
c. They do not exhibit any chromatic aberrations. Apart from these, reflex gratings can also be used as they ensure greater resolution, linear dispersion and resistance to water.

Sample holder   Narrow sample holders (0.1–1 mm) are used to avoid the absorption of solvent used. Nowadays demountable sample holders are used. This permits the variation in path lengths. For better results, the sample concentration is maintained between 0.1–10%.

Apparatus detectors   Thermal detectors (thermocouple, bolometer, thermistor) and photodetectors are the two types of detectors used commercially to evaluate the absorption spectra. Crystal detectors such as pyroelectric triglycine sulphate, barium titanate and lithium niobate can also be used. However the cost of these detectors restricts their use.

Amplifiers   Rectification of signal is essential as the radiation emitted in the infrared region by the materials other than the sample may interfere. Therefore amplifiers are used to provide alternating current amplification.

1.2.3 Sample Preparation

The sample may be either in the form of solids, liquids or gases and their processing differs accordingly.

Gaseous samples   The sample holder is a glass or metal cylinder (10 cm long) with mirrors that allow the radiation to pass through the sample. The sample should be made free of water vapour, subjected to a partial pressure of 5 to 50 mm Hg and finally let into the sample holder where absorption takes place.

Liquid samples   Since both the liquid and solid samples have to be dissolved in solvents, selection of solvent is the most important criterion in infrared spectrometry. This is due to the fact that each solvent has its own absorption ability which would give a value greater than the original value of the sample. Therefore carbon tetrachloride that allows only a limited number of substances to dissolve is preferable. It has a suitable wavelength in the range of 2.5–7.5 μ. Carbon disulphide that is efficient within the range of 7.5–16 μ can also be used. Several solvents like alcohol, acetone, water, chloroform and cyclohexane can also be used depending on the liquid samples. Using liquids as such without solvents will give more accurate results.

Solid samples   For soluble solid samples, suitable solvent can be used. However insoluble samples should be handled in a different way. A known amount of powdered potassium bromide mixed with the thoroughly homogenized sample is subjected to high pressure by careful insertion in between a die. Now the sample is tightly held in a fine plate, which can be readily used. Another method is to finely grind 1 mg of the sample with a drop of nujol. The function of the liquid paraffin nujol is to act as a mulling agent. The mull thus formed is pressed between plates made of sodium chloride. The thin layer produced can be inserted into a spectrophotometer. The concentration of the sample should not exceed 10%. Sample holders are usually removable and in the case of fixed sample holders, the samples are filled with a syringe. They are equipped with sodium chloride windows and exposure to moisture may fog them. To tackle this problem buffer powder is used. The sample holders are held in containers made of sodium chloride or potassium bromide.

1.2.4 Procedure

After processing the sample, it is filled in the sample holder in required concentrations. Suitable light rays, after filtration by monochromator, are allowed to pass through the sample. The sample absorbs the radiation in the infrared region and exhibits absorption curve as bands. This is then measured by a detector to give the exact concentration of the compound.

1.2.5 Applications

It is widely used in the quantitative identification of several organic compounds such as alkanes (branched, unbranched); alkynes; aromatic hydrocarbons; alcohols and phenols; aldehydes and ketones; ethers; esters; acid halides; acid anhydrides; amides; amines; and carboxylic acid. Even simple organic molecules exhibit complex spectra with a number of maxima and minima. This gives an accurate picture of the molecules and helps in comparative studies. However this instrument has its own drawbacks.

    a.   It is a complex spectra when compared to the ultraviolet and visible spectra.

    b.   It is less sensitive and therefore needs a large sample size (10^{-4} to 10^{-5} moles).

    c.   When a sample with a mixture of compounds is analysed, the spectral formation of the specific compound to be identified is influenced by the spectral formation of interfering compounds. Therefore for the quantification to be more accurate, the compound of interest should be isolated, which is a lengthy procedure.

EMISSION METHODS

Under controlled conditions, metallic/nonmetallic elements, when exposed to a powerful light source, emit radiation at some wavelength, which is a characteristic property of that compound. This helps in their qualitative analysis. Also, it is interesting to note that the intensity of the emitted radiation is directly proportional to the quantity of the compound in the sample. This favours its quantitative analysis also.

1.3 ATOMIC ABSORPTION SPECTROPHOTOMETER (AAS)

Atomic absorption spectrophotometer is more popular due to its versatility in measuring elements like copper, iron, magnesium, nickel and zinc at extremely low concentrations (mg/l). Moreover it takes into account the presence of unexcited free atoms in the flame. This favours the estimation of elements like Zn and Mg that do not get excited in flames as easily and therefore cannot be estimated in flame photometer. Nowadays, the instrument is even more refined in having a small cylindrical graphite furnace tube instead of an atomized burner. This favours the estimation of elements even in trace levels/amounts (mg/l). The characteristic feature of this graphite tube is that it can be adjusted and programmed to a wide range of temperatures. Also it measures the selected wavelength emitted by the cathode lamp as soon as it passes through the open ends of the horizontal cylinder.

1.3.1 Principle

When a sample is converted into its vapour form (the neutral atoms) and exposed to a light source, it absorbs the radiation in the ultraviolet/visible region. The amount of absorption is measured at a specific wavelength which is characteristic for each element. Since the radiation absorbed is directly proportional to the concentration of the element, comparison of the evaluated absorbance with the reference solution of known composition gives the quantitative analysis of the specific element.

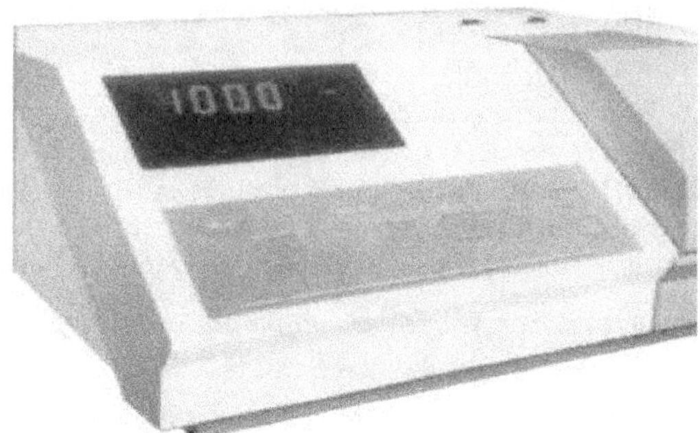

Figure 1.2   Atomic absorption spectrophotometer

1.3.2 Components of the Instrument

Even though the mechanism of AAS is based on absorption of radiation, it is included under emission spectroscopy because it is similar to the flame photometer in many respects. The components of AAS are

1.   Source/flame unit
2.   Monochromator/prism for dispersion and isolation of emission
3.   Sample container
4.   Detector
5.   Amplifiers

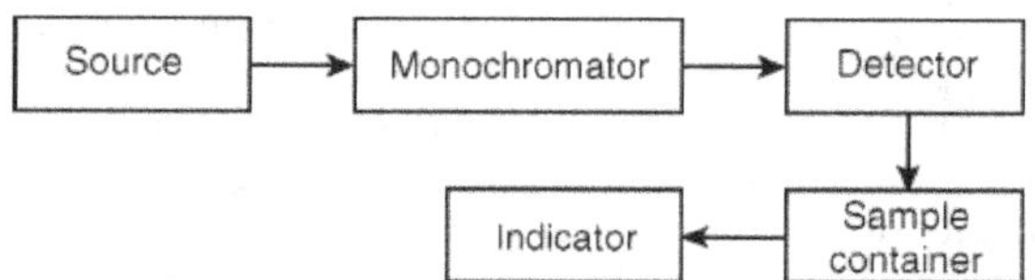

Flame   Initially it was difficult to have different light sources for different elements. However, nowadays hollow cathode lamps that are able to emit light for a variety of elements are available. Flame is released under controlled conditions and the flame requirement differs between elements. It should be noted that the maximum radiation absorption is accomplished by adjusting the burner light. An exhaust hood is equipped over the burner which immediately removes the combustion products formed due to the sample introduction in flame. As soon as the sample reaches the flame it is vaporized for the elimination of solvent and subsequently atomized for radiation absorption. Initially the sample in the minute droplet form is introduced at the base region of the flame. Here the vaporization of solvent takes place leaving behind the sample in dry form. The solid form of sample reaches the inner core of the flame where it is converted to the atomic state. It is here that the atoms get excited and absorb the radiation from the light source. In the reaction zone of the flame, the atoms are converted to their respective oxides and are readily ejected from the outer mantle of the flame. However, the entire portion of the sample to be processed is exposed to flame in a stepwise process and the successful sample conversion depends on several factors such as the droplet size of the sample particles, amount of oxidant added to the flame, number of species that form oxides in the reaction zone and the flame type that ensures maximum absorption. For the conversion of sample solution into element vapour, it is sprayed into the flame. This is because, elements in the gaseous atomic form get excited at high temperature so that their outermost electrons are shifted to the higher energy orbitals. These transitions occur in the ultraviolet and visible regions and this results in the development of well-defined narrow lines. Here the radiation is provided by a hollow cathode source and it is warmed up for about 15 minutes prior to the operation. Selection of flame type depends on the type of the metal to be analysed and its composition. Commercially air coal gas, air propane and air acetylene flame types are available.

Atomic vapour forming devices   Commercially available atomic vapour forming devices are discussed below.

Oven	Vaporization of the sample is accomplished by exposing it under very high temperature.
Electric arc and spark	A high voltage alternating current applied to the sample readily atomizes it.
Sputtering device	Positively charged ions of an inert gas when allowed to bombard against the sample in the cathode atomize the sample.
Non-flame atomizer	Electrically heating devices replace the flame to atomize the sample.

Monochromators and filters   Monochromators should be efficient enough to isolate atomic spectral lines of the element to be determined from the undesirable spectrum of liquid source.

Optical system, detectors, indicators   To detect the lines in the ultraviolet/visible region, photomultiplier tubes are used and the chart recorders are used as readout devices.

1.3.3 Sample Preparation

Sample in solution form is the basic requirement of atomic absorption spectrophotometry. Preparation procedures for a variety of samples are given below.

Aqueous solution	Diluted with water and then used.
Non-aqueous solutions	Sample should be dissolved in suitable solvents like methyl isobutyl ketone and then used.
Sample for biological source	The sample is digested using triple acid digestion method. Then it is diluted with distilled water and used.
Metals/Alloys	Dissolved in acid alkali and then diluted with water.
Trace elements	Concentration of the sample is done by extraction with chelating agent.

1.3.4 Procedure

The light source is a characteristic of atomic absorption spectroscopy. The method designed is based on the fact that each element has a characteristic wavelength for them to absorb. A light source emitting the light with a wavelength specific for the element of interest is passed through the flame. Now the intensity of the light is measured with and without the sample. Depending on the concentration of the element in the sample, the intensity of light is decreased and this difference is taken into account for the determination of element concentration. Now the sample is introduced into the cylinder through a side hole and the temperature programme is started. At the initial temperature, 100°C, the water particles of the sample are vaporized leaving behind the inorganic metal salts. Then with the progressive increase in the temperature (100°C), the cations are volatilized to occupy the cylindrical space. At this stage, the cations of interest absorb the radiation of their characteristic wavelength emitted from the cathode tube. The detector does the measurement of light intensity and the concentration of the element is evaluated.

1.3.5 Application

1.　Atomic absorption spectrophotometer is widely used in the determination of about sixty metals, metallic ores and minerals.

2.　It is used to determine the presence of the metals in plant, soil and soil extracts and this is very essential in the field of agronomic research.

3.　It is used to determine the heavy metal concentration in biological samples like urine, blood and tissues and hence this method is employed to check out the water quality by evaluating the presence of heavy metals; in the field of medicine to evaluate occupational-exposure-related diseases, for example, the presence of lead in traffic policemen, the presence of chromium, zinc and cadmium in industrial workers, etc. This helps to identify the cause of the disease.

4.　It is also used in environmental pollution control, i.e., the estimation of heavy metals in water, soil and plants, and helps to check out the pollution status of the environment in concern. Also this method is employed to estimate the heavy metal concentration in water so that its quality is explored.

5.　For the treatment of industrial effluents, the determination of heavy metals gives an idea about suitable treatment methods and if the concentration is exceedingly high it can be recovered.

Advantages

　i.　Highly sensitive method that can determine even trace levels of the individual elements.

　ii.　Wide range of elements can be analysed using AAS.

　iii.　The operation is not influenced by flame temperature.

　iv.　Completion of the analysis within a short duration and accurate results are obtained even for minute quantities of sample.

Disadvantages

　i.　Requirement of different lamp sources for every element.

　ii.　Those elements, which give rise to oxides in the flame cannot be detected.

　iii.　Non-metals cannot be detected.

　iv.　When samples in aqueous form are analysed, they may give irregular signals due to the anion in the sample.

1.4 EMISSION SPECTROSCOPY

Invention of emission spectroscopy has given a new dimension to the waste water analysis. Other than the basic components like prism and detector with amplifiers, it employs high-energy excitation sources like a-c arc; d-c arc or a high voltage spark that can excite all metallic and nonmetallic elements. Both solid and liquid samples including sludges and complex wastes can be analysed. Nowadays, use of inductively coupled plasma sources has popularized emission spectroscopy in waste water analytical laboratories as it is widely accepted as a standard method for several metal pollutants.

1.4.1 Principle

On exposure to a thermal or electrical source, the atoms of the given sample excite to give characteristic radiations in the UV and visible regions. This can be explained by the fact that as a result of excitation, atoms of the molecule acquire a high energy state and while returning to the ground state, they emit spectral lines at discrete wavelength of light. Since the wavelength of the emitted spectral line is inversely proportional to the energy difference between the high and low energy level, the amount of energy involved in the transition can be calculated. This aids in identifying the elements, as each element has characteristic energy level or spectral line.

1.4.2 Instrument

The components of the instrument include

1.   Radiation source
2.   Slit and a collimator lens
3.   Prism
4.   Photomultipliers

The radiation source vaporizes the sample which dissociates its molecules into ions. The ions are then excited. To accomplish this process, suitable light sources should be used and the light source to be used in this instrument should be able to meet the following criteria.

1.   The excitation should be constant over a period of time.
2.   Sample vaporization should be uniform and reproducible.
3.   It should excite the sample to give sufficient light intensity.
4.   It should excite a wide range of elements.

In this regard several light sources such as flame, a-c arc, d-c arc and a-c spark are used. Light emitted by the radiation source reaches the slit which filters the selective radiation. Two metallic blades are held by a micrometric screw from the slit. The function of the collimator lens is to focus the radiation to the prism. Resolution, based on the wavelength of the radiation, takes place in the prism. The emerging spectral line is then recorded by photomultipliers.

1.4.3 Sample Preparation

The instrument requires the sample concentration in minute amount. Sample in the form of metals, powders, solutions and gases can be analysed by this method. However powdered samples are more preferred and they should be homogenized before use.

Pelletized samples are ground by pelletized graphites. Following this, they are kept under a hydraulic press which passes the sample into a pellet. Metal samples are directly inserted into the radiation source.

1.4.4 Procedure

Allow a stream of argon gas to pass through three quartz tubes arranged concentrically. A water-cooled induction coil encircles the quartz tubes powered by a radio frequency generator. This creates a strong magnetic field inside. Now a high voltage spark excites the argon so that its ions and electrons exhibit a spiral flow pattern within the tubes, under the influence of the magnetic field. This ultimately results in the collision and hindrance in the movement of ions within the tubes. As a result the temperature is elevated to 4000 K and it may increase up to 8000 K which ensures complete classification of molecules giving no place for the other elements to interface. The sample is introduced into the central tube but above the argon flow. The element to be analysed emits their characteristic radiation which is isolated for intensity measurements. It covers a wavelength range from 180 to 900 nm.

1.4.5 Applications

1.   It is used to determine the qualitative and quantitative analysis of metals and metalloids.
2.   Metals and metalloids in all the three states, gas, liquid and solid, can be determined.
3.   It is used to determine the isotopic ratios.
4.   It is an accurate method for samples of low concentrations.
5.   It is used to semiquantitatively analyse metals that do not have suitable standards.

Advantages

i.   Even in mixture, the determination of the metals to be analysed can be done with great accuracy due to the minimum interference by other compounds in the mixture.

ii.   The method produces individual spectral lines for different compounds under the same excitation condition and therefore different compounds can be identified at the same time.

iii.   Many samples can be handled using the same light source.

iv.   It is a universally accepted technique.

Disadvantages

i.   Instrument operation and maintenance costs are very high to the extent that it cannot be adopted in small-scale laboratories.

ii.   Skilled personnel is required to operate the instrument.

iii.   Gives less accurate results compared to that obtained in atomic absorption spectroscopy.

1.5 FLAME PHOTOMETRY

This method is used to measure the amount of alkali and alkaline earth metals like sodium, potassium and calcium in the sample. The minimum size of the sample should be 1 mg/l. Even below this concentration, the determination can be performed but the accuracy will be lesser.

1.5.1 Principle

Variation in the radiation of sodium lamp with the amount of sample introduced, is the principle behind flame photometry. Each metal has a specific emission spectrum and based on its composition, the metal exhibits itself in various wavelengths. To determine the exact amount of metals in a mixture of solution, the intensity of the isolated radiation is measured. Under controlled conditions when an element is introduced into the flame it gets vaporized and excited to emit a characteristic radiation. The sample dissolved in suitable solvent is introduced into the sodium flame. Depending on the concentration of the sample, the flame emits radiation which enters a dispersing device for isolation. The intensity of the isolated radiation is measured in a phototube and the quantitative analysis of the sample is done by comparing the results with that of a solution whose concentration and composition is already known. The instrument finds great use in the determination of sodium and potassium in biological samples.

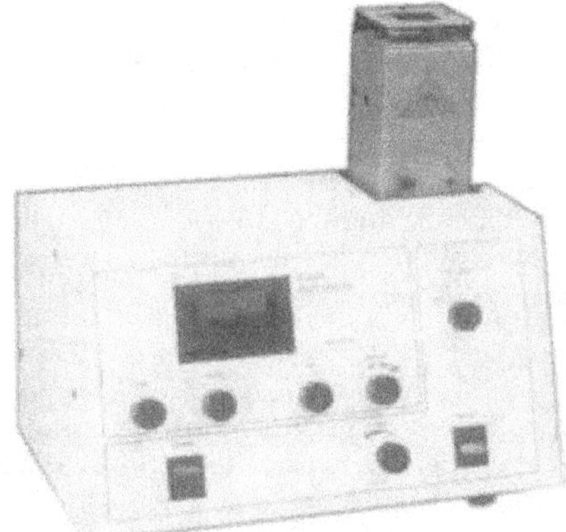

Figure 1.3   Flame photometer

1.5.2 Instrument

The instrument (Figure 1.3) consists of the following components.

1.   Pressure regulators and flow meters
2.   Atomizing device
3.   Optical system
4.   Photosensitive detectors

Pressure regulators and flow meters   They are installed in line from the gas cylinders to the burners and their function is to check the flow rate and pressure of fuel and oxidant. Since the pressure to be applied differs between the fuel 101 b/in^2 and oxidant 201 b/in^2, double diaphragm pressure regulators are used. Under constant pressure and flow rate they are allowed to pass through separate chemicals in the burner. At the flame end they are mixed. Due to a vacuum force, the sample, in limited quantities, is exposed to the plasma. Electrically heated argon gas gives rise to highly energetic plasma, which is more advantageous than other sources of flame. This is because they possess the ability to dissociate a wide range of elements.

Atomizers and burners　Since an oxygen acetylene flame can be used, the flame produced by the mixture of fuel and air concentrates the source of emitted radiation in a limited space, and the temperature of the flame ranges from 2700–3000°C.

Optical and electronic system　Emission lines or bands produced by the flame should be properly isolated. This can be successfully done when a flame photometer is equipped with monochromators. This type of photometer greatly resembles the atomic absorption spectrophotometer and is called as flame spectrophotometer. The monochromators used prefer the wavelengths which efficiently separate the easily excited emission lines with their resolving power.

Detectors　Detectors that can emit flame emission between 800 and 1000 μm are not available. Vacuum phototubes or photomultiplier tubes are used as detectors which give maximum signal even for weak flame emission.

1.5.3 Sample Preparation

Dissolve 0.23 g sodium carbonate in a little hydrochloric acid. Dilute this solution to 1 l using distilled water. This is the sodium stock solution containing 0.1 g Na in 1000 ml solution. Prepare several dilutions of sodium standard solution with their concentration ranging from 4 ppm and 40 ppm at appropriate intervals. Now measure the flame emission of these solutions many times. Using the emission values, draw a standard graph. Dilute the sample so that its concentration lies within the concentration range of the calibration curve.

1.5.4 Procedure

The liquid sample in appropriate amount is sprayed into the flame. For flaming, fuel gases like acetylene, hydrogen or natural gas and oxidants such as tank oxygen are used. This ensures the burning of flame at a very high temperature, which is very essential to excite more atoms. Flaming of sample evaporates the water particles from the sample, leaving behind the inorganic salts as minute particles. After decomposition into atoms or radicals, they are vaporized. The thermal energy of the flame excites the core electrons of the metallic atoms in the sample and raise them to the high energy orbitals. These electrons after reversal to their original state emit radiant energy which when passed through a prism separates into the different wavelengths. Now the desired region is isolated and a photocell and an amplifier measure their intensity.

1.5.5 Quantitative Analysis

It is a highly sensitive method for certain elements. The major disadvantage of this method is that it requires a suitable solvent for dissolution. Moreover, only fewer elements can be excited in a flame. This restricts the technique to have a wide range of applications. The sample is quantitatively analysed by correlating the sample concentration with its emission intensity and the concentration <1 ppm gives better results. The sample should contain the element to be analysed as a major constituent. The composition of the unknown sample should be similar to that of the sample composition. Flame photometry has been widely used in the analysis of sodium, potassium and other metals. To analyse the Na and K content in blood, urine and other biological fluids, internal standard method is employed using known quality of lithium.

1.5.6 Application

The various applications of a flame photometer are listed in Table 1.2.

Table 1.2 Application of flame emission spectroscopy in the identification of compounds.

Materials	Compounds that can be analysed
Water	Calcium, iron, magnesium, silicon, sodium and potassium, nickel and strontium
Glasses	Sodium, potassium, aluminum and boron
Cement	Sodium, potassium, manganese and lithium
Petroleum products	Tetraethyl lead; manganese in gasoline stocks
Metallurgical products	Aluminum, barium, calcium, copper, iron, lead, lithium, magnesium, manganese, potassium, rubidium and sodium
Rare earths	All rare earths except calcium, rhodium, ruthenium, scandium, silver, tellurium, tin, thorium, yttrium
Biological fluids, urine, blood	Sodium and potassium
Plant and soil	Calcium, magnesium, sodium and potassium

Disadvantages The formation of molecular species, incomplete vaporization and incomplete excitation are some of the processes that restrict the emission intensity of dissociated atoms and therefore this technique is selectively used for the determination of alkali and alkaline earth metals.

DISPERSION AND SCATTERING

The interaction of the sample with light depends on the presence of suspended particulates in it. That is, in the presence of suspended solids, it scatters light, whereas clear samples transmit light. This helps in the identification of turbidity in the sample and based on this property, two types of instruments are in practice.

1. Turbidimetry
2. Nephelometry

In the former, amount of transmittance is recorded and the latter takes into consideration the amount of light scattered by the suspended particulates. Using the same principle, instruments are designed to measure the scattered light. Fluorimetry and phosphorimetry are the typical examples of this type.

1.6 FLUORIMETRY

Fluorometers are the instruments that measure fluorescence. When equipped with two monochromators, they are called spectrofluorometers.

1.6.1 Principle

Molecules excited by electromagnetic radiation re-emit radiation of same or different wavelength. As a result of the excitation they acquire vibrational as well as electronic energy and collide. Due to this collision they drop to lower vibrational states. The process continues till it reaches an excited singlet electronic level, emits photon and immediately reaches the ground level. This phenomenon is called fluorescence. The molecule fluoresces till the radiation continues and spectra with many lines are exhibited in the visible region. When a system is subjected to a light ray, based on the intensity of the light energy, two processes may occur.

1. Ejection of electron—Photoelectric effect
2. Shifting the electron to a higher energy level—fluorescence

With sufficient energy in the light source the electrons will be completely removed from the system thus ionizing it. However if the energy intensity is low it will shift the electron to a higher energy level. These electrons emit light energy when they return to the ground state. If the emission is immediate, the phenomenon is called fluorescence.

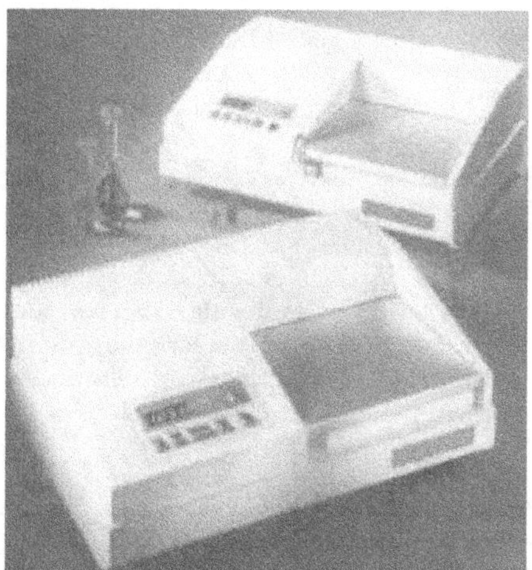

Figure 1.4   Fluorimeter

1.6.2 Sample Preparation

The sample is dissolved in a suitable solvent. Methyl alcohol, ethyl alcohol and dimethyl formamide are some of the solvents used. These solvents should be redistilled and the middle portion of the distillate should be used. The solvent after redistillation should not be stored in bottles with rubber cork, stoppers, grease, filter paper, etc. Also, the solution should not be exposed to UV radiation for a long period of time.

1.6.3 Instrument

This method is used to detect the concentration of certain organic and inorganic compounds that have fluorescent ability, i.e., they absorb radiant energy of one wavelength and emit it at a different wavelength. The instrument termed as fluorometer is similar to that of spectrophotometer and has the following components.

1. Light source
2. Condensing lens
3. Primary filter
4. Sample holder
5. Secondary filter
6. Phototube

Light source, condensing lens, primary filter and the sample holder are in a horizontal line whereas the secondary filter and phototube are at right angles to the light source. UV or visible radiation is the light source, which flows through the primary filter to the sample holder. The prime function of the primary filter is to filter only the desired wavelength. The fluorescent material in the sample absorbs the radiation and in turn emits the visible light which reaches the second radiation and allows only the wavelength emitted by fluorescent materials in the sample whose intensity is measured in the phototube. They are useful in the detection of biochemical compounds (ATP and F430). It is highly applicable in water quality studies. Here high fluorescent dyes like rhodamine B and pontacyl pink B are added to the water and their subsequent movement is detected by fluoroscopy.

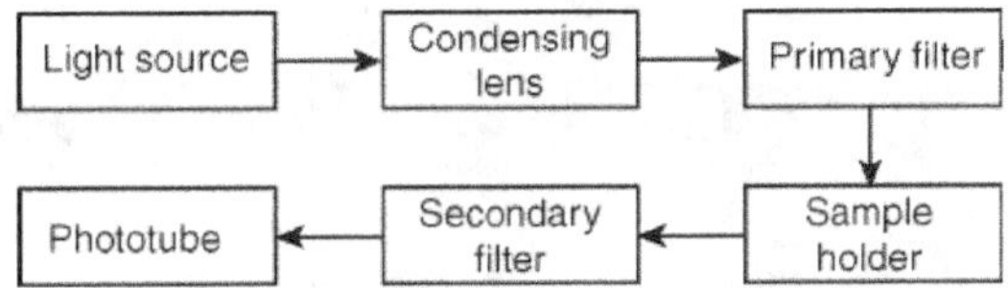

Source Different sources of radiation used in fluorimetry are mercury vapour lamps, xenon arc lamp and tungsten lamp. Of these, mercury vapour lamps are considered to be more advantageous than other sources due to the followings reasons.

a. Easy operation at both high and low pressure. The lamp emits radiation at 350 mμ. In the presence of apt filters high pressure lamps produce lines at several mμ (365, 398, 436, 546, 579, 690, 734) whereas low pressure lamps produce lines at 254 mμ.
b. Widely acceptable
c. Sharp line limitation
d. Maximum intensity

Nowadays, xenon lamps have replaced mercury lamps as they can give high intensity radiation. This is made possible by allowing the current to pass through a xenon atmosphere. Tungsten lamp is highly preferable for substances with excitation band beyond 450 mμ.

Condensing lens The heat of the lamp should not in any way affect the sample or phototube. Therefore a condensing lens is placed between the sample and light source.

Filters and monochromators The primary filter is kept between the light source and sample and its main function is to transmit the ultraviolet radiation alone to reach the sample whereas the second filter placed between the sample and phototube absorbs it efficiently.

Table 1.3   Compounds that favour and inhibit fluorescence

Compounds that favour fluorescence	Compounds that inhibit fluorescence
Molecules with conjugate double bonds, aliphatic, salicylic carbon structures.	Molecules that do not absorb ultraviolet radiation
Compounds with $-NH_2$, $-OH$ groups	Compounds with $COOH$, NO_2 groups
Heterocyclics with fused rings, quinoline, isoquinoline and aromatic hydrocarbons in unsubstituted form	Heterocyclics such as pyridine, furan, thiophene and pyrrole
Compounds with rigid structures	Aromatic ring compounds substituted with $100H$ or $> C = O$ groups

Figure 1.4a   Spectrofluorometer

Fluorometers equipped with two monochromators are called spectrofluorometers and the components of the instrument include:

1. Excitation source
2. Monochromator I
3. Monochromator II
4. Sample
5. Phototube
6. Microphotometer
7. Recorder

1.6.4 Factors Influencing Fluorescence

Increased polarity of the solvent enhances fluorescence. In the presence of dissolved oxygen, emission intensity of the fluorescent solution gets reduced, an action mediated by the photochemically induced oxidation of the fluorescent material. Moreover the paramagnetic properties of oxygen extinguish the fluorescence. Change in pH brings about a marked effect on fluorescent compounds. This is due to the fact that the wavelength and emission intensity of the compounds vary between the pH. Wavelengths lower than 280 nm do not fluoresce the compounds to be analysed, since wavelength <250 nm is sufficient enough to deactivate the excited molecules by dissociation process. Therefore selection of transition type that enhances the fluorescent process rather than the deactivation process is necessary.

1.6.5 Applications

Fluorometry finds application in the following.

1. It is more sensitive than spectrophotometer and so can be used for low concentration ranges.

2. It can be used to measure a broad spectrum of minerals and inorganic solid-state phosphors.

3. It can be used to measure organic and organo-metallic compounds that have the ability to absorb radiation in the UV range.

4. It can be used in the quantitative analysis of several biological compounds including thiamine, riboflavin, steroids, enzymes, coenzymes and plant products such as chlorophyll, alkaloids, fluoronoids and rotenone.

5. Fluorometric analysis can be used in the determination of organic compounds such as anthranilic acid; aromatic polycyclic hydrocarbons; indole; naphthols; certain nerve gases; proteins; salicylic acid; uric acid; tryptophan and skatole.

6. Furthermore it can be used to determine the chemical structure of the compound, the isomeric form in which it exists, nature of bond formation and rate of reaction.

7. It can be used in the quantitative analysis of rare earth metals (bismuth, aluminum in steel and beryllium in silicates).

Advantages

i. It can measure even minute concentrations of the sample (1.0 µg/ml).

ii. It is highly specific.

iii. It is an advanced method.

iv. It gives accurate results.

Disadvantages

i. The intensity of radiation is based on the pH of the sample and therefore continuous monitoring of pH is very essential.

ii. At times, the UV light may induce photochemical destruction in the fluorescent molecule, which in turn affects the intensity of radiation.

iii. If the sample contains dissolved oxygen, it will interfere with the fluorescent molecule to reduce its intensity of fluorescence.

iv. Other interfering compounds are iodide and nitrogen oxides.

v. Its determination is limited to low concentrations and to fluorescence-emitting compounds.

Phosphorimetry

A large number of organic compounds when cooled with liquid nitrogen exhibit phosphorescence. This is because only at this temperature, the excitation of ions from excited singlet to metastable triplet state takes place. Collision of molecules as a result of electromagnetic radiation absorption leads to its transfer to lower vibration states. The process comes to an end when the molecule moves to an excited triplet electronic level and subsequently emits photons to come back to its ground state. Even after the radiation is stopped, luminescence exists for a considerable period of time. Solvents used in this method should be a mixture of ethyl ether, isopentane and ethyl alcohol in the ratio of 5 : 5 : 2. Since these compounds have the ability to fluoresce too, flash tube sources can be used. Electronic timing circuit kept in line with this tube induces the flash to give a high voltage negative pulse at different time intervals. Fluorimetry is more widely used than phosphorimetry.

ELECTRICAL METHODS

The interaction between the ionic species of the compound and the electric current applied are measured. This method of analysis is of two types.

1.   potentiometric analysis
2.   polarographic analysis

1.7 POTENTIOMETRIC ANALYSIS

Glass and reference electrodes are introduced into a sample solution and a direct current is allowed to flow through the solution. Based on the hydrogen ion concentration of the sample, a suitable voltage is passed between the electrodes and the measurement of electric potential therefore gives the pH of the sample. pH meter comes under this category.

1.7.1 Principle

The immersion of electrodes in the solution favours the direct current flow through the solution which in turn induces a chemical change at the electrodes. This interdependence clearly reveals the fact that the nature of the chemical change is determined by several factors such as composition of the solution, the material from which the electrode is made and the strength of the electromotive force.

To make this description more clear, take two platinum electrodes connected by an external metallic conductor and immerse it in a solution of hydrochloric acid. One end acts as cathode and the other end acts as anode. Apply electric current (1.3 volts) so that the electrons get transferred from anodic end to the cathodic end. This ensures the flow of current through the solution and it gives negative charge at the cathode and positive charge at the anode. Now notice the chemical change at the electrodes due to the passage of electric current. Hydrochloric acid dissociates into H^+ and Cl^-. When the H^+ reaches the cathode, the cathode due to its reducing property donates an electron and reduces the H^+ to hydrogen gas.

$$H^+ + e^- \rightarrow \tfrac{1}{2}H_2 \text{ (gas)}$$

The anodic end, due to its oxidizing ability accepts an electron form Cl^- ion and converts it to chlorine gas.

$$Cl^- \rightarrow \tfrac{1}{2}Cl_2 \text{ (gas)} + e^-$$

It is this electron that is carried by the driving force of the battery from the anode to the cathode so as to be utilized by the H^+ ions. Therefore the net chemical charge observed is

$$H^+ + Cl^- \rightarrow \tfrac{1}{2}H_2(g) + \tfrac{1}{2}Cl_2(g)$$

Even though the flow of electrons determines the chemical change at the electrode, the type of electrode used influences the reaction. For instance, a single electrode is termed as half cell whereas when 2 half cells are combined together it is called as electrochemical cell. In electrochemical cell, the metallic conductor carries the electron between the two electrodes so as to produce a chemical change. Now when the half cells are subjected to an electric current, emf is developed which when measured gives the free energy of the solution. Therefore the electric energy utilized to give one mole of chemical change is expelled as *ZFE*,

where,

Z = Number of electron equivalents per mole

F = Faraday per equivalent

E = Emf of cell in volts

If the reaction is in progress, the electromotive force is said to be positive and therefore the relation between free energy and electrical energy is

$$\Delta G = -ZFE$$

The emf is said to achieve zero when the electrochemical cell has reached an equilibrium state.

1.7.2. Instrument

Potentiometric analysis makes use of the concentration-dependent relative potential of an electrode in a solution and helps to measure the activity of a specific ion. The instrumental set-up involves the following

1.   Special electrode
2.   Calomel electrode
3.   Potential measuring device

The special electrode is developed specifically for the component to be analysed and is therefore used to measure the concentration of the unknown solution component; it should be calibrated with the standard of known concentration and the solution to be analysed should be free from interfering compounds. Based on the ions to be evaluated, different electrodes are used and a few of them are discussed below.

1.7.3 Membrane Electrodes

Membrane electrodes play a key role in water quality measurements. The principle behind the membrane electrode is that when a thin glass membrane is kept between two solutions of two different hydrogen ion concentrations, a potential is developed. Nowadays, many membranes that favour selective adsorption of ions are available. This includes K^+, Na^+, F^-, Ca^{2+}, NO_3^-, and NH_4^+. In general, membrane electrodes are divided into three types.

1.   Glass electrode
2.   Liquid membrane electrode
3.   Crystalline membrane electrode

The mechanism behind all the three types are similar. However glass electrode is more widely accepted than the other two.

Glass electrode The universally accepted standard procedure for pH is based on glass electrode. It is successfully used in coloured solutions, oxidizing media, reducing media and colloidal systems and the results obtained are satisfactory. In earlier days, the major drawback encountered in the glass electrode was that it established a potential with Na^+ in alkaline solution and this was later solved. However this is taken into account for the development of glass membrane electrodes specific to sodium. The composition and thickness of both the glass and calomel electrodes are the same except for the sensitive part of the glass electrode where an opening is made for direct electrical connection with the surrounding fluid. Basically the glass electrode is filled with an electrolyte, an acid solution of predetermined strength. Therefore the hydrogen ion activity in the solution is in relation to the electrolyte activity within the electrode and based on these, a single electrode potential is established. Since the potential developed with different ions depends on the composition of glass, different types of glass electrodes are presently in use to measure the concentration of several ions like Na^+, K^+, NH_4^+ and Ag^+. It should be remembered that prior to their analysis, it is very essential to evaluate the presence of cations and other interfering ions in water.

Liquid membrane electrode Liquid membrane electrodes were recently developed to evaluate the activities of many polyvalent cations (Ca^{2+} and Mg^{2+}) and few anions. Here the glass membrane in the glass electrode is replaced by an insoluble organic compound (liquid) that is kept in support, between the porous glass and plastic discs. Based on the ions of interest, the functional group of the organic compound is selected, i.e., these functional groups should have greater affinity for the ions of interest. This allows the exchange of specific ions with that of the ions present in water–liquid exchange interface and ultimately develops a potential.

Crystalline membrane electrode The electrode filled with the reference solution is impregnated by a solid membrane. The reference solution selectively absorbs anions and the anion to be analysed is present in the solid membrane as sparingly soluble salt. When the electrode is placed in solution, the anion of interest from the solution is exchanged with the ions of the solid membrane. This establishes a potential which in turn is measured to evaluate the ionic activity. This method is suitable for the measurement of fluoride in water. Lanthanum fluoride crystals, the constituent of solid membrane is pretreated with rare earths before use. This enhances its electrical conductivity and efficiency.

1.7.4 Metal Electrode in Contact with Slightly Soluble Salt

Calomel electrode used as reference electrode comes under this category. Here the metal electrode is kept in contact with the slightly soluble salt, mercury, which in turn, is in contact with a KCl solution. Since it is stable and can be handled with ease, it is highly preferable in electrometric pH determinations, oxidation–reduction measurements and electrochemical analysis. The reaction of the electrode is represented as

$$Hg + Cl^- \rightarrow \tfrac{1}{2}\, Hg_2Cl_2 + e^-$$

Moreover, concentration of KCl solution in the calomel electrode determines its potential development. Therefore to acquire difference in potentials, different concentrations such as normal (–0.281 volts), one-tenth normal (–0.334 volts) and saturated (–0.242 volts) are prepared. Lack of knowledge about the concentration of KCl will misinterpret the data. However, the saturated type of calomel electrode is extensively used.

1.7.5 Oxidation–Reduction Electrode

This includes platinum wire, a nonreactive electrode kept immersed in a solution of ferrous (reduced) and ferric (oxidized) chloride ions. Also a calomel reference electrode is kept immersed as a standard. They are extensively used as stoichiometric end point indicators in oxidation–reduction titrations and to measure relative oxidation–reduction potential in biological systems. However, their commercial use in complex waste water systems is not so successful. Probably this may be due to the presence of oxidants and reductants in large amounts. But still a general view about the oxidizing or reducing condition of the aquatic system can be evaluated. The reaction at this electrode is represented as

$$Fe^{2+} \rightarrow Fe^{3+} + e^-$$

1.7.6 Metal Electrode

When a metal is placed in a solution of its own ions, reversible electron transfer between the metal and its ions occurs and based on this principle, metal electrodes are used. Its usage is restricted to metals that are able to establish reproducible potentials. Typical examples are silver, copper, mercury, lead and cadmium. On the other hand, oxide-coated metals such as iron and nickel cannot be used.

1.7.7 Gas Electrode

The gas inside the electrode is protected by a strip of nonreactive metal, usually platinum or gold. Several gas electrodes are available of which hydrogen electrode gains more importance. In this electrode, a platinum sheet is coated with platinum black and then bathed in a stream of hydrogen gas at 1 atm pressure. When immersed in a solution, a potential is developed.

1.7.8 Fluoride-Specific Electrodes

These electrodes cannot be involved in the direct measurement of concentration of compounds. Instead it measures its ionic activities which are directly proportional to the ionic strength of that particular compound. Therefore to measure the concentrations of compounds using electrodes, it should be calibrated in a standard solution of similar ionic strength. Since the electrodes do not take into account species complexes, they must be broken up prior to the electrode use. For example fluoride electrode measures only the ionic form but leaves behind the portion of fluoride that complexes with cations such as A1 (III), Fe (III) and Si (IV). Probably adjustment of the pH of the solution or addition of materials that compete with the fluoride to complex with cations may serve best to ensure the measurement of the entire concentration. For routine analysis of water quality, ion-specific electrodes can be used.

1.8 POLAROGRAPHIC ANALYSIS

When a measured voltage is applied to an electrode kept in a sample solution, the current flows throughout the solution and the flow is directly proportional to the composition of the sample. Therefore this method is used for analytical purposes and the analytical procedures are called voltammetry. Using this principle, several methods are in practice of which two methods are discussed below:

1. Anodic stripping voltammetry
2. Membrane probe

1.8.1 Anodic Stripping Voltammetry

Anodic stripping voltammetry is superior to direct polarographic analytical methods, due to its high degree of sensitivity in measuring metal pollutants like lead and cadmium in water. The instrumental set-up is similar to that of polarography and in addition, it consists of a hanging drop mercury electrode or thin mercury film electrode.

The instrumental set-up involves

a.   electrochemical cell

b.   Voltmeter—to measure the exact voltage applied

c.   Microammeter—to measure the flow of current through the solution.

As an initial step, a minimum volt far less than the halfway potential of the metal to be analysed is applied between the two electrodes. This favours a hanging drop of mercury to form and within a specific period, the metal of interest is deposited on the drop. Now the applied volt is gradually decreased in such a way that the decrease is noted in the voltmeter. As a result, an oxidizing potential is established which ultimately strips off the metal from the electrode. Since the current flow at the time of stripping is directly proportional to the concentration of the metal, the results obtained are more accurate and therefore it is widely accepted as a standard procedure for metal analysis.

1.8.2 Membrane Probe

Membrane probes are extensively used to determine the concentration of gaseous or non-ionized molecules in the water. Measurements of dissolved oxygen and ammonia are accepted as standard procedures. O_2 electrode system consists of gold or platinum as cathode and silver as anode. Potassium chloride is used as electrolyte. Polyethylene or teflon permeable to gases separates the cell from the solution to be analysed. The purpose of using the teflon membrane is to protect the cell from interfering compounds, contamination of which would limit its functional life. 0.5 to 0.8 volt supplied across the anode and cathode is sufficient enough to reduce the O_2 passing through the membrane. This creates the current to flow and its magnitude equals the O_2 concentration in the sample. Gold or platinum is inert until O_2 is reduced here. Another development of this method is that an externally applied voltage is not required. Here lead cathode acts as an inert electrode, silver as anode and potassium hydroxide as electrolyte. As a suitable combination, they develop a galvanic cell which has a potential sufficient enough to bring about O_2 reduction.

CHROMATOGRAPHIC METHODS

Separation of individual components in a mixture is the basic principle behind chromatographic methods. In the presence of two phases, stationary phase and moving phase, the sample components are allowed to separate and based on their relative affinity for partitioning they get migrated along the moving phase. Solids or liquids can be used as stationary phase whereas liquids or gases can be used as mobile phase. If the mobile phase is gas, it is called as gas chromatography and if it is liquid the method is termed as liquid chromatography. Names of some methods are still more descriptive and self explanatory. For example in gas liquid chromatography, the mobile and immobile phases are gas and liquid respectively. In paper chromatography the mobile phase is liquid and the immobile phase is paper. These instruments are widely used in the field of engineering and science.

1.9 GAS CHROMATOGRAPHY

In recent years, gas chromatography has become inevitable for a variety of routine analyses in water pollution studies. It is a widely used separation technique in which the mobile phase is gas and the stationary phase is a solid or liquid. If the stationary phase is a solid (diatomaceous earth of kieselguhr) the technique is termed as gas solid chromatography (GSC) and if the stationary phase is a nonvolatile liquid (granular silica/alumina of carbon), it is termed as gas liquid chromatography (GLC).

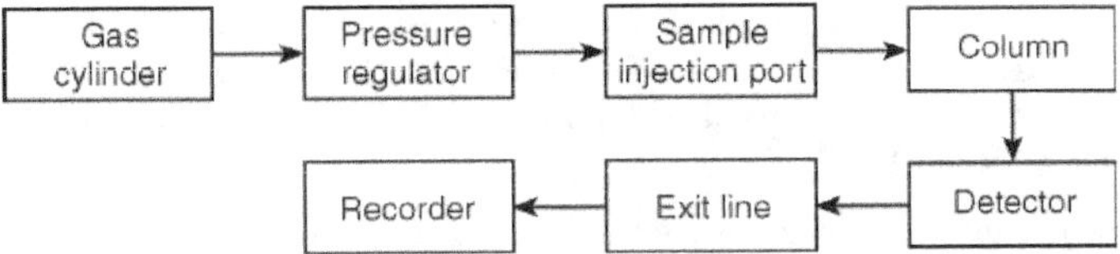

1.9.1 Principle

The partition between the gaseous mobile phase and stationary solid/liquid phase enables the separation and fractionation of the vaporized sample. When a mixture of components in a liquid sample is analysed under gas chromatography, individual components are identified, quantified and then measured after they have vaporized into gaseous phase and therefore this technique is used for the quantitative and qualitative analysis of the sample.

1.9.2 Instrument

The various components of gas chromatography involve

1. Gas cylinder
2. Pressure regulator
3. Sample injection port
4. Column
5. Exit line and
6. Recorder

Figure 1.5   Gas chromatography

Gas cylinder　The gas cylinder with a reducing valve is filled with carrier gas (hydrogen, helium or nitrogen). Helium or nitrogen is used as carrier gas in gas chromatography. Their ability to remain inert, easy availability, inexpensiveness and their ability to indirectly induce the detector to exhibit good response make their application successful in GC. The experiment is initiated by allowing the carrier gas to pass into the column.

Pressure regulator　The pressure regulator controls the temperature and flow rate of the carrier gas. Temperature of the sample should be appreciably high to maintain it in the gaseous state throughout its movement in the column. This is because only the sample in gaseous form can be detected by gas chromatography. The procedure should not be continued without the vaporization of samples. This is because the compounds to be analysed, if present in the solid or liquid state, will close the sampling port rather than move along the port.

Sample injection port　The sample to be analysed is injected into the sampling port using a syringe where it is immediately converted to gaseous components by flash evaporation. While drawing the liquid samples with a syringe, gas bubbles should not be entertained and it should be immediately injected into the gas stream. Syringes with capacities of 0.5 to 10 ml are used. As far as solid samples are concerned, they should be dissolved in suitable solvents and then drawn as solution.

Columns　In general, chromatographic columns are packed with an inert solid protected by nonvolatile liquids like silicone oil and polyethylene glycol. Their length varies within 1–10 m. Nowadays, capillary columns with 0.2 to 0.4 mm diameter and 20 to 30 m lengths are available. Since they give better resolution of peaks they are considered to be advantageous over other columns. Moreover they are provided with a temperature adjuster in which the required temperature is programmed for a specific time period. This gains the solubility of different components of the sample within a single column of materials. Moreover the temperature should not in any way degrade the sample. Carrier gas takes the gaseous components of the sample to the column. Now based on their solubility the components get partitioned between the stationary liquid phase and moving gaseous phase of the column. Naturally components solubilized in the liquid phase exhibit a slow rate of movement. Therefore by suitable selection of stationary phases and column lengths, the materials to be analysed can be separated. This enables the gas components to emerge out from the column at different times.

In the columns, different constituents of the sample, based on their interaction with the packed material, get separated. Column packing therefore plays an important role in solute separation. Usually packed columns are copper or stainless steel tubes with two different types of packing such as solid substrate and liquid coating on an inert solid. For packing, a known weight of the liquid phase is mixed thoroughly with an inert dried solid in an open container. However this can be achieved by dissolving the liquid phase in a suitable solvent and subsequent addition of solid. Solids acquire the liquid coat when the solvent is vaporized. Then it is packed in the column. See to it that the column is tightly packed and now the column is loosely folded. For solid substrate packing, plug one end of the column with a glass wool and fill it with the desired solid. The solid is filled into the column through a funnel. After packing is over, close the other end also and fold it.

Silicious material is widely used as the solid inert support. Some of the characteristic features of silica are as follows—high porosity, large surface area, excellent mechanical strength, uniform particle size, nonreactivity even at high temperatures and easy integration with liquid phase. It is these characters that make it suitable to act as a support. Similarly liquids, to function as stationary liquid phase, have to meet several requirements. They are

a. Suitability of the solvent for the compound to be analysed

b. Difference in dissolution ability between the compounds

c. Nonreactive nature

d. Stability even at high temperatures

e. Low volatility and high boiling point (200°C)

Since the selection of the liquid phase depends on the compound to be analysed it cannot be generalized and different liquids are employed based on their application.

- Water and methyl alcohol mixture → semipolar liquid phase
- Hydrocarbon mixture → nonpolar liquid phase
- Mixture of pentane, butane and propane → nonpolar liquid phase
- Alcohol mixture → polar liquid phase

The efficiency of a column to bring out high resolution depends on several factors such as

a. Particle size of the stationary phase

b. Surface area of the stationary phase

c. Rate at which the carrier gas is passed through the column

d. Structural similarity of the liquid stationary phase with the components to be separated

e. Concentration of the stationary phase

f. Column dimensions

g. Column temperature

Particle size and the surface area determine the distribution of sample components in the stationary phase. Moderate speed is required for the passage of carrier gas through the column because slow movement of the carrier gas gives rise to broad peaks and with speedy flow, the peaks will not be resolved. Similarly, excessive concentration tails the peak and therefore a liquid phase with a concentration of about 1–15% is found to be reasonable for columns with 1–10 m length and 0.25 inch diameter. The operable temperature use of the column should be adjusted in such a way that it keeps the sample in vaporized form but does not disturb the stationary phase.

Detectors Emerged out gas components are detected and recorded as peaks. Each specific compound or a mixture of compounds that exhibits same flow rate are represented as 'peak' and the area covered by each peak is directly proportional to its concentration in the sample. Another factor that helps to identify a compound is that each of them has a characteristic retention time, i.e., the time taken for them to emerge out from the column, which in turn depends on the solubility of that component in the liquid phase.

Detection of compounds flowing out of the column is the most important step for an analyst. Detectors of gas chromatography are selected in such a way that they render more accuracy in result interpretation. This is possible when it possesses certain characteristics like suitability for the compound to be analysed; stability even at high temperatures; easy operation; specific response to compounds of interest; production of output signal that is directly proportional to the concentration of the compound and height of accuracy irrespective of sample concentration. The detectors should take into consideration the physical or chemical property of the emerging gaseous components so as to convert them into electrical signal. The different detectors that possess these characteristics include:

1. Thermal conductivity detectors
2. Flame ionization detectors
3. Photo-ionization detector
4. Electrical conductivity detectors

Thermal conductivity reactors　　This detector is based on the characteristic thermal conductivity of different gaseous components. When a heated wire is placed in a stream of gases coming out from the chromatographic column, the temperature of the wire fluctuates according to the thermal conductivity of the gaseous component flowing over it. This electrical response is recorded. These detectors are highly applicable for the analysis of gases emerging from anaerobic digestion process.

Flame ionization detector　　Since this detector does not respond to the water vapour, samples can be directly injected. They are useful in analysing organic compounds. Several hundred volts are applied across the burner and a collector electrode so that the organic component when eluted from the chromatographic column is burnt to yield ions and electrons. Due to the volt applied, they carry current and the measurement of current carried by them gives a clear picture of the specific compound present in the sample. Organic acids and petroleum hydrocarbons can be separated and measured with high degree of accuracy.

Photo-ionization detector　　This works on the principle of flame ionization detector, but instead of flame, an intense beam of ultraviolet radiation (105–150 nm) is used to ionize the molecules. This detector is suitable for the analysis of small organic molecules with double band.

Electrical conductivity detectors　　These detectors are also represented as electron capture or coulometric detectors. A radioactive source such as ^{63}Ni or tritium are used. The ability of the compounds to capture the beta particles emitted by these sources is measured. Certain compounds with halogen atoms or polar functional groups have the ability to absorb these β emissions and therefore can be identified by these detectors.

Gas chromatography has been developed with a lot of sophistication. For instance, detectors that can measure gas density, change in potential at electrodes and mass spectrometry are being used today.

1.9.3 Procedure

In this technique, the sample in vaporized form is introduced into the column head. Sample components distribute themselves in the stationary phase and the mobile phase so as to maintain equilibrium. Inert gas forced through the column transfers the sample components to the column end. However the rate of their movement depends on their solubility in the stationary phase. Sample components that are insoluble in the liquid phase move rapidly than the soluble components in the solid phase. The former is evaluated by noting down the time duration for peak formation whereas the latter is determined by measuring the areas covered by individual peaks.

1.9.4 Applications

1. It is widely used to detect, quantify and measure a wide range of volatile organic compounds.

2. Good precision in results is acquired. Samples with varied size (from µg to 11) can be used. To increase the sensitivity, passing inert nitrogen gas through the sample vaporizes the volatile organic compounds in liquid samples. The vapour is transferred to a packed column where the volatile organics are absorbed. The column is now connected to the GC and then the procedure described above is followed. This eliminates the interfering compounds and water from the compound of interest.

3. Qualitative analysis of the sample component can be done by measuring the retention time (t_R) or retention volume (t_V). Time taken by the sample to travel from the column head to its end and appear as peak in the detector is defined as retention time and the volume of the gas required to transfer the sample throughout the column is called as retention volume (t_V). In this technique, the t_R is calculated by injecting the sample into the column with the stationary phase of different polarity. Again the sample is injected into the column but with a different stationary phase and (t_R) is recorded. A standard graph is drawn in a logarithm scale plotting the t_R of one column with the t_R of another column. This yields a straight line since the intercept of the plot varies between compounds with different polar characteristics. Thus it is possible to separate compounds between homologous series.

Table 1.4   Application of gas chromatography

Products	Compounds to be quantified
Petroleum Products	Gasoline, wax, LPG, sulphate and nitrogen compounds
Food products	Milk, bread and crushed fruit
Alcoholic beverages	Bourbon, Vermauth and rum
Plastics	Plastics identification, esters in acrylic copolymers; long chain alcohol esters in acrylic copolymers, styrene monomers in styrene, plastics and vinyl acetate in copolymers
Detergents and soap	Liquid detergents, ethyl alcohol
Water	Soap, fatty acid and pesticides
Rubber	Amine antioxidants, copolymer composition
Other items	Water in butane gas, creams, emulsions ointments and pastes
Pesticides	Determination of organic phosphates
Biological samples	Blood oestrogen, vanillmandelic acid, homovanillic acid

1.9.5 Quantitative Analysis

Even though peak areas and peak heights are the two parameters taken into consideration for quantitative analysis, samples with short retention time exhibit narrow peak areas whose measurement is difficult. However the peaks obtained are tall enough to be measured for the determination of sample concentration. Samples with long retention time have broader peaks where the area included along with height of the peak is influenced by several factors such as column length, porosity of the material used in packing, column operable temperature and the flow rate of the carrier gas.

Given in Table 1.4 are few examples in which gas chromatography is employed.

1.10 HIGH-PERFORMANCE LIQUID CHROMATOGRAPHY (HPLC)

High-performance liquid chromatography is advantageous over gas chromatography in that it can analyse thermally unstable compounds and nonvolatile species such as amino acids, proteins, nucleic acids, hydrocarbons, fatty acids, carbohydrates, phenols, pesticides, antibiotics, metal organic species and a wide variety of inorganic substances.

1.10.1 Principle

This method favours the separation of thermally unstable, nonvolatile species in a mixture. The liquid mobile phase drives the sample along the stationary phase packed in a column. The sample thus eluted can be detected at the end of the column.

1.10.2 Instrument

The experimental set-up of HPLC involves:

1. Solvent delivery system
2. Sample injection system
3. Chromatographic column
4. Detector and Recorder

Figure 1.6　High-performance liquid chromatography

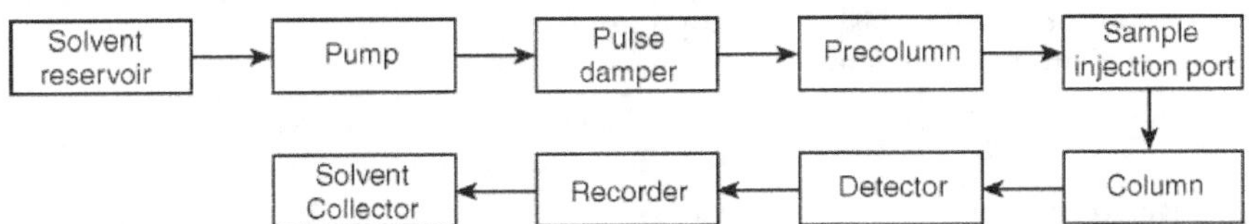

Solvent delivery system   The mobile phase when pumped under high pressure pushes the sample through the column. Before use, the mobile phase should be deaerated using a vacuum pump. Selection of solvent to be used as a mobile phase is based on its polarity, its eluting power, boiling point, viscosity, detector compatibility, flammability and toxicity. Above all, the solvent should not interact with the sample.

Sample injection system   Sample introduction can be done by three methods. Method I injects the specific volume of sample by a fixed volume loop injector. Method II injects the sample by injection valve and the valve aids in adjusting the sample volume and method III injects the sample through a syringe chromatographic column.

Columns   Analytical columns are short (10–30 cm long) metallic tubes kept stretched with still more limited inner diameter (1+10 mm). About 3 to10 mm of this area is uniformly packed with inert materials such as silica. The material used to coat silica is obviously an organic phase of required characteristics that facilitate the maximum partitioning of compounds between mobile and stationary phase. Based on the polarity between the mobile and stationary phases, HPLC is of two types:

1. Normal phase chromatography
2. Reverse phase chromatography

Normal phase chromatography   Here the low-polarity-bearing components emerge out of the column first as the mobile phase is a nonpolar solvent (hexane) and the stationary phase is a high-polar material (water).

Reverse phase chromatography   High-polarity-bearing components emerge out of the column first as the mobile phase is a polar solvent (methanol/water) and the stationary phase is a nonpolar material (hydrocarbon).

Again the selection of detectors depends on characteristics of the mobile phase like its density, dielectric constant and refractive index. Also the properties of the compound to be analysed such as its ability to fluoresce, to absorb UV radiation and to conduct electric current should be considered. The main drawback of this instrument is that it cannot identify two elements when they emerge from the compound at the same time. Using different columns, solvents or detectors the problem can be solved. But it is highly expensive and time-consuming.

Types of column packing   The columns are constructed with heavy glass or stainless steel to bear high pressure. Care should be taken for the sample volume and concentration to be within the capacity of columns. Three different types of packing are used in HPLC—a) Porous polymeric beds, b) Porous layer beds and c) Totally porous silicate particles. A provision is made at the end to collect the waste solvent. The selection of solvent as a carrier, column packing and detector depends on the compound to be analysed. Column heating is not required, however a temperature controller that maintains uniform temperature throughout is essential to control the retention time of respective compounds.

Porous polymeric beds composition/components	Application	Modification
Styrene divinyl benzene copolymers	Ion exchange and size exclusion chromatography	Replaced by silica-based packings
Thin shell of silica on glass beds	Analytical	Totally porous micro-particulate packings.
Silica particles less than 10 μm	Analytical	Widely used

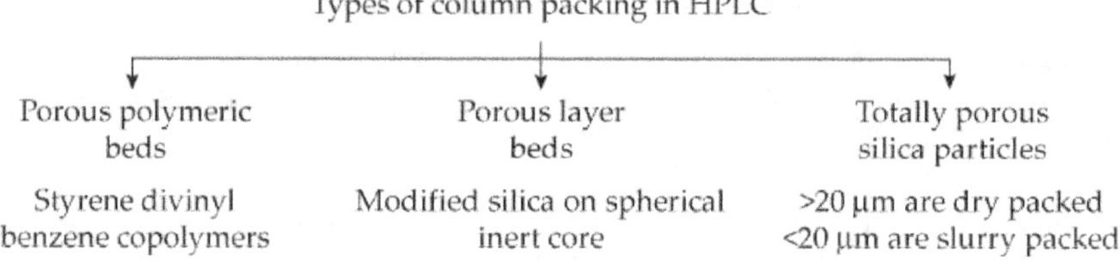

Application: Ion exchange size exclusion chromatography.

Detectors Two types of detectors are recognized a) bulk property detectors b) solute property detectors. The former takes into consideration the overall changes in the physical property of the mobile phase whereas the latter detects the physical property of the solute alone and is therefore considered to be more sensitive even for minute sample concentrations (nanograms). Based on the response of detectors to samples, they can again be classified into two types: a) universal and b) selective. Universal detectors can be applied to detect the sample as a whole whereas selective detectors are specific to certain components of solute.

1.10.3 Sample Preparation

Dissolve the sample in a suitable solvent. See to it that the sample solution is not clear. Then transfer a measured volume of sample into the sample injection system.

1.10.4 Procedure

The solvent to be analysed is degassed and filtered as a preliminary step. Then it is filled in the solvent reservoirs. The solvent is now pumped into a column system at 400 atm. pressure and then passed to a precolumn through a pulse damper, which regulates the flow rate of the solvent. Precolumn termed as guard column removes the particulates and contaminants from the solvent and the filtration process serves two purposes a) to make the solvent fit as a carrier phase by eliminating the interfering compounds and b) to protect the analytical column and increase its functional life. Individual organic compounds such as hexane, methanol, water (isocratic elution) or mixture of compounds (gradient elution) are used as solvents and their polarity is not usually considered. Better separation efficiency is observed in solvents mixture as they exhibit different polarity. The sample is injected into the sampling port and the solvent pushes the sample into the analytical column where the separation of components occurs. The detector at the other end of the column identifies and records the results so as to quantify the separated component.

1.10.5 Applications

1. It is used in the quantitative analysis of inorganic compounds that can be dissolved in solvents used in HPLC columns.

2. It is very useful in the field of forensic toxicology where a large number of intoxicants are needed to be detected. They include:

   Medicament—hypnotics

   Additive drugs—heroin, opium alkaloids

   Plant protection agents—cholinesterase-inhibitors, CO_2, alcohol

   Inorganic agents—cyanide, fluoride

   Industrial chemicals—formic acid

   Since most of the poisons are polar compounds and susceptible to heat, HPLC serves as the best option to identify them.

3. Since it can be used in the identification of lipstick smears in cloths, glasses and papers, it plays an important role in crime investigations.

4. Also it can be used in the analysis of different explosives such as TNT, TETRYC, RDS, HMX, PETN, EGDN and NG.

5. Application of HPLC in quantifying lipids in food products has great importance in nutritional science.

6. It is used to separate steroid hormones such as oestrone and androgen, a useful technique in the medical field.

7. It finds wide application in the separation and purification of nucleic acids.

8. It is used in the separation of oxide variety of antibiotics such as amoxycillin; ampicillin; anisomycin; chloramphenicol; cycloheximide; erythromycin; levorin; neomycin; penicillin; streptomycin; trichloromycin and vancomycin.

9. It is used in the analysis and separation of amino acids, proteins and carbohydrates.

10. Various preservatives and antioxidants used in the food products can be easily quantified by HPLC. e.g. sorbic acid, benzoic acid, PHB, methyl ester, PHB ethyl ester, PHB-propyl ester, biphenylol, biphenyl.

11. It is used to determine both water-soluble and water-insoluble vitamins.

Advantages

    i. High-performance liquid chromatography is more advantageous than other analytical instruments as it guarantees efficient and rapid separation of samples.

    ii. It is used successfully for the quantitative analysis of complex mixtures and the reported results are found to be more accurate.

    iii. It aids in the analysis of multiple components simultaneously.

    iv. It can be used for a wide variety of thermally stable and high molecular weight compounds.

    v. The acquired results are more reliable.

    vi. Excellent resolution ability.

    vii. Even complex compounds can be separated and then quantified.

    viii. Quantitative measurements are highly reliable.

ix. Successive use of column for different samples.

x. Accuracy of automated data processor.

xi. Applicability for a wide range of samples (volatile and nonvolatile components can be analysed).

xii. Choice for the selection of solvents and columns specific for samples is more.

xiii. Easy operation.

xiv. Simple isolation procedure for separated components.

xv. Determination of several components in a single analysis.

xvi. Completion of analyses within a short duration of time.

1.11 ION CHROMATOGRAPHY

This can be used to measure any compound that exists in an ionized form including organic species, inorganic species, anions and cations. For the removal of hardness-impacting multivalent cations such as Ca^{2+} and Mg^{2+} in water, natural ion exchangers like clays and zeolites were used in the past. However with the development of synthetic ion exchange resins the problem is successfully tackled. Ion chromatography is useful in analysing anions like chloride, bromide, nitrite, nitrate, phosphate and sulphate with a high degree of accuracy and within a limited time period. Also these ions can be quantified in a single analysis.

1.11.1 Principle

Synthetic ion exchange resins are high molecular weight polymeric materials with ionic functional groups. Based on the nature of the functional group, they exchange cations or

Table 1.5   Application of Ion exchange resins

Ion exchange resins	Examples	Applications
strongly basic	quaternary ammonium, polystyrene (optimum pH range: 0–12)	fractionation of anions, halogens, alkaloids, fatty acids
weakly basic	phenol formaldehyde and polyamine polystyrene resin (optimum pH range: 0–9)	fractionation of anions, vitamins, amino acids
weakly acidic	sulphonated polystyrene resins (optimum pH range 1–14)	fractionation of cations, biochemical separation of lanthanides, vitamins, peptides, aminoacids
strongly acidic	carboxylic polymethacrylate (optimum pH range 5–14)	fractionation of cations, inorganic compounds, lanthanides, vitamins, amino acids and peptides

anions. In this regard ion exchange resins are divided into three types a) Cation exchange resins b) Anion exchange resins c) Amphoteric electrolytes.

Cation exchange resins   They are aromatic hydrocarbon polymers with acid residues $(-SO_3H, -COOH, -SH, -OH)$. The hydrogen ion of this acid residue removes cations from the solution.

$$[R] \, H + Me^+ \rightleftharpoons [R] \, Me + M^+$$

Anion exchange resins   They are polyamine aromatic compounds with amino groups linked to acid or water molecules $(-NH_3OH, -NH_2OH, -NHOH, NHOH)$. Hydroxyl ions of this residue are exchanged with the hydroxyl ions of anions in the solution.

$$[R] \, OH + H^- \rightleftharpoons R \, [A] + OH^-$$

Amphoteric electrolytes   These sorbents can exchange both anions and cations. Four types of ion exchange resins are recognized a) strongly basic, b) weakly basic, c) strongly acidic and d) weakly acidic.

When a solution of sodium chloride is passed through a cation exchange resin, it retains the sodium ions to neutralize their negative charge. Now when the hard water is passed through the resin filled with sodium ions, multivalent cations due to their affinity to resins compete with the sodium ions and displace them so that the water is made free of Ca^{2+} and Mg^{2+}.

1.11.2 Instrument

The instrumental set-up for ion chromatography is similar to that of high-performance liquid chromatography except for the analytical column that is packed with ion exchange resin.

Column   The packed column should allow the free flow of liquid passing through. The downward movement of the mobile phase facilitates its contact with ionic exchangers in the column. The solvent coming out from the column has exchanged ions and therefore during preparation of the column there should not be any air bubbles. The height and diameter of the column should be in the ratio 10 : 1 or 100 : 1. This is denoted as the critical length beyond which the flow of liquid is hindered. Slurry is added at intervals and every time after addition, the resin is allowed to settle. The addition is continued till the resins are in equilibrium with the solvent. The solvent is allowed to pass through the column and check that the solvent flows freely through the entire column and finally settles at the bottom of the resin bed. Now the sample to be analysed is carefully transferred to the column by using micropipettes. If the difference is minimum, a suppressor column with a second ion exchange resin can be kept after the analytical column. This converts the ions in the solvent into a component of limited ionization. However the ion to be analysed will not be affected in any way. This ultimately increases the conductivity difference between the two, and based on this the ions of interest can be quantified.

Resins   Ion exchange resins have the tendency to swell when immersed in water. The process behind is hydration, which allows the free diffusion of small ions through the ion resins. To prevent this, cross linking agents are added whose function is to connect the chains of the resin and this provides strength to it. Divinyl benzene is used as a cross linking agent. Fine particle size (50–100 mesh) is usually preferred. It can successfully extract cations from biochemical compounds, transition elements, amino acids, antibiotics and organic bases.

Separation is influenced by factors like pH of the solution, nature and concentration of the ion to be separated, ability of ions to undergo hydration, chemical nature of the ion exchange used, speed of the solution through the solvent, dimensions of the column and temperature of the solution to be analysed. pH influences ion exchangers with weak acidic groups, i.e., alkaline pH favours the sorption capacity of ion exchanger. Ions adsorbed by the ion exchangers from the solution can be successfully removed by immersing them in a solution with higher ion exchanged constant, and the adsorbed ions are desorbed into the solution. Solution containing high concentration of lower ion exchange constant gives better results. For this purpose HCl ions are used. The effective separation of ions can also be accomplished by another method where the ions adsorbed on the exchange resins form a complex with a compound in the desorbing solution and the charge of the complex form is opposite to that of the ion of concern. For example ion exchanges containing cations of rare earth elements when immersed in a citrate solution form a complex ion $M(C_6H_5O_7)_2^{3-}$. An increase in the surface area of the ion exchanger and the temperature of the solution bring out effective separation. The elute can be analysed by several methods like spectrophotometric, polarographic, conductometric and radiochemical methods.

Spectrophotometric method	Spectrophotometers are used to determine the separated components
Polarographic method	The solution is subjected to a current of constant potential and the areas covered by the recorded curve reveal the concentration of solute in the solution
Conductometric method	EC value of the solution is recorded, which in turn indicates the concentration of solute in the solution
Radiochemical method	Radioactive isotopes are used as tracers for fission products, which can be recorded by Geiger-Müller counter.

1.11.3 Procedure

Ion exchange chromatography is otherwise called liquid chromatography. Here the sample to be analysed is added onto the column containing anionic or cationic exchangers. These ions are exchanged with the ions in the mobile phase. Later they are eluted to the bottom of the tube where they can be collected and determined. The sample is allowed to pass through the analytical column packed with ion exchange resin. Naturally the resins retain the ion of the sample. To displace these ions, a solvent with ions having greater affinity for resins than the ions of interest is passed through the column. The displaced ions emerge out of the column. Based on their difference in relative affinities for the resin they come out of the column at different times. Then by using the conductivity detectors, conductivity difference between the emerged out ions and those left behind in the solvent is evaluated.

1.11.4 Application

1. It is used in the removal of minerals in water, a process called as demineralization. Water, when passed through a cation exchanger followed by an anion exchanger, will get rid of cations and anions other than H^+ and OH^-. Thus we get pure water, a basic need for analytical work.

2. Similarly it is used in the removal of hardness-causing cations from the water and thus makes it soft.

3. It is used to evaluate the total cation content of the sample. Also the specific cations can be separated by this method.

4. It is used to determine trace amounts of ion concentration in a diluted solution.

5. Separation of alloys and high alloy steels is possible by this method.

6. Ion exchange chromatography is the most convenient and efficient method to separate rare earth metals and to acquire metals in pure form.

7. It is a rapid and accurate method for isolation of transuranic elements.

8. It is used in the production of radioactive metals such as uranium and plutonium and therefore widely used in the field of radiochemistry.

9. It is used to separate
   a. isotopes of several metals (boron, beryllium, calcium, cobalt and uranium)
   b. non-amphoteric metals from amphoteric metals
   c. two substances of closely related characteristics for separation of rare earth and transuranium elements
   d. ions in groups
   e. transition elements

10. It is used to concentrate and purify samples.

1.12 PAPER CHROMATOGRAPHY

Out of the various chromatographic procedures, paper chromatography is the simplest one.

1.12.1 Principle

Here, allowing the solvent to flow on the filter paper does the quantitative analysis of unknown substance and the mixture of substances present in the solvent is separated based on their differential migration on the filter paper. The two solvents used in this experiment are immiscible or partially miscible and therefore have different partition coefficients. This enables their effective separation.

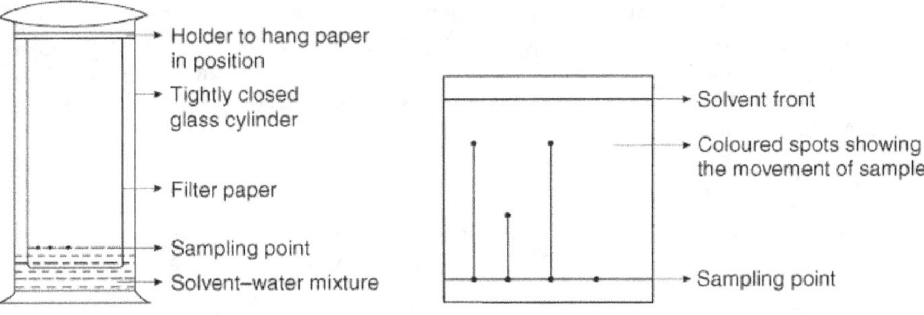

Figure 1.7a   Paper chromatography   **Figure 1.7b   Paper Chromatogram**

1.12.2 Types of Paper Chromatography

Two types of paper chromatography are adopted.

Paper partition chromatography   Here two solvents, one as mobile phase (organic phase) and another as immobile phase (aqueous phase) are used. Filter paper is used as an inert support.

Paper adsorption chromatography   Here the filter paper is impregnated with adsorbents like silica or alumina and the mobile organic phase is used to flow over the modified filter paper.

1.12.3 Apparatus

The whole set-up of paper chromatography includes:

Whatman filter paper

Paper holder

Glass chamber

The container used to run the chromatogram can be made up of glass/polyethylene/ stainless steel/porcelain. It should be noted that the material which the container is made of should not be soluble in the solvent. The inside portion of the container should be well-saturated with the solvent vapour. The paper is made to hang inside the glass chamber and the holder prevents the movement of paper from its original position. A drop of substance to be quantified is kept at the lower side of the paper. The point at which the sample is placed is marked and the developer moves in the upward direction due to capillary forces (ascending movement). The paper is placed erect in the mobile organic phase in such a way that the sample point is just above the level of organic phase. This ensures the mobile phase to move against the gravitational force. Substances migrate as a result of flow of mobile phase called developer and the process of movement is called development. As the solvent moves upwards, the rate of movement decreases during which the partition equilibrium of both the organic and aqueous phase occurs. Similarly if the mobile phase exhibits downward movement, the technique is called descending movement which is caused by the combined effect of capillary and gravitational force. When the mobile phase moves from a central portion in an outward direction, it is called disc development. A circular paper with sample applied in its central spot is used. The maximum distance reached by the solvent is marked and the point is represented as "solvent front". Resolution is the degree of component separation and Resolution Front (RF) is evaluated by comparing the rate of the distance covered by the specific component with the rate of the distance covered by the solvent front. Distance travelled by the individual constituents of the sample is determined by physical and chemical methods.

1.12.4 Physical Methods

Ultraviolet light source can be used to detect the invisible sample spots. Also, radioactive substances can be used and detection of radioactivity is carried out by Geiger counter.

1.12.5 Chemical Methods

Several chemicals called as locating agents are used to identify the sample migration in the filter paper. These chemicals with their functional groups combine with the sample constituents and exhibit colour development which is visible to naked eye.

Solid reagents	Potassium chromate
Liquid reagents	water, methyl, ethyl and n-butyl alcohol
Gas reagents	Hydrogen sulphide

Application of the locating reagents can be done either by spraying method or by dipping method. The solvent used in dipping method should not solubilize the substances in the paper. Moreover, they should be dried from the chromatogram after use. The chromatographic separation may be one-dimensional or two dimensional.

1.12.6 Sample preparation

Liquid samples can be directly applied onto the paper. However solid samples, plant and animal extracts should be dissolved in suitable solvents and then used. Care should be taken while loading the sample (i.e.) the sample size should be minimum. For better resolution use rectangular shaped Whatman filter paper.

Aqueous phase	(i)	Hydrophilic stationary phase – methanol, formamide, glycol, glycerol
	(ii)	Hydrophobic stationary phase – kerosene, aromatic and aliphatic hydrocarbons, dimethyl formamide.
Mobile phase	(i)	Isopropanol, ammonia, water
	(ii)	N-butanol, acetic acid, water
	(iii)	Water, phenol
	(iv)	Kerosene, 70% isopropanol
	(v)	Dimethyl formamide, cyclohexane
	(vi)	Formamide, benzene
	(vii)	Formamide, benzene, cyclohexane

One Dimensional Chromatography

1.12.7 Procedure

A line is drawn one or two cm above the base of the filter paper strip. The sample is loaded using a capillary tube, at the centre of the line. This clearly marks the position at which the loading is done. The filter paper strip is about 15–30 cm long and a few centimetres wide. Allow the sample spot to air-dry. Now place the filter paper erect in a container filled with organic mobile phase. Tightly close the container. Leave it undisturbed until the solvent traverses the entire length of the paper. The upward movement of the solvent carries with it the sample constituents and based on their partition coefficients, their rate of movement differs. Now carefully remove the paper and dry it in an oven or in a hot plate. Note down the position of the solvent front. The paper is now termed as paper chromatogram.

1.12.8 Calculation

$$\left.\begin{array}{l}\text{Rate of movement of individual} \\ \text{components in the filter paper}\end{array}\right\} RF = \frac{\text{Distance covered by the component}}{\text{Distance covered by the solvent}}$$

Two Dimensional Chromatography

Sample loading is done at the corner of a square sheet of filter paper. Chromatography is performed following the procedure described above. After the position of solvent front is marked, the filter paper is allowed to dry. Then the paper is rotated to 90° and the procedure is repeated again with a different solvent.

1.12.9 Applications

1. It is used for the quantitative analysis of organic and inorganic pollutants.
2. It is used to ascertain the purity of compounds separated by distillation process.
3. It is used to assess the individual amino acids and sugars in a mixture.
4. It is used to analyse the organic compounds in a mixture.

1.13 THIN LAYER CHROMATOGRAPHY (TLC)

1.13.1 Principle

Here separation of sample constituents occur on a thin layer of adsorbent and hence the name thin layer chromatography. The layer is supported on a glass plate and the whole set-up is prepared by spreading of adsorbent in aqueous slurry from over the glass plate. To facilitate uniform spreading, the adsorbent is homogenized before use. To stabilize the layer on the glass plate, it is heated in a hot-air oven for several hours. It is more preferable than other chromatographic methods as it is cheap, easy to operate and ensures high degree of resolution.

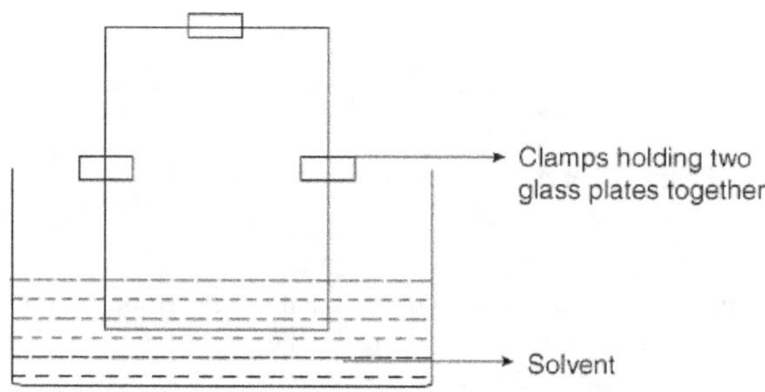

Figure 1.8   Thin layer chromatography

1.13.2 Apparatus

The apparatus of TLC includes the following components:

1. Adsorbent
2. Glass plates
3. Plate holder
4. Tank for the experiment to run

Coating of adsorbent layer over the plate is the preliminary step in the thin layer chromatographic method. In general, two layers are recognized a) Loose layers and b) Solid layers. Loose layers can be obtained either by spraying or dipping the solvent in the plate whereas solid layers are prepared by using machined applicators. Layers formed vary in size from 0.25 mm to 2.0 mm. Addition of calcium sulphate to the adsorbent helps it to bind strongly to the plate. Silica gel is the most commonly used adsorbent in TLC. Along with calcium sulphate, 2 ultraviolet indicators are added with the adsorbent. However it should be noted that the samples to be characterized should not in any way react with these individual components. The aqueous slurry of the adsorbent is prepared by grinding 25 g of silica in a mortar with 50 ml distilled water. After 10 minutes of air-drying, it is kept in a hot-air oven for two hours and then applied on the glass plates. Several solvents such as petroleum ether, carbon tetrachloride, benzene dichloromethane, chloroform and diethyl ether can be used in TLC. However a mixture of solvents is found to give better results—hexane (90 parts); diethyl ether (9 parts); acetic acid (1 part). Iodine vapour and sulphuric acid are the two detecting agents used to locate the partitioned components.

1.13.3 Sample Preparation

The sample, small in size (0.1%), is applied on the plates by using capillaries, micropipettes and microsyringes. Allow the sample to dry. The sample loading is done on one side of the plate, 2 cm above the base. The sample spot should be restricted to a limited space (0.5 cm) and its diffusion is inhibited when the sampling is performed in nitrogen-saturated atmosphere.

1.13.4 Procedure

Glass plates of the size of about 20 × 20 cm is sufficient enough to develop 10 to 20 sample spots loaded in a row. Remove the coating material on the edge of the plate before development. While loading the samples, cover the layer with transparent templates. The migrated sample spots can be visualized under fluorescent lamp. Phosphorus-incorporated adsorbent layer, when viewed under ultraviolet light source, shows the sample spots by fluorescence. Detective agents used in paper chromatography prove to give equally good results in TLC. Qualitative analysis of the sample is done by comparing the ratio of the distance covered by the sample and the distance covered by the solvent front. Quantitative analysis of the sample is done by two methods. In method I, the sample spot is removed and then diluted in a suitable solvent for further estimation. The selection of the method depends on the physical and chemical properties of that specific solute. In method II, area covered by the sample spot is measured and the logic behind is that the square root of the sample spot area is directly proportional to the logarithm of the substance concentration. The time taken for the solvent to travel along the glass plate is rapid. Cover the inner portion of the container with filter paper so as to saturate the container atmosphere with solvent vapours. As soon as the solvent travels 10–12 cm above the sample loading position, the plate is removed.

1.13.5 Application

It is used to

1. identify components in drugs and biochemical products.
2. identify and isolate plant extracts.
3. detect minute quantities of pollutants in water.

4. ascertain the purity of a natural product or drug.

5. characterize compounds like acids, alcohols, glycols, amides, alkaloids, vitamins, amino acids, antibiotics and foodstuff.

6. characterize organic compounds such as citric and tartaric acid.

7. isolate purine alkaloids that helps to evaluate their toxic properties.

8. isolate and characterize a wide array of chemical compounds.

9. ascertain the industrial processes performing purification processes such as distillation and therefore plays an important role in quality control.

Table 1.6   Differences between TLC and other chromatographic methods

TLC	Other chromatographic methods
Minimum sample requirement	Sample size is comparatively high
Very short running period	Development process requires some time
Accurate	Accurate only for specific compounds
Acidic detecting agents can be used	Acidic detecting agents destroy paper
Plates can be exposed to hot-air oven for several hours	Exposure to hot-air oven above the prescribed limit destroys the paper
Corrosive reagents can be used to coat glass plates	Corrosive agents damage the paper
Capacity is more	Capacity is limited

1.14 ADSORPTION CHROMATOGRAPHY

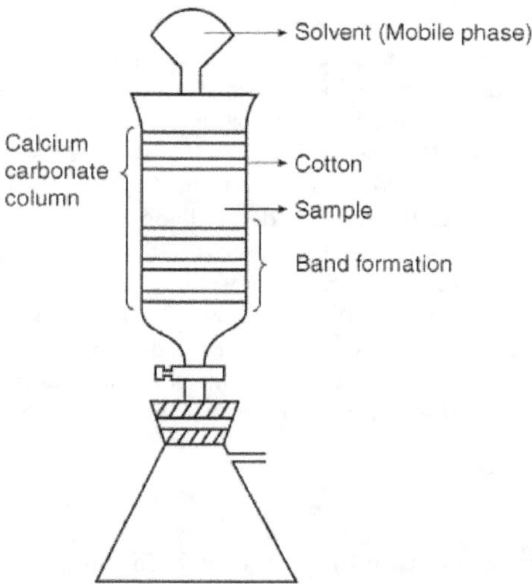

Figure 1.9   Adsorption chromatography

1.14.1 Principle

It consists of a stationary solid phase and mobile liquid or gaseous phase and the separation of individual components in a mixture is based on a) their affinity towards the stationary phase b) the surface area of the stationary phase and c) the distribution coefficient of a substance between the two phases of a system. Distribution coefficient is calculated by comparing the ratio of amount of solute per unit of stationary phase with the ratio of amount of solute per unit of mobile phase and its value is influenced by the temperature and concentration of the substance to be analysed.

1.14.2 Apparatus

The chromatographic assembly consists of a cylindrical glass tube mounted vertically by a shaft. The bottom of the tube has a porous septum and the tube is inserted into a tightly stoppered conical flask for the collection of separated components whereas the top of the tube is closed with a hollow cork in which a small funnel is positioned vertically. It is through this funnel that the solvent is allowed to pass through. The tube is lengthy and can withhold considerable amount of adsorbent. However the diameter is restricted and thus makes the tube appear like a rod with uniform cross-sectional area. The ratio of column length to its diameter should be >10. Apart from glass, stainless steel, steel teflon and polythene columns can also be used.

The column is packed with suitable adsorbents. In dry packing method, before filling, the porous septum is plugged with cotton. Then the adsorbent is packed inside the column. To ensure tight packing, the tube is tapped slightly until the column reaches the marked level. In another method, (wet packing) the porous septum is plugged with cotton. Loading of adsorbent in small quantities after the passage through the column allows them to settle down against the viscosity of the solvent introduced. However in both the methods, air bubbles should not be entertained. Adsorbent in this method (substance which act as) should posses the following characteristics.

a. It should not react with the sample mixture.

b. It should not be dissolved in either the sample solution or in solvent used for elution process.

c. The particle size of the adsorbent should be moderate (100–200 mesh range) and neither the big-sized nor the fine-sized particles should be used.

d. It should have low viscosity and proper elution strength.

e. It is better to use colourless adsorbent.

f. Surface energy and surface area of adsorbent should be more.

g. Based on the force by which they attract the ions the absorbents are classified into 3 classes and according to the nature of the substance to be analysed the absorbents are used.

Class I	Adsorbents with strong force: alumina, charcoal
Class II	Adsorbents with intermediate force: calcium hydroxide, magnesium hydroxide
Class III	Adsorbent with weak force: sucrose, starch

The solvent otherwise called as eluting agent helps to separate sample components in bands and furthermore favours their elution. Viscosity, purity, stability and solubility in sample and efficiency to elute the components are the main properties taken into consideration for solvent selection. Commonly used solvents are petroleum ether and chloroform. Sample and column packing ratios ranging from 1 : 20 to 1 : 100 is highly preferable and the chromatographic operation is performed at room temperatures. For speedy elution, the temperature of the column can be increased. The column should not be too tightly packed as it may inhibit the flow of solvent.

1.14.3 Procedure

The glass rod is filled with adsorbent following wet or dry packing method. Then the sample mixture is poured over the packed column. Depending on their attraction force with the stationary phase the sample components will fix themselves/adsorb on the column surface. Now the adsorbent is flooded with the mobile solvent phase, which desorbs the sample components. Loosely held components desorb first to reach the base of the glass tube. Strongly held components occupy the position above the first layer. These components form distinct bands along the length of the tube. After band formation, the sample components are separated by two methods

1. Extrusion method
2. Liquid chromatography method

1.14.4 Extrusion Method

The column is carefully removed from the glass tube and laid over the table. The column is positioned in the horizontal position. The distinct layers of bands are separated by knife. The separated bands are then extracted with solvent and the extracted substances are quantified by suitable method.

1.14.5 Liquid Chromatography

The column is continuously washed with solvent which brings down the components to the lower end of the column. Since the solutes are washed out at different speeds they can be collected separately in a conical flask fitted at its bottom. For the collection of the next component, change the container. Accurate results in column chromatography depend on selection of adsorbent, migration, speed of the solvent in the column, particle size of adsorbent, reactive nature of solvent and column dimension.

1.14.6 Applications

This method is widely used in the separation of

1. organic compounds (urinary 17 ketosteroids, 17 ketosteroids, gluconoids, plasma cortisol, etc.)
2. related compounds/stereoisomers in a mixture
3. nonpolar/fairly polar organic compounds
4. polycyclic aromatic compounds, phenol and amines

1.15 PARTITION CHROMATOGRAPHY

1.15.1 Principle

In this method, separation of individual components in a mixture is based on their differential adsorbent speed over the silica gel layer.

Differences between adsorption and partition chromatography	
Adsorption chromatography	**Partition Chromatography**
Easy to operate	Attention is required during operation
It can be used to separate large-sized samples	It is used to separate trace amounts
It is used to separate closely related compounds	Separation of homologous series
Used for nonpolar/less polar molecules	Used for polar molecules
Low resolving power	High resolving power

1.15.2 Apparatus

The apparatus used in this method is similar to that of the adsorption chromatography. However, the adsorbent layer is replaced by a cellulose powder or silica gel complexed with water molecule (stationary liquid phase). The mobile phase used may be gas or liquid. Preparation of gel for packing involves the following steps.

a. Washing with acid
b. Washing with water
c. Drying at 110°C in a hot-air oven
d. Addition of 50% water/alcohol

   The powder thus obtained is dry or it can be made into an aqueous slurry.

e. Saturation with methyl alcohol when combined with petroleum ether. Now it is ready to be filled in the tube.

1.15.3 Procedure

A glass tube with a porous plate at its base is stuffed with homogenized silica gel. After setting, a small disc that covers the entire top surface layer is placed on it. This prevents the direct contact of mobile phase with the column so that the packed column is not disturbed by the irrigation of solvent. The acid mixture to be analysed is mixed with petroleum ether and then added to the column. Individual acids based on their adsorbent ability to the column, exhibit different distribution ratios, i.e., acids with smaller ratios move faster than acids with bigger ratios. They are then separated into distinct band layers. Identification of coloured compounds is easier than colourless compounds. However, the colourless compounds can be detected by exposing under ultraviolet lamp. If the column is packed in a glass tube the layers should be extruded out and then laid horizontal under the lamp. Radio isotopic methods can also be used to detect the colourless compounds wherein the radioactivity is detected by the Geiger-Müller counter. On the other hand, incorporation of an indicator like bromocresol green with silica gel will give colour to the separated components. Nature of components such as the change in viscosity, density and refractive index can also be used as tools in detecting fraction of chromatography.

1.15.4 Applications

Partition chromatography finds wide application in

1. the separation of closely related compounds.
2. isolation of nonpolar/fairly polar organic molecules.
3. isolation of polycyclic aromatic compounds, phenols, amines.
4. the purification process of organic compounds. Here the impurities are successfully adsorbed onto the stationary phase thus leaving behind the pure form.
5. determining the identity of two sample constituents in a mixture.
6. separating urinary 17-ketosteroids and plasma cortisol and therefore is greatly useful in medical laboratories.

1.16 GEL CHROMATOGRAPHY

1.16.1 Principle

Gel chromatography is called by several names such as gel permeation chromatography, exclusion chromatography and molecular sieve chromatography. Here separation is effected by the molecular size of the solute particles in the solution.

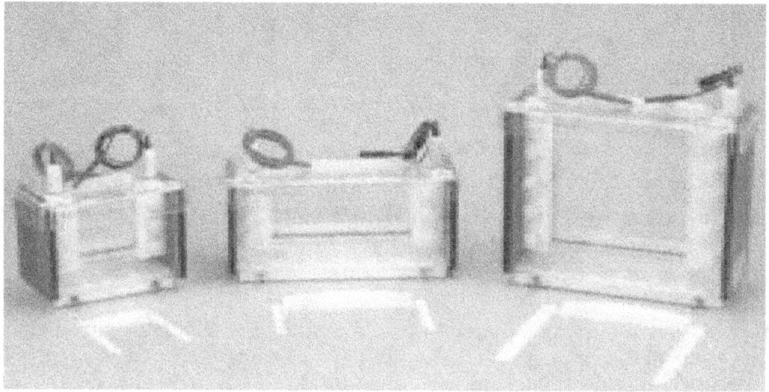

Figure 1.10   Gel chromatography

1.16.2 Apparatus

Natural and synthetic zeolites are used as molecular sieve. These substances are characterized by permanent cavities and channels and it is these structural formations that make them act as molecular sieves. Also the presence of M, Al, Si and O in the zeolite accompanied by their surface area favours the adsorption of solute. Efficiency of molecular sieves depends on the size of the cavities. Three-dimensional network of cross-linked polymer chains are used as solid support. Epichlorhydrin cross-linked with dextrin is a commercially used polymer. Slight modification in the amount of epichlorohydrin can bring out resins of different pore sizes. Polyacrylamide cross-linked with methylene bis acrylamide is widely used in gel chromatography. When they contact the solvent, these gels called as sephadex swell and this ultimately increases its pore size. Molecules larger than the pore space do not enter into the gel and therefore exhibit a free flow movement throughout the column whereas the small-sized particles are held onto the surface of the gel. Therefore solute particles based on their molecular weight can be separated. Usage of sephadex with different size range facilitates the separation of solute with specific molecule size. Stationary phase solute is applied over the entire gel phase and the solvent passing through the column is the mobile phase.

Gel is prepared by either dry method or wet method. In the dry method the silica gel is allowed to swell by mixing a known quantity of dry silica powder with solvent. Leave it undisturbed until the silica gel is in equilibrium with the solvent. In wet method, swelling of gel is done by boiling them at 100°C in a water bath. After cooling, it is packed in the column. Before packing the silica gel into the column, the column is filled with the solvent which allows the gel to precipitate in a slow manner and continuous addition of gel tightly packs the column. However uniform packing of column is done by the following method. Similar to the dry method, the silica gel to be packed is mixed with the solvent and allowed to swell. Heating the silica gel–solvent mixture favours the swelling process. After swelling, it is cooled to reduce the temperature of the gel. Since the gel may contain air bubbles, they cannot be directly used for packing. The problem is solved by removing the supernatant liquid almost to half the volume of settled gel. Now the gel suspension or slurry formation can be accomplished by the addition of solvent to the gel. Continuous stirring prevents the ingress of air. This slurry is used to pack the addition of column and while packing, the slurry is loaded in a single step. Addition of the solvent to the column should be slow and steady so as to maintain the gel layer.

If the gel is hard (agarose gels) prior to heating, it should be mixed with suitable buffer solution. For efficient separation, the column should be sufficiently long (100 mm). Diameter of column for better resolution varies from 10–12 mm (small samples) to 20–30 mm (large samples).

1.16.3 Procedure

The sample is loaded at the top of the gel bed using plugger. Sufficient time is given to allow the sample to percolate through the bed. Avoid direct contact of the mobile phase with the column so that the gel bed will not be disturbed. After some time, the solute particles based on their molecular weight are separated into distinct band layers. Detection is done by exposing them under ultraviolet lamp. If the gel bed is packed in a glass tube, extrude the layers and then lay it horizontal under lamp. Radioisotopic methods can also be used for detection.

1.16.4 Applications

1. It is used to separate organic (sugars, polypeptides, proteins) and inorganic materials (polyethylene, polystyrenes).
2. It is used to determine the molecular weight of molecules.
3. It is used to characterize the complex nature of highly polymerized molecules.

OTHER INSTRUMENTAL METHODS

1.17 MASS SPECTROMETRY

The instrument is applicable in tracer experiments with stable isotopes (15N) and for evaluating the age of water in natural resource studies. It is possible by determining the isotope ratios for water.

1.17.1 Principle

By using suitable devices, the substance to be analysed is vaporized and the gaseous components when bombarded against the fast-moving electrons are converted to positive ions. Now the gas stream with positive ions is subjected to an electric field which ultimately accelerates the positive ions. They are further separated according to their charge-to-mass ratio and particles of different masses are identified by detectors and are then measured quantitatively and qualitatively.

1.17.2 Instrument

The components of the instrument include

1.　Sample loading system
2.　Ion source
3.　Ion separator
4.　Ion collector
5.　Measuring system

The sample loading system is operated under vacuum. This prevents the loss of ions from the sample. An ion source held in line with the sample loading system ionizes the sample and accelerates towards the ion separator. Selection of sample loading system depends on the nature of the sample such as its melting and boiling points. According to the physical state of the sample, the ionization method differs.

Gaseous samples　　Initially gaseous samples are loaded in a reservoir and then transferred to an evacuated glass or metal bulb chamber which is connected to the ion source through a pinhole or molecular leak. High pressure in the reservoir favours the passive transfer of sample to the vacuum chamber. Gas inlet systems are maintained either at high or low temperatures. This expands the gaseous volume.

Liquid samples　　Liquids are heated at a high temperature and then evaporated. The vaporized form of the sample is allowed to pass through an evacuated reservoir. Since room temperature in maintained in the reservoir, the vapour expands. In the case of volatile liquids, it can be handled similar to gaseous samples.

Solids　　A stainless steel probe with a cup at its tip is used for loading samples. The sample loaded in the cup is directly introduced into a spark source where the sample is ionized.

In the ion separators, the ions are separated as series of beams and the ion separation is based on their difference in charge-to-mass ratio. Separated beams are then collected and measured by the ion collector and measuring system. Several devices such as pen recorder, tape recorder and oscilloscope are used. The instrument possessing the photographic recorder is called mass spectrograph.

1.17.3 Sample preparation

Low concentration of sample gives reliable results.

1.17.4 Procedure

Mass spectrometry in conjunction with gas chromatography helps to identify different compounds that are eluted from the column at the same time. This is made possible by bombarding the organic molecules by fast-moving electrons. This will break the organic molecules into a number of charged fragments. Actually the bombardment is based on the fact that each organic molecule bombards in a specific pattern so that their fragmentation in unique. Since the charge-to-mass ratio is the quantity of each fragment, it is measured for each fragment and then compared with known materials so that the organic composition can be identified and quantified.

1.17.5 Applications

1. It is used to analyse synthetic gas mixtures, individual gas constants and noble gases.
2. It is used to evaluate the chemical composition of organic compounds.
3. It is used to evaluate the age of rocks by isotopic measurements.
4. It is used to determine isotopic creation of a molecule.
5. It is used to characterize high molecular weight polymeric materials.
6. It is used to characterize and quantify the individual components of petroleum products such as gasoline, fuel oil, paraffin, olefin, alcohols and ketones.
7. It is used to evaluate the reaction between ions and unionized gases.
8. It is used to measure the ionization potential of the molecule.
9. It is used to determine the molecular structure.
10. It is used to identify individual components in industrial effluents.

EXERCISES

1. UV spectrophotometry is preferable for _____________ compounds.
2. Electrodes are used to measure _____________.
3. Crystalline membrane electrode is used to evaluate _____________ in the sample.
4. Membrane electrodes play a key role in _____________.
5. Membrane probes are extensively used to determine _____________.
6. Emission spectroscopy is highly applicable in the quantitative and qualitative analysis of _____________ and _____________.
7. Mass spectrometry is used to measure _____________ of a molecule.
8. Ion chromatography is used to measure _____________.
9. Flame photometry is used to measure _____________.
10. What are the basic criteria for the energy source in UV spectrophotometry?
11. What is the function of monochromator in UV spectrophotometry?
12. What is the mechanism behind photoelectric devices?

13. List out the changes that take place when a molecule absorbs radiation.

14. How will you dilute the sample in UV spectrophotometry?

15. Write down the applications of UV spectrophotometer in the field of medicine.

16. Name the instrument used to identify pesticides and complex organic chemicals in the sample?

17. In what way is IR spectrophotometry advantageous over other spectrophotometric methods?

18. Name the light source acting as energy source in IR spectrophotometry?

19. What are the components of IR spectrophotometer?

20. Which monochromators are highly preferred — prism/grating?

21. How will you prevent the stretching of monochromators in an infrared spectrophotometer?

22. How will you prepare solid and liquid samples for analysis using infrared spectrophotometer?

23. What is the function of nujol in IR spectrophotometry?

24. How will you measure the species complexes using electrodes?

25. What is the principle behind liquid membrane electrode?

26. What is the major drawback associated with the application of glass electrode?

27. What happens when a metal is placed in a solution of its own ions?

28. What is the nonreactive metal present within the electrode?

29. What is a half cell?

30. Name the method that is widely accepted as a standard procedure for metal analysis.

31. In membrane probe, which is the metal used as an inert cathode?

32. What are the components involved in the instrumental set-up of polarographic analysis?

33. List out the high excitation sources involved in emission spectroscopy.

34. What is the major disadvantage associated with emission spectroscopy?

35. What is the function of primary filter in fluorimetry?

36. Differentiate phosphorescence and fluorescence.

37. What are the different sources of radiation in photoluminescent chemical systems?

38. Differentiate fluorometers from spectrofluorometers.

39. How will you prepare samples in fluorometric method?

40. How will you process metal sample in HAS method?

41. Discuss the advantages of AAS method.

42. What is the role of ion exchange resins in ion chromatography method?

43. Give examples for weakly basic and strongly basic ion exchange resins.

44. What are the two types of paper chromatography?

45. How will you prepare samples in paper chromatographic method?

46. How will you pack a column in adsorption chromatography?

47. What is an eluting agent?

48. What is the principle behind partition chromatography?

49. Explain with examples cation and anion exchange resins.
50. Which is the substance used as a solid support in gel chromatography?
51. How will you prepare gel in gel chromatographic method?
52. Explain the principle of electrophoresis.
53. What are the components of high pressure liquid chromatographic instrument?
54. Discuss the two types of HPLC.
55. How will you inject the sample in AAS method?
56. Explain the significance of HPLC method in the field of forensic toxicology.
57. Discuss the types of column packing in HPLC.
58. What is the principle of gas chromatography?
59. Give an account of different types of detectors used in the gas chromatographic method.
60. Name the gas used as a carrier in GC method?
61. Why should the samples be maintained in vaporized state in GC?
62. Write down the characteristic features of silica in gas chromatography.
63. On what basis are the detectors in gas chromatography chosen?
64. Name a few petroleum products that can be quantified in gas chromatography.
65. What are the two factors taken into consideration for quantitative analysis in gas chromatography?
66. What are the components of a fluorimeter?
67. What is the role of optical and electronic system in flame photometer?
68. Name the different metallurgical products that can be determined by flame photometer.

（2）

WATER POLLUTION

2.1 INTRODUCTION

Water is the essential constituent of any form of life. On an average, a human being consumes about 2 litres of water every day and it accounts for about 70% of the weight of human body. Not surprisingly, the unique properties of water make us feel that it is especially designed for the living organisms and no other liquid can absolutely replace it. Water moves between the various compartments of the earth, (lithosphere, atmosphere and biosphere) thus forming a perpetual global circulation. This movement is termed as water/hydrological cycle. Solar energy evaporates the vast expanse of water from oceans, rivers and lakes. The vapour goes into the atmosphere to form clouds. Wind currents drift the thick dark water-bearing clouds to the land. When these drifts are dashed by mountain ranges, the clouds move to a higher altitude where condensation takes place to make it fall as rain. A part of initial rain is converted into vapour, a part gets soaked in the soil to form various forms of underground water (shallow well, deep well, spring, tube well, artesian well, infiltration galley) and when the soil gets saturated, no more percolation takes place and now the water takes the various forms of surface water (streams, rivers, ponds, pools). It is estimated that the quantity of freshwater resource is 1500 million cubic kms out of which 84.4 million cubic kms is in available form.

The water of the earth's surface constitute the hydrosphere, about 97% of which is trapped in ice glaciers and the remaining 1% is freshwater. The unique properties of water include:

- high melting and boiling point
- high vapour pressure and specific heat capacity
- high enthalpy of melting and vaporization
- low thermal conductivity
- excellent solvent

It is these properties of water that make it a deserving candidate to play a pivotal role in climate regulation, weathering process, biochemical reactions in cell and in the nutrient cycling between biotic and abiotic components of the ecosystem.

Man requires water for a variety of purposes including irrigation, industries, livestock management, thermal power generation, fisheries, navigation and recreational activities. The agricultural sector is the biggest consumer of freshwater (76%) followed by power generation (6.2%), industries (5.7%) and domestic and livestock management (4.3%). In general, water is considered to be fit for drinking only when it possesses the following characteristics.

- It should be colourless, odourless and tasteless in nature.
- It should be free from turbidity and suspended impurities.
- It should be free from pathogenic microorganisms.
- It should possess a pH from 7 to 8.5.
- It's hardness value should be from 50–100 ppm.
- It should have an aesthetic value.
- It should be noncorrosive and should be free from hazardous substances.

The question whether rivers, the direct source of drinking water, meet all this criteria remains unanswered because of the fact that all major Indian rivers such as Ganga, Yamuna, Tapti, Narmada, Sone, Channbal, Daha, Damodar, Krishna, Cauvery, Brahmaputra and Mahi are severely polluted. The pollutants which make the river waters unfit and unsafe for drinking and bathing come from three important sources:

1. Sewage discharge into the river.
2. Highly toxic industrial effluents discharged into the river without prior treatment by industries.
3. Surface run-off from agricultural land with chemical fertilizers, pesticides, insecticides and manures.

The impulsive shift of the human community from the traditional lifestyle to modernization and urbanization involved the destruction of valuable nonrenewable natural resources and disintegration of the environment. For the past three decades, we have been living in the illusion that we are making great progresses due to industrial developments, but these have only caused irreparable damage to the environment. Any amount of effort in the form of expertise, know-how, planning, designing implementation and operation put forth to interfere with the beaming pollution increase has proved futile. Hence billions of gallons of wastes discharged from cities, housing settlements, industries and agricultural processes are poured into freshwater bodies everyday (Figure 2.1). In the northern part of our country, the situation is more pathetic as most of the rivers have been turned into sewers. The problems are aggravated due to the persistent nondegradable inorganic pollutants and hazardous heavy metals. This is due to the fact that our environmental law plays "hide and seek" with environmental quality even with massive ecological damage cost. The situation may get worse in future as the global population continues to shoot up thus demanding more water for domestic and industrial uses, agricultural production and hydropower generation which may result in the generation of contaminants in quantities far exceeding the ability of their management. Besides, the sophisticated lifestyle of people has increased the per capita demand and reckless over-consumption and misuse of water. This contributes significantly to the wastage and degeneration of our freshwater bodies. Everywhere more water is used than is necessary, probably because of the attitude that it is available in plenty. This leads to over-consumption and wastage and a decrease in the amount of water available per head to an alarming level. At present, the demand for underground water has gone up considerably. While water drawn up from the subsurface layers goes on increasing every year, the recharging of underground water has greatly slowed down.

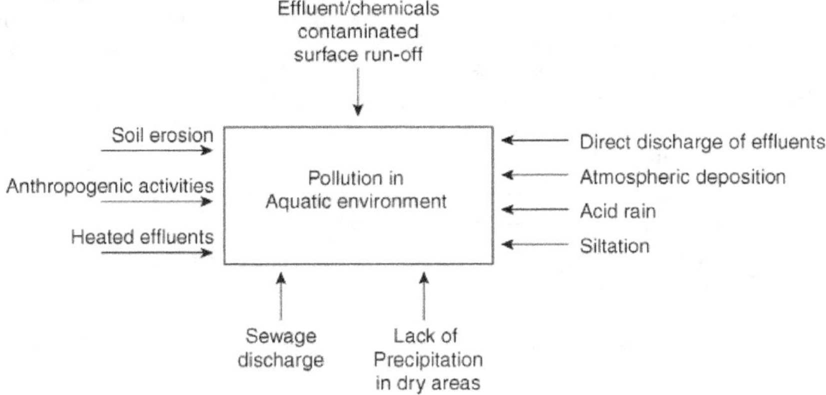

Figure 2.1   Sources of water pollution

2.2 WATER QUALITY

Water quality is a broad term and can be interpreted in different ways according to its intended use—drinking, irrigation, industries, power generation and recreation. Presumably, the variation in the water quality requirement among users does not allow it to get grouped under common standards. Therefore based on their use and quality demand, water sources are classified into five major types.

Class I	Potable water source after disinfection
Class II	Water source fit for bathing
Class III	Potable water source suitable after conventional treatment and classification
Class IV	Water for wildlife and fisheries
Class V	Water for irrigation, industrial use and for treated effluent discharge

These classes have separable standards and water quality of each class is required to meet its own standards. Thus water quality considered as "too good" in class V may be rejected as "too bad" by class II or I. Therefore the standards established for each class take into account the special constraints on water quality imposed by that use. Adoption of water quality criteria needs acquiring of knowledge from many segments of science like chemistry, bacteriology, biology and toxicology. However, addition of every new chemical compound in the industrial processes and its subsequent discharge into water necessitates the publishing of a new standard. Also when a scientific research reveals an unknown dimension of a known chemical, the standards have to be revised again. The ancient human community had recognized water as the absolute necessity of life. Since relatively little was known about the diseases and absolutely nothing was known about the chemical contaminants, visual appearance and palatability were the only two factors taken into consideration to maintain water quality. It is surprising to know that several treatment methods like boiling, filtering, sedimentation and salt application were used to improve the water quality. Later, with the increasing knowledge about microbe-related waterborne diseases and the hazardous impact of heavy metals on environment and biotic life, the standards for drinking water was published on October 21, 1914. About fifteen renowned persons in the field of hydrology were selected as members of the advisory committee to prepare the standards. The main features of the standards include the following.

- It stressed on the quality of water in terms of bacteriology and it commended procedures to enumerate the coliforms.

- Physical and chemical analysis was not specified as they could not specify parameters to be analysed for maintaining water quality.

- The report was the output of discussions by the committee members regarding the selection of a few simple examinations to determine the quality of drinking water.

2.2.1 Water Quality Standards of 1925

In the standards published in 1925, the committee had begun a defined approach towards its goal. According to it,

- Water used for drinking should be free from pollution and if found to be contaminated should be treated before use.

- Water quality assessment should strictly follow the bacteriological examination.

- Water quality assessment should involve the measurement of heavy metals, lead, copper or zinc and if present in quantities above the specified standard, it should be rejected.

- Water should be clear and pleasant to taste. Also it should be free from colour and odour.

2.2.2 Revised Water Quality Standard of 1925

The revised edition of the standard published in the year 1925 was divided into two volumes, one specifying standards for contaminants and the other suggesting the procedure for its analysis. A gist of the contents is given below.

- Regarding the bacteriological examination of water, the report specified the minimum number of samples to be analysed in proportion to the population and the time duration of sampling. Also it strove to frame a supervising committee which had every right to inspect the microbiology lab at the time of examination.

- It established maximum acceptable limits for heavy metals like lead, fluoride, arsenic and selenium, and chemicals causing hazardous effects like salts of barium and hexavalent chromium should not be allowed into the water. Periodical examination for the presence of these chemicals is necessary.

- It also established maximum acceptable limits for copper, iron, manganese, magnesium, zinc, chloride, sulphate, phenolic compounds, total solids and alkalinity.

2.2.3 Water Quality Standards of 1942

The salient features of the 1942 standards include:

* Maximum permissible concentration was mentioned for more chemicals.
* With the knowledge about the deleterious physiological effects of heavy metals, their maximum permissible concentration was minimized e.g. lead.

The 1946 standards was published with slight changes in the 1942 standards.

2.2.4 Water Quality Standards of 1962

The main objective of 1962 standards was to project the health and promote the well-being of individuals and the community as well. The important features of this standards are as follows:

* Distribution of potable water to be done under the supervision of qualified personnel.
* Introduction of alternative bacteriological method to enumerate microorganisms (membrane filter technique).
* Addition of new compounds like alkyl benzene sulphonate (ABS), barium, cadmium, carbon-chloroform extract, cyanide, nitrate and silver and their permissible limits.
* Acceptable limit for fluoride in fluoride-treated water.
* Inclusion of radioactivity.
* Maximum permissible limits for pesticides like aldrin (0.017 mgl^{-1}); chlordane (0.003 mgl^{-1}); DDT (0.042 mgl^{-1}); Dieldrin (0.017 mgl^{-1}); Endrin (0.001 mgl^{-1}); Heptachlor (0.018 mgl^{-1}) heptachlor epoxide (0.018 mgl^{-1}) Lindane (0.056 mgl^{-1}); and methoxychlor (0.035 mgl^{-1}).

These standards were followed by the one drafted by the World Health Organization and individual countries. Even though it gained acceptance, it was feared that a maximum permissible value would allow the water chemist to take full advantage to degrade the water quality to that level. However it was understood that the maximum permissible limit allows minimum degradation and definitely can be tolerated by the aquatic environment without any quality change. The quality standards for water and effluents are listed in Table 2.1.

Table 2.1   Water and effluent quality standards

Parameter	Surface water	Public sewer	Drinking water	
			Desirable limit	MAL*
Turbidity	-	-	5.0	25.0
Colour	-	-	5.0	50
PH	5.5–9.0	5.5–9.0	7–8.5	6.5–9.2
Total solids	-	-	500	1500
Total hardness	1000	1000	100	500
Chlorides	1000	1000	200	600
Sulphates	2.0	15	200	400

Table 2.1 Contd...

Parameter	Surface water	Public sewer	Drinking water	
			Desirable limit	MAL*
Fluorides	-	-	1.0	1.5
Nitrates	-	-	45	45
Calcium	-	-	75	200
Magnesium	-	-	30	150
Iron	-	-	0.1	1.0
Manganese	-	-	0.05	0.5
Copper	3.0	3.0	0.05	1.0
Zinc	5.0	15	5.0	15.0
Phenol	1.0	15	0.001	0.002
Mineral oil	10	20	0.01	0.30
Heavy metals				
Total chromium	2.0	2.0	-	0.01
Cyanide			-	0.05
Lead	0.1	1.0	-	0.10
Selenium	0.05	0.05	-	0.01
Cadmium	2.0	1.0	-	0.01
Mercury	0.01	0.01	-	0.001
PCB			-	0.2
TSS	100	600	-	-
TDS	2100	2100	-	-
Temperature	<40	<45	-	-
Total residual chlorine	1.0	-	-	-
AN	50	50	-	-
TKN	100	-	-	-
BOD	30	350	-	-
COD	250	-	-	-
Arsenic	0.2	0.2	-	-
Nickel	3.0	3.0	-	-
Boron	2.0	2.0	-	-
Hoxavalent chromium	0.1	2.0	-	-

* Maximum acceptable limit

2.3 POLLUTION-INDUCED CHANGES IN AQUATIC SYSTEM

Animal and plant communities respond to the pollutants in different ways such as elimination–invasion, tolerant–sensitive, and increase–decrease. Their individual response leads to adverse effects. Altogether they act as indicators of any change in the aquatic system. But how they are considered on a chemical surveillance programme still remains a question. In some cases, these organisms are used in the biomonitoring system using very few organisms as test species. This is solely because of the lack of knowledge about the response of different organisms to a wide range of pollutants. For instance, macro-invertebrates show noticeable changes even to a mild load of pollution but not many of them have been studied in this regard so far. Most of the industries situated near the bank of the rivers are wise enough to discharge their effluents during night time. When they are analysed by the quality assessment crew during daytime, the pollutant concentration becomes too low to be detected. Under these conditions, when the tissues of an organism is tested, concentration of the same pollutant will be found to be accumulated, far exceeding the acceptable limits. Therefore it is very clear that the pollution assessment of aquatic system should be viewed from the biological side too. In the freshwater ecosystem, the following pollutants and pollution-causing processes are encountered.

- a.   Organic pollutants
- b.   Eutrophication
- c.   Acidification
- d.   Thermal pollutants
- e.   Oil pollution
- f.   Industrial wastes

2.3.1 Organic Pollution

Research work on the impact of oxygen-demanding wastes on the freshwater ecosystem has confirmed their significant effect on indigenous flora and fauna. The main sources of oxygen-demanding wastes are sewage discharge, agricultural run-off, spillage, industrial wastes rich in inorganic content such as that of distilleries, textile industry, pulp and paper industry, breweries, slaughterhouse waste, and food processing plants. Basically nature's inbuilt mechanism is such, that any environment can repair its damage to some extent. Rivers too are not an exception and with the aid of diversified microbial species, they can decompose the organic matter and bring the environment back to normalcy, a process termed as self-purification. The only requirement is oxygen while the food for microbes is already available in the form of pollutant. As soon as the pollutants are discharged, they enhance the proliferation of microorganisms. This is a positive sign as the microbial members will begin to digest the pollutants. This continues to happen until the microbial number is directly proportional to the amount of available oxygen and only when the former outweighs the latter does the real problem arise.

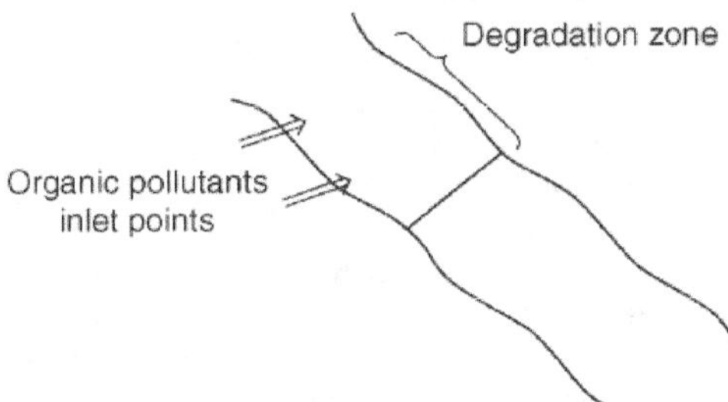

Figure 2.2   Riverine area showing organic inlet points

2.3.2 Eutrophication

Eutrophication is the ageing process of any water body. This is more common in lakes where excessive addition of nutrients to the water in due course of time creates conditions unsuitable for aquatic life. When this process happens naturally it is termed as natural eutrophication. However it takes a very long time period for the process to occur and obviously it is the initial step in the transformation of an aquatic ecosystem to a terrestrial one, which later undergoes a series of succession and finally develops into a forest, the climax community. The real problem arises when anthropogenic activities induce nutrient addition, speeding up the ageing of water bodies—the process rightly termed as induced or artificial eutrophication.

Nitrogen and phosphorus are the two major nutrients implicated in eutrophication and their important sources are phosphate-rich detergents, agricultural run-off, leaching of artificial fertilizers, washing of manure from biocide manufacturing plants, topsoil erosion and its leaching due to deforestation and nitrogen content in rain. It is estimated that about 50% of nitrogen and 90% of phosphorus applied to the crops find their way to the nearby water bodies. Of the two nutrients, phosphorus is the growth-limiting factor for plants and animals and naturally when added in excess, it stimulates the proliferation of aquatic algae. Even though the process of eutrophication is common in both static and running water bodies, the latter gets replenished downstream due to the availability of oxygen which is the critical factor that supports aquatic life. Steps involved in the eutrophication process are

1. Stimulation of algal growth
2. Decrease in species diversity
3. Change in species dominance
4. Aerobic microbial proliferation
5. Depletion of O_2
6. Facultative microbial proliferation
7. Anaerobic microbial proliferation followed by its dominance

Stimulation of algal growth   In the presence of excess nitrogen and phosphorus, aquatic algae grow fast. Diatoms are the initiators. Soon after this, green algae (Chlorophyta) and blue-green algae (Cyanophyta or Cyanobacteria) take over. They proliferate in an exuberant manner and in a short period of time, form a dense mat preventing the diffusion of atmospheric oxygen. Submerged vegetation and aquatic animals are the prime victims as they crave for oxygen. Floating macrophytes and rooted plants pose a different problem. Growth of epiphytic algae on their leaf surface reduces their light uptake and inhibits their growth and reproduction.

Decrease in species diversity and change in species dominance   Sensitive species are killed. Zooplankton are the first among them. Even though oxygen scarcity kills them, a maximum of their population are lost by predation. This is because submerged plants act as excellent shelter for them and without these plants they are quite exposed to the predators. On the other hand, their reduction rate still enhances the phytoplankton population which again intensifies the oxygen demand problem. Invertebrate communities such as stoneflies and mayflies are completely lost. Fish, blessed creatures of aquatic ecosystem, can easily avoid the pollution incidents. However, in static waters with continuous pollutant discharge, they too are victimized. Loss or dominance of specific fish species depends on two important factors. (a) their tolerance to the toxic/organic contaminants (b) the availability of prey. It is reported that the fish, 3-spined stickleback (*Gastrosteus aculeatus*), is found to be more tolerant followed by salmonids and white fish. These fishes are replaced by cyprinids. Further, the migration of sensitive species such as salmon and sea trout is prevented.

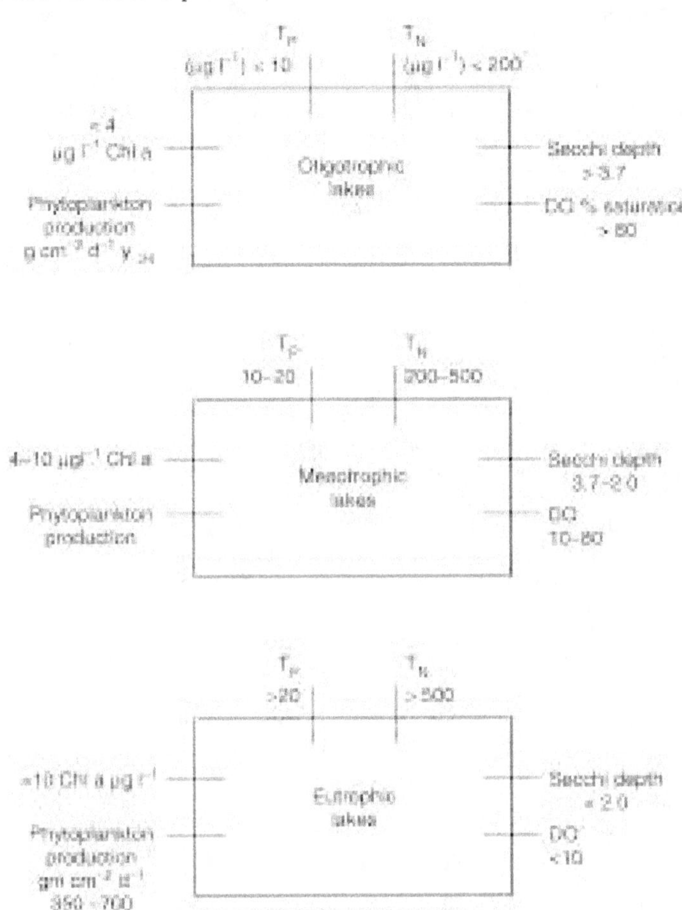

Figure 2.3   Nutrient productivity in different lakes

Depletion of O_2 O_2 depletion is the result of direct and indirect effects. As described above, inhibition of photosynthesis by algal bloom is the direct effect. Death induced by oxygen scarcity and decomposition of plants and animals is the indirect effect. Altogether a slow transformation from aerobic to anaerobic condition occurs.

Transition of microbial communities   Based on the availability of oxygen, the microbial community takes it lead. Under normal conditions, aerobic microorganisms predominate. The facultative microbes soon replace them. Even at this stage, anaerobic microorganisms begin to colonize and finally in the complete absence of oxygen, they are the only biological components remaining in the aquatic system. In running waters, deterioration of total aquatic environment does not occur. Instead, all the processes mentioned above will be restricted to the zone where the pollutant is discharged. However downstream, biological communities diversify. But still, certain species sensitive even to the mildest organic pollution may not recolonize.

2.3.3 Acidification

Catchment areas with igneous and granite rocks and uplands afforested with coniferous plantations during floods get eroded, reach the water body and lower the pH. However the susceptibility of any water body to these inflows depends on the buffering capacity of underlying rocks and soil type. For example chalk or limestone release the bicarbonate anions (HCO_3^-) in abundance which subsequently can neutralize the incoming hydrogen (H^+) ions. Rivers embedded with rocks of this type can tolerate acid pollutants and maintain their pH.

$$H^+ + HCO_3 \rightarrow H_2O + CO_2$$

At times, when the acidic inflow goes beyond the limit, the pH drops below 5.0 and naturally when the inflow is controlled or minimized, the original pH is recovered. But nowadays, with increased pace of industrialization and urbanization, the discharge of acid pollutants has geared up from a variety of sources such as burning of fossil fuels, exhaust from vehicles and power stations. Oxides of sulphur and nitrogen from these sources reach the atmosphere under moist conditions and become acidic, the main acidifying species being sulphurous acid, sulphuric acid and nitric and. They return to earth as wet deposition or acid rain. Falling of acid rain directly into water and washing of the acidified soil into the water makes it acidic. Apart from this, the direct discharge of effluents from acid mines contaminate the aquatic system. Thus acid rain and acid mine drainage are the two main sources for the acidification of waters. Also, acidification occurs under natural condition.

To evaluate the impact of acidification, a lake in Canada was artificially acidified in the year 1975. Over an eight-year period, the pH dropped from 6.8 to 5.0. With the gradual decrease in pH, there was observed a slow but steady change in species diversity and species abundance.

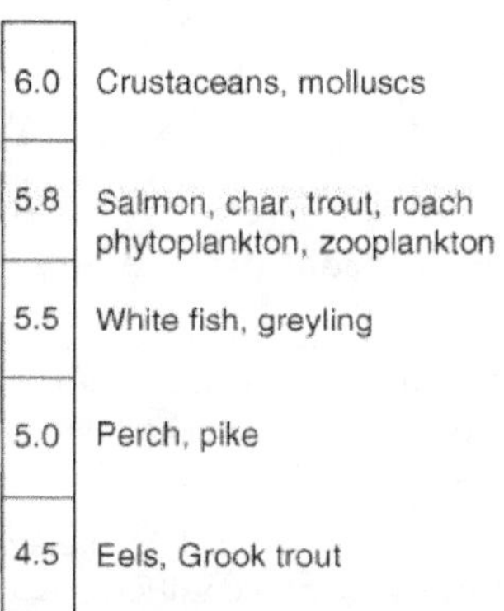

Figure 2.4 Lethal pH to aquatic organisms

Interestingly, primary producers and decomposers were found to remain unaffected even at low pH thus showing a regular flow of nutrients within the system. Several research studies conducted to analyse the acid-induced biotic community change in the aquatic environment exhibited a general pattern and the observed change associated with decrease in pH is summarized below.

- Reduction in number of most sensitive species followed by their elimination.
- Dominance by acid-tolerant species.
- Behavioural changes in tolerant species (for example cannibalism is exhibited by trout).
- Resorption of calcium from the exoskeleton of molluscs.
- Loss of reproductive ability in fishes.
- Mortality of fishes.

Fishes, the active members of the aquatic system, are the primary victims in acidified waters. Considerable work has been done to trace out the mechanism behind fish mortality. According to these studies, due to the osmoregulatory process, fish maintain high levels of sodium and chloride ions in their blood plasma. To replace these ions that have been lost in the urine, they are actively transported from the water to the cells. In the presence of calcium ions in water, concentration-gradient-based outward movement of sodium and chloride ions and inward movement of hydrogen ions are prevented. But under acidified conditions, these ions are excessively lost thus allowing more water to enter into the cells. To prevent the bursting of the swelled cell, potassium ions may be lost and finally they die. Added effect is the aluminium toxicity i.e., at low pH, sediment-bound aluminium gets mobilized and mixes with water. The presence of aluminium in water is a serious threat as it interrupts the calcium regulation of gill permeability which in turn leads to the egress of sodium, plugs the gills with mucus and blocks active exchange of gases and resorbs the calcium from the skeleton of young fish thus preventing its growth and development. If this is the case for a considerable period of time the fish population will very soon be driven to extinction. Lowering of sensitive species will duly affect other organisms at various trophic levels, as they are closely interconnected by a food web. Birds that depend on (aquatic) fish will be starved to death.

2.3.4 Thermal Pollution

Discharge of heated water from the thermal power plants is the major source of thermal pollution. Except for a very high temperature they do not have noticeable chemical contaminants. However the impact they cause on the ecology of the receiving stream is beyond imagination. Some of the harmful effects of thermal pollutants are

- dramatic drop in dissolved oxygen content
- change in the physicochemical properties of water
- enhancement of toxicity of hazardous substances
- interference with biological activities such as physiology, metabolism and biochemical processes
- inhibition of reproductive ability of aquatic organisms
- victimization of aquatic organisms to diseases
- change in the basal metabolic rate (BMR)
- proliferation of alien species
- exuberant growth of algal population

- shortened lifespan
- biochemical oxygen demand
- mortality of sensitive species
- changes in species distribution
- depletion of bottom dwellers
- behavioural changes in fishes

Another deleterious effect caused by thermal effluents is that certain pathogens that cannot sustain under normal temperature will thrive best when temperature is enhanced and attack susceptible organisms.

2.3.5 Oil Pollution

Crude oil, being the complex mixture of thousands of different organic molecules mainly hydrocarbons (aromatics, alkenes and cyclohexanes), when refined by fractional distillation process gives rise to valuable products such as petrol, diesel oil and tar. They reach the aquatic environment through road wash off, discharge of engine oil, from boats and irrigation pumps and accidental spillage from storage tanks. Unfortunately oil pollution in freshwater environment is given least attention even though its effect on the freshwater aquatic environment is severe. It is rather difficult to define the toxicity of crude oil to aquatic organisms. This may be due to the fact that monohydric aromatic compounds are toxic, and of the several components in oil, PCB is found to bioaccumulate in the target tissues and cause deleterious effects. The presence of surface active agents in oil still more intensify their effect on animals i.e., these enhance the permeability of toxic compounds into the biological membranes. Addition of emulsifiers and dispersants to clean up the spillage are themselves more toxic. Apart from their direct effect on organisms, they indirectly change the community structure and the organisms of the aquatic ecosystem. Oil slime on the feathers of birds reduces their buoyancy and insulation capacity.

2.3.6 Industrial Wastes

Industries are the prime sources of continuous discharge of toxic chemicals into the aquatic environment. Industrial chemicals may be in the form of metals (zinc, copper, lead, mercury, nickel, chromium, lead); organic compounds (pesticides, herbicides, polychlorinated biphenyls, phenols); gases (chlorine, ammonia); anions (cyanide, sulphide, sulphite); acids and alkalis. Clearly speaking, these substances have a variety of applications and hence their usage in industrial processing cannot be prevented. The problem really occurs when they react with several components of the environment like soil, air and water. It should be noted that the real impact of these compounds in completely biotic and abiotic environments is not assessed yet. This is because, the toxic effect is different in different habitats. Similarly the tolerance ability of organisms varies with species and even within the individuals of the same species. Also the compounds categorized under a single group exhibit different toxic effects. Above all, effluents consist of a complex mixture of hazardous substances and when they react with each other, they show three different types of effects.

1. *Combined effect*   The combination of two compounds may result in an additive effect, i.e., $1 + 1 = 2$, e.g. zinc and copper; ammonia and zinc.
2. *Antagonistic effect*   The combination of two compounds results in an overall toxicity that is less than their individual effects i.e., $1 + 1 > = 2$ e.g. calcium antagonizes lead and aluminium.

3. *Synergistic effect*   The combination of two compounds results in an overall toxicity that is greater than their individual effects, i.e., 1 + 1 > = 2, e.g. mixture of nickel and chromium.

Aquatic toxicity tests aim for the evaluation of maximum acceptable concentration for each compound to be released into the aquatic environment. Also the lethal concentration for these compounds is framed out by plotting concentration against the maximum survival period so that the concentration at which 50% of the population survives is taken into consideration and depicted as the lethal concentration of that particular compound (LC 50). As far as the effect of toxic compounds is concerned, they are manifested at several levels like biochemical, physiological and behavioural levels and in the life cycle. Along with this they may be carcinogenic (form cancer), teratogenic (affect youth) or mutagenic (affect at genetic level). From this, it can be understood that those organisms that survive at the lethal concentration cannot be expected to have nil effect. Unfortunately, sub-acute lethal tests and chronic toxicity tests that would give a clear picture about the toxic impact are not performed on a routine basis. Time and financial constraints restrict their practice. Moreover, toxicity tests carried out for a specific organism cannot be generalized and all these factors compel a project manager to keep these tests aside and go in for the estimation of heavy metal concentration alone.

Factors that facilitate organisms to develop tolerance to pollutants are (a) genetic make-up and (b) continuous exposure. It is reported that the algae living in mine-waste-contaminated streams are highly tolerant and when analysed, the adaptation has been shown to be genetically determined. Similarly acute toxicity tests conducted using isopods (*Asellus aquaticus* and *Aselies meridians*) showed a tolerance limit higher than those collected from uncontaminated sediments. Evaluation of the mechanism behind revealed that these organisms when exposed to heavy metals for a long time stimulate the synthesis of metal-binding proteins which can subsequently bind with metals and make them inactive. However the situation is different in the case of large mammals. Results of a research study reported that eels were affected more as they themselves carry a major sink of heavy metals. The flesh and tissues of these organisms had cadmium and mercury levels far exceeding the standard limit. Another characteristic feature of these heavy metals is their ability to bioaccumulate or biomagnify. Bioaccumulation process affects only the exposed organism but still it needs to be checked as there is a possibility for the sensitive species to get completely eliminated. Biomagnification on the other hand is an environmental threat as it affects the organisms at several trophic levels from the producer to the top carnivore.

Table 2.2   Dominant species prevailing in different zones of the freshwater environment

Degradation zone	Sewage fungus, Protozoa, Tubicid worms (*Tubifex tubifex, Limnodrillus hoffmeisteri*)
Decomposed Zone	Midge larva (*Chironomous riparius*), isopod crustacean (*Asellus aquaticus*), Cladophora, Molluscs, Leech, Predatory alder fly fish (*Gasterosteus aculeatus*)
Recovered zone	Diversified species members

With the slow depletion of oxygen, almost all the organisms including zooplankton, invertebrates and fishes strive hard to thrive. However certain members like algae, protozoans, bacteria and specialists (those organisms that live only under specific conditions) proliferate and bring about an imbalance in the whole ecosystem. However in downstream areas with sufficient aeration, the organic content is decomposed and the water is purified.

2.4 WATER QUALITY ASSESSMENT

It is a tedious procedure involving several steps. A wise approach is needed to conduct an assessment of water quality.

2.4.1 Role of Chemical Analysis in Water Quality Assessment

Even in the 1950s, adoption of chemical methods to determine the water quality was in practice. The parameters considered were ammonia, nitrogen, nitrates, nitrites and albuminoidal nitrogen, and the different forms of nitrogen, when reduced were taken as indicators of recent pollution. However the routine analysis of these parameters could not prevent the sudden outbreak of waterborne diseases. The evaluation of the positive correlation between the microorganisms and diseases and formulation of bacteriological procedures by the scientific community did take some time. With the increased knowledge about microorganisms, bacteriological examinations were considered inevitable to determine the water quality, but since it demanded considerable time, it could not be performed on a daily basis and even today, the method remains as a last option. As an alternative, chlorine was used for the purpose of disinfection. Enumeration of microbial forms in a water body is the first step. Application of chlorine at different concentrations and measuring the residual chlorine at the outlet is the next step. Now comparison of bacterial count with the applied chlorine concentration and residual chlorine content gives the amount of chlorine to be used for disinfections. This ensures satisfactory control over the coliform species.

But the question of health hazards associated with metal pollutants remained unanswered. Only with the development of toxicology, a separate branch of biology, was the impact of these compounds on the environment and the organisms understood. Toxicological testing methods using animals at different trophic levels were evaluated. As a result, "limits of discharge" for chemical compounds was framed out and subsequently included in drinking water standards. It was in the year 1925 that acceptable limits of three heavy metals—copper, lead and zinc were included in the drinking water standards. Further in the revised edition of 1946, new chemicals were added. In the 1962 standards, the previously published acceptable limits of these chemicals were further reduced. Any amendment in the permissible levels depends on the toxicity of a particular compound. This can be understood only by extensive research in all aspects such as their action on organisms (physiological, biochemical, genetic, behavioural and pathological changes), on the environment (change in structure and function of air, water and soil ecosystem) and on other compounds (additive, synergetic, antagonistic). Therefore with every new finding, the permissible limits have been revised. For instance, tolerance level of lead in the 1925 standard was 0.1 mg/l which was later revised to 0.05 mg/l in 1962. This is due to the fact that lead is a cumulative poison and finds its way to be deposited in the target tissues, kidney, bone and brain. Thus the revision of tolerance level of metal pollutants seems logical.

The chemical analysis also aims at maintaining the aesthetic quality of water. This includes colour, turbidity, smell and taste. Usually people do not bother about the presence of chemicals in water unless it appears totally unappealing. Suitable analytical methods are adopted to measure these parameters deciding the aesthetic quality of water, and even a narrow range relaxation is not permitted as the end users can accept or reject the water only based on these qualities.

2.4.2 Role of Bacteriological Examinations in Water Quality Assessment

Only recently has the term 'coliform group' and its bacterial examination gained importance. In the past, these organisms had been referred to as *B.coli*, *Bacterium coli*, *Bacillus colon* and *Escherichia coli*. It was Escherish in the year 1882 who isolated these microorganisms from the faeces of cholera victims and was surprised to see the same species in the intestinal tract of healthy individuals. From then, the presence of these coliforms in the environment has been considered as an absolute index of faecal contamination. This is because it is estimated that about one-third to one-fifth of the weight of an average individual's faeces is nothing but coliforms. Even though the causative agent for waterborne diseases was recognized much earlier, standard procedures to enumerate them was formulated only after 10 years. Since the number of microorganisms is considerable even in mild pollution, the results acquired are satisfactory. However the drawback of this procedure is that the specific pathogens cannot be evaluated by this method.

The various steps involved in the assessment of water quality are

- Identification of water resource in terms of quality
- Selection of environmental indicators
- Search of existing source of information
- Sample collection
- Laboratory analysis
- Data reduction by statistical analysis
- Interpretation of data

2.4.3 Identification of Water Resource in Terms of Quality

Water quality criteria are the required levels of specific concentration of constituents to allure its suitability for specific uses. It is interesting to note that these standards are different for different specific uses. Water is used for a variety of purposes like domestic supply, industrial supply, stock and wild life watering, aquatic life, recreation, power and navigation. Of these, domestic water supply demands highest quality. This does not mean that water for other applications can tolerate exceedingly high level of impurities. Therefore based on the specific use of water, the objective of the analysis should be framed out followed by the analysis of required parameters. Water quality standards are established by the respective state governments indicating the acceptable concentrations of various constituents in water. Similarly effluent guidelines for specific industries have been developed by the United States Environmental Protection Agency (USEPA) and state Pollution Control Boards (PCB). Since these regulations are continuously revised, the most recent standards should be taken into consideration.

2.4.4 Selection of Environmental Indicators

Quality assessment in the first step gives a clear picture about the essential constituents in water and their acceptable concentration. Also the presence of alien substances can be identified. Analysis of these parameters and their quantitative assessment evaluates the status of water quality. Source of water pollutants may be sewage, sullage, industrial effluents, agrochemicals, surface run-off and erosion. Therefore parameters to be analysed for water quality assessment cannot be generalized. The specific environmental indicators for different pollutants are given in Table 2.2.

Table 2.3   Environmental indicators for different pollutants

Pollution Source	Environmental Indicators
Sewage	pH, temperature, colour, odour, density, viscosity, turbidity, electric conductivity, redox potential, total dissolved solids, total suspended solids, BOD, COD, TOC, hardness, chlorides, hydrogen sulphide, organic nitrogen, ammoniacal nitrogen, nitrate nitrogen, nitrite nitrogen, phosphate, orthophosphate, fluoride, total coliforms, faecal coliforms, residual chlorine, detergents.
Industrial effluents	pH, temperature, colour, odour, density, viscosity, turbidity, electric conductivity, redox potential, total dissolved solids, total suspended solids, volatile solids, BOD, COD, TOC, hardness, calcium, magnesium, acidity, alkalinity, chlorides, sulphates, total nitrogen, phosphate.
	Iron, manganese, silica, fluoride, Cd, Cu, Hg, Pb, Ar, Se, Zn, Ni, Cr, Ag, CN, sulphide, surfactants, trace organics, potential carcinogens.
Agrochemicals	Pesticides, weedicides, fungicides, algaecides, fertilizers, heavy metals, suspended particulates, trace metals, BOD, COD, TOC.
Surface run-off	Inorganic pollutants, suspended particulates, dissolved substances.

2.4.5 Search of Existing Source of Information

Since the quality of a water resource is considerably influenced by the surrounding environment, a thorough investigation of the water resource and its surrounding is very much essential for successful assessment of water quality. It involves field visit and its survey; documentation of reports; investigation of study area; referring to previous reports; close co-ordination with biological and socio-economic groups; existing water use; existing treatment facilities and quality of raw and treated water. However this is only the initial step and would lead to more detailed information.

Field visit   A complete picture of the sampling area will be acquired only if it is visited by the key members of the assessment team. If the assessment area is quite large, for instance, a running stream giving way to inputs from a number of sources, extensive field surveys can be conducted to adequately define the environment to be assessed.

Documentation of reports   The field visit should be appropriately documented with photographs. Also a report should be prepared illustrating the investigated site, time and date along with the information gathered during the visit. Specific informations from this report can be incorporated at the back of the photograph too. It is very important to mention the direction at which the photograph is taken.

Investigation of study area   The nature of the study area can be easily assessed by collecting information about meteorology, hydrology, water quality and sewage and effluent inlets. These information can be collected from the PWD department.

Referring to previous reports   Referring to previous reports is an easy but valuable method to get first-hand information about the study area, i.e., environmental impact assessment studies carried out in previous years throw light on each and every aspect to be assessed. These reports can be totally relied upon as they would have dealt with this problem taking into consideration both the biotic and abiotic components.

Coordination with scientific groups   Verification of collected information is the next step. Data confirmation approval by the scientific committee helps in proceeding further. Otherwise individual departments of educational institutes can be approached for information concerning soils, geology, hydrology and water quality. University environmental science departments may furnish the evaluation of details pertaining to the study area.

Existing water use   Running waters are usually distributed for domestic water supply and no doubt, they are the major sinks for any kind of wastes. However based on their location, the specific pollutants predominate. For instance if it is nearby an industrial area, considerable amount will be reformed back as effluents. Thus, evaluation of existing water use gives a clear picture about the parameter to be analysed.

Evaluation of existing treatment facilities   The treatment methods designed for water purification are based on the composition of waste water to be treated. In the case of water with complicated industrial effluents, advanced tertiary treatment methods should be adopted whereas minimum amount of sewage mix in water requires basic disinfection methods. Therefore it is an indirect measure to evaluate the water quality.

Comparison of raw and treated water   Difference between the pollutant levels before and after treatment reveals the efficiency of the treatment method in practice and also helps us to switch over to another option if necessary.

2.4.6 Sample Collection

The success of the water quality surveys, depends to a large extent on the sample collection and method of transportation. It is very important that trained personnel are employed for field work. Along with the technical skill, they should be aware of what they are doing. Only a properly planned and well-conceived sampling programme will give accurate results. Otherwise the observed results may be too high or too low compared to their actual level in the environment. The overall sampling programme involves

- Site selection
- Parameter-specific sampling method
- Time and frequency of sampling

- Data collection
- Sample handling prior to analysis
- Quality control

Site selection The selected sampling site should be able to define exactly the water quality change. If the area to be assessed is very large, sampling points at regular intervals are chosen. This will facilitate observation of even minute variations. Always select the sites in such a way that the sampling is made easier. Having a topographic study of that area or referring to previous reports will help in this regard. Besides location, the number of sampling points to be selected totally depends on the nature of the study area with respect to water flow, velocity, depth, breadth, sewage and effluent inlet points. Based on this, water resources can be classified into the following three types.

Narrow water resource In a fast-moving, shallow and narrow water resource, pollutants will be completely mixed and therefore sampling points can be restricted to a minimum, i.e., one sampling point for each location.

Wide water resource Multiple sampling is required for wide water resource (rivers, canals, lakes and streams). Also sampling should be taken at each cross-section along the stream. For waters with depth of 100 to 1000 ft, three sampling points at equal distance across the cross-section are suggested whereas for rivers greater than 1000 ft depth, five equally spaced sampling points are needed.

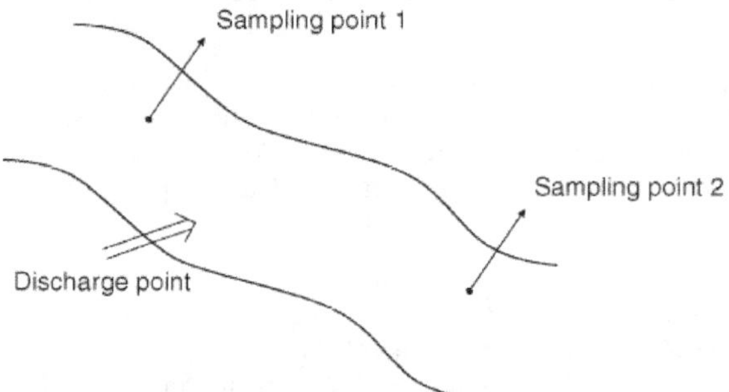

Figure 2.5 Sampling points in riverine system

Stratified water resource In stratified waters the upper layer (epilimnion) is warmer and the lower layer (hypolimnion) is cooler. The zone of transition is termed as mesolimnion on thermocline. Here two samples one for each layer should be taken. Take care to collect samples from the middle point of each layer.

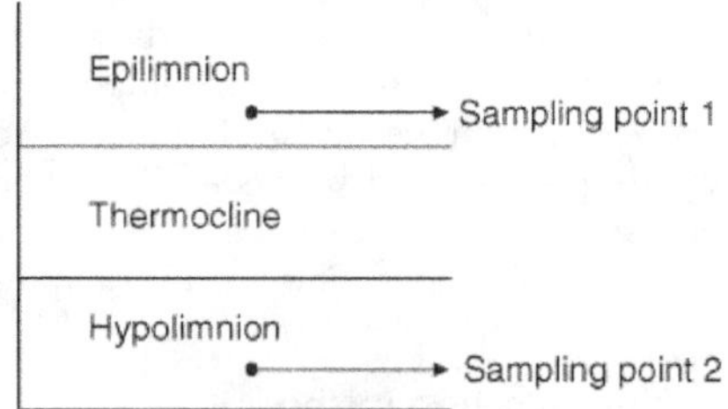

Figure 2.6 Sampling points in stratified water

Parameter-specific sampling   Water quality parameters, based on their susceptibility to change with time and environmental conditions, are of two types.

1.   Conservative materials
2.   Nonconservative materials

Conservative materials   With respect to time, conservative materials do not change in their form or concentration. Chlorides, total dissolved solids and sulphates are some of the parameters that come under this category. Therefore taking samples to analyse these parameters is quite easy. Two samples, one above and one below the discharge point is sufficient. To get a three-dimensional view in this aspect, samples can be taken at cross-sections.

Nonconservative materials   Parameters like, DO, BOD and temperature are subjected to change and hence called as nonconservative materials. Dilution of water decreases the temperature but increases the dissolved oxygen content. Similarly BOD undergoes decomposition to get reduced. Under these circumstances, it is very essential to evaluate the point at which the change takes place. Therefore five sampling points should be selected and they include:

a.   unpolluted zone (control) – Initiation point (A)
b.   Degradation zone 1
c.   Decomposition zone 2
d.   Recovery zone 3
e.   Unpolluted zone – End point (B)

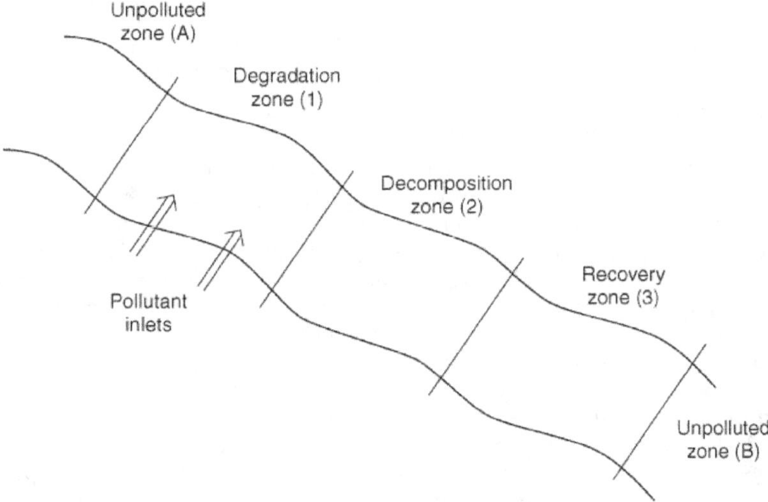

Figure 2.7a   Zone classification in rivers based on the concentration of contaminants

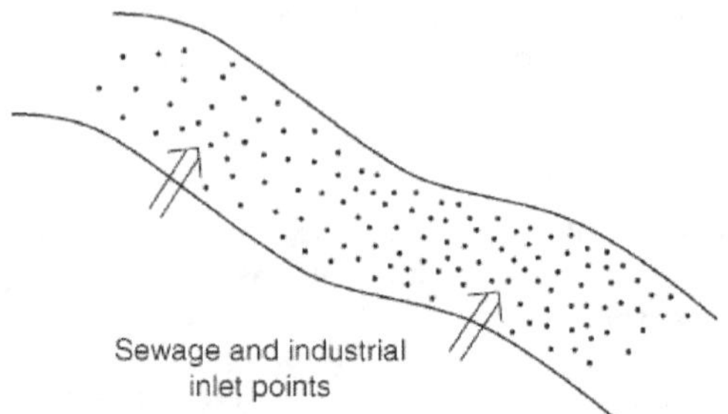

Figure 2.7b　Composite sampling in polluted waters

Unpolluted zones "A and B" can be kept as control. Samples taken at equal distance between the two controls are designated as sampling points 1, 2 and 3. The point at which the pollutant is discharged is the degradation zone. Two to three samples can be taken at cross-section. Next is the decomposition zone followed by recovery zone. Since the sample collection method differs between these parameters, a decision has to be made as to which water quality parameters to investigate. It depends on the objective of the assessment study. This will clearly reveal the indicator parameter to be analysed and based on this the sampling programme can be planned.

Objective of assessment study	Parameters to be analysed
Thermal discharge	Temperature, DO BOD
Organic waste discharge	Temperature, DO, BOD, SPC
Sanitary waste discharge	Total coliform, faecal coliform, DO, BOD
Industrial discharge	TDS, acidity, alkalinity pH, metal compounds TOC, COD, EC, redox political
Nutrients	Nitrogen phosphorus

Time and frequency of sampling　Based on the duration of assessment study, sampling is performed. In case of short-term projects, sampling can be done on any one day and therefore two important things are to be considered in this regard. One is the sampling of the study area and it should be completed on that day itself. If the distance between the sampling points is quite large, note down the time of sampling for each point so that the sampling can be continued on the next day at the same time. Secondly do not go in for sampling when the weather condition is in contrast with the prevailing season. For example, sampling on a cloudy day in summer will misinterpret the results. To have diurnal variations, one day of round-the-clock sampling can be conducted. To acquire sufficient information, increase the size of samples. For long-term assessment studies, conduct the sampling programme on a routine basis, weekly/monthly, seasonally. If the assessment study involves the impact of environmental conditions like precipitation, storm, floods and erosion, a special sampling programme can be designed besides routine sampling procedures.

Data collection　Again the nature of the study area determines the sampling type. In general two types of sampling methods are adopted.

1. Grab samples
2. Composite samples

Grab samples are taken instantly at a particular place. They represent the existing conditions of that particular place. For water resources that are less prone to any change, grab samples are sufficient. However certain water bodies are subjected to rapid variations due to continuous inflow of domestic and industrial waste waters. Therefore pollutant concentration taken in the remaining portions would lead to a remark as "too good". Similarly samples taken near the inlet points show it as "too bad". Both the remarks do not give a clear picture of the water body. In these situations, multiple grab samples termed as composite samples should be preferred, since series of samples taken over a period of time better define the existing conditions. Also to know the temporal variations, samples at cross-section can be taken.

Sample handling prior to analysis   The project manager should involve himself in the field visit to see that a carefully designed sampling programme is implemented. Field technicians should maintain a field note book with all the information properly documented. The note book should be submitted to the supervisor and the basic data regarding the associated parameters should be collected and used for data interpretation for weather conditions, hydrology and geology of the study area.

The following precautions are to be taken during sampling.

- Always use clean bottles for sampling.
- Before filling the sample, rinse the bottle at least two times. Do it leisurely so that the inside of the bottle including sides, top (lid) and bottom is completely rinsed.
- Take care to collect water samples which are devoid of either debris or sediments.
- Immediately after sampling, label the sample bottle with information like sample number, name of the place, date and time of sampling, parameters to be analysed, name of the fixatives if used and expiry date.
- While sampling, immerse the sample bottles in water in such a way that the sample overflows the bottle and then carefully close the lid.
- Do not entertain air bubbles while sampling.
- Since certain parameters need to be immediately analysed, perform them in the field itself, e.g., pH and temperature. Use automated analyser for these parameters.
- Since the water quality is assessed by several parameters, sufficient volume of sample is needed. Usually 2 litres of sample is preferred.
- Take samples in separate bottles for physical, chemical, bacteriological and microscopical analyses.
- Perform the analysis soon after sampling and if not possible, use appropriate fixatives and store it for later use.

Sampling devices   Sampling of top layer can be done manually. Polyethylene or polyvinyl chloride sampling bottles are preferred. However for the samples to be taken from a certain depth, suitable devices are needed. For streams with minimum flow, kemmerer sampler can be used. Its large size prevents its usage in swift flow streams. If the water body is too deep, samples can be pumped from preselected depths using peristaltic pumps or a vacuum pump. Core samplers are available for obtaining samples from bottom sediments. Also dredges are used for taking bottom samples. The different types of dredges and their uses are given in Table 2.4.

Table 2.4 Types of dredges

Dredge type	Nature of sediment	Water resource
Ekinan dredge	soft	shallow-water bodies
Ponar grab dredge	hard (clay, gravel)	deep waters (lakes, sea)
Peterson grab dredge	soft	deep waters (lakes, sea)

Preservation techniques Soon after sampling, analysis should be done immediately. However practical difficulties such as increasing sample number and distance between the sampling point and laboratories do not always permit immediate analysis. In this case, the samples should be refrigerated, so that biochemical or biological reactions cannot occur. Also, the samples should be fixed using preservatives and the preservative used should not interfere with the analysis. Parameters that require similar preservation techniques are grouped under the same class as given below.

Group I (residue, specific conductance, alkalinity, nitrate, nitrite)

Sample containers Polyethylene, glass

Required sample volume 100 ml

Storage procedure Refrigerate at 4°C (for nitrate and nitrite, use H_2SO_4 as preservative and adjust the pH to less than 2).

Group II (COD, chloride, colour, organic carbon, phosphate and sulphate)

Sample containers Polyethylene, glass

Required sample volume 50 ml

Storage procedure Refrigerate at 4°C (for COD and organic content use H_2SO_4 as preservative and adjust the pH to less than 2).

Group III (cyanides, ammonia, kjeldahl, phenol, coliform bacteria)

Sample containers Polyethylene, glass

Required Sample volume 500 ml

Storage procedure Refrigerate at 4°C.

Fixative Add NaOH and adjust the pH to 12 for cyanides. Add H_2SO_4 and adjust the pH to 2 for ammonia, Kjeldahl, phenol and coliform bacteria.

Group IV (DO, temperature)

Parameters to be determined on site.

Group V (BOD, pesticides)

Sample containers Polyethylene, glass – BOD glass

Storage procedure Refrigerate at 4°C.

The maximum holding time for various parameters are given below.

1. 24 hours for acidity, alkalinity, colour, cyanides, ammonia, nitrate, organic carbon phenol, phosphate, pesticide, specific conductance
2. 6 hours for BOD
3. 36 hours for coliform bacteria
4. 7 days for COD, chloride, hardness, Kjeldahl residue, sulphate, turbidity
5. 6 months for metals

Quality control   The main objective of the water quality assessment is to get accurate results. Now after careful sampling, the total pressure lies on the analytical part. 100% accuracy cannot be expected, as human error is likely to occur. However if the analytical work is performed with utmost care, defined results that go in hand with the existing status can be produced. It is the responsibility of the supervisor to get the work properly done. This can be accomplished by collecting duplicate samples and conducting analysis of the same parameter by two or three individuals. Comparing these results obviously gives clear-cut data that can be totally relied on. Also the working ability of lab technicians can be monitored. Using spiked samples is another option. Dissolution of a known amount of the pollutant to be analysed in the sample is called as spiked sample. Estimation of the specific pollutant in the sample reveals the efficiency of the analytical procedure to recover the devolved pollutant. Based on this, the same procedure can be adopted for real samples. All the analytical procedures should be performed under the direct supervision of the project chief.

2.4.7 Laboratory Analysis

Collected samples are taken to the laboratory for analysis. The analyst with sound practical and theoretical knowledge can provide results that may be interpreted in an accurate and wise manner.

2.4.8 Data Reduction by Statistical Analysis

Large data complicate the data interpretation, and the significance of the results obtained can be gained only when it is statistically analysed. The advantages of statistical analysis are as follows:

a. Data from various sources can be compared.
b. Gives an overall view about the data.
c. Different parameters can be compared from the same point of view.
d. Based on the objective, different methods can be employed.

2.4.9 Interpretation of Data

Of all the components discussed, this is more important because of the fact that it is the final result of the assessment process. Before interpretation, the analyst should have a clear knowledge of what is needed, should be able to coordinate various aspects of the assessment and then interpret in a wise manner.

2.5 MEASURES TO BE TAKEN FOR LABORATORY SAFETY AND HEALTH

- Only trained personnel should be involved in analysis. They should be well-informed about the reactive nature of the chemicals, protective measures to be taken while handling hazardous chemicals and should be trained to provide basic first aid to themselves and to their colleagues when the situation demands.

- Sterilize the solutions that may change its composition by the action of microorganisms. This is done because certain solutions in the presence of suitable microorganisms are prone to either oxidation or reduction and when these solutions are used as such, the result may mislead the analyst.

- Since pH plays a major role in analytical chemistry, check the accuracy of the pH meter frequently. Also use freshly prepared solutions of standard pH. Use specific containers for different solutions. The containers should not in any way react with the solution to change its form.

- Based on their reactivity, store the reagents in appropriate containers.

- Label the date of preparation and expiry for all reagents and following the expiry date, discard them without delay.

- Highly reactive chemicals and acids should be kept in a separate rack.

- To avoid spilling, always place them in lower racks at reachable distance.

- Always keep the lab clean.

- Maintain a separate register to document the daily activities.

- After use, put off the instruments, disconnect the electric connection, wipe it clean taking care not to leave any spilled chemicals/reagents/solution inside the laboratory.

- Periodical servicing of all the instruments is very much necessary.

- Only trained personnel should be allowed to handle the instruments.

- At the end of the day's work, check out whether the electric connection, gas flame and water connection have been turned off.

- Dispose all the unwanted things (wrappers, cotton, rubber band, etc.) properly.

- Do not discard highly reactive chemicals and acid in the water sink.

- Wear gloves, apron and face mask compulsorily as soon as you enter into the laboratory.

- Even for non-microbial analysis, clean all the glassware before use. Sterilization of all the glassware once a month is advised.

- Use separate spatula for different chemicals or else clean it thoroughly before using it for the other chemicals.

- Routine health check up for all the lab workers is necessary.

- Follow standard procedures for all analyses.

- Provide sufficient ventilation in the laboratory and do not keep any instrument that blocks the airflow.

- See to it that at least two people are inside the lab at a time.

- After handling chemicals, wash hands immediately.

- Do not use the lab refrigerators to store food, beverages or drinking water.

- Do not eat inside the lab.

- Create a working environment and do not sit and chat inside the lab.

- Do not perform mouth suction for pipetting solutions and never place the used pipettes on table tops to avoid dripping of pipetted solutions.

- Take care to choose outfits that are neither too tight nor too loose. At the end of the day's work immediately remove gloves, apron and face mask before leaving the lab.

- Try to wind up the analysis before the end of the day unless the procedure demands overnight incubation. If not possible, document the day's work (solutions prepared, steps completed) in a chart so that it can be continued next day without any confusion.

- Carcinogenic substances should be handled with extreme care. Given here are few examples that are harmful when improperly used.

  Organic acid   -   irritation

  Pesticides   -   poisonous

  Dyes   -   aerosol

- Do not build two laboratories adjacent to each other with a common ventilation and exhaust duct.

- Do not use the working platform as a writing table. This may lead to the direct contact of skin with spilled chemicals.

- Do not smell or taste chemicals at any cost.

2.6 EFFLUENT ANALYSIS

The main purpose of effluent analysis is to evaluate better-performing, advanced treatment methods with the aim of reuse, recovery of valuable products or safe disposal. On the other hand, this helps to evaluate the efficiency of the treatment system by analysing the pollutant level remaining in the treated effluent. Above all, the quality of water should be known prior to its release into the environment. To fulfil these needs, suitable analytical methods for pollutant characterization should be followed. The methods used should be able to furnish reliable information and sensitive results. Measurement of pollutants by automated techniques minimizes human error and gives a full picture of effluent quality in terms of its physical, chemical and biological characteristics. However it is the duty of the environmentalist to go in for suitable methods after thorough diagnosis of several factors such as aim of analysis, accuracy of results, interference compounds in the effluent and method of sampling and transportation.

Analysis of industrial waste water is a problematic one. This is due to the fact that the composition of effluent differs with different industries and even between the different processing units of the same industry. They possess a bad colour, offensive odour, turbidity, acid or alkaline pH, solids in different forms, hazardous substances, toxic impurities, metallic pollutants such as arsenic, barium, cadmium, chromium, copper, lead, iron, manganese, selenium, silver and zinc, and anions like F^-, NO_3^-, PO_4^{3-}, SO_4^2 and Cn^-. The compounds of the effluent from different processing units, when mixed to form a combined effluent, may react with each other and form still more complex mixtures.

For selection of analytical procedures it is essential to group the industrial effluents based on their composition. The predominating pollutant content can be taken into consideration, e.g. effluents rich in organic content and effluents rich in mineral content. However for the selection of the analytical method for the same parameter, identification of interfering compounds is very much essential. Only the method that is able to eliminate interfering compounds and give accurate results regarding pollutant concentration should be chosen.

Effluent analysis can be broadly classified into four types. They include

a.　analysis of physical pollutants
b.　analysis of organic pollutants
c.　analysis of metallic pollutants
d.　analysis of anion and dissolved gases

2.6.1 Measurement of Physical Characteristics

This involves the measurement of colour, temperature, odour, electrical conductivity, turbulence, oil and immiscible liquids, density and viscosity. The various physical characteristics and their concentration are given in Table 2.5

Table 2.5　List of physical characteristics to be evaluated for effluent quality

Characteristics	Index
Colour	Colour-producing compounds
Odour	Volatile organic matter and biological materials such as algal and actinomycetes.
Oil and immiscible liquids	Volatile and nonvolatile oily matter
Volatile and dissolved solids	Presence of solids
Electrical conductivity	Ionic contents
Turbidity	Colloidal and suspended particulate matter
Density	Dissolved materials
Viscosity	Liquid flow resistance

2.6.2 Measurement of Organic Pollutants

This involves the analysis of carbon, hydrogen, nitrogen, sulphur, phosphorus and oxygen content of the effluent. Based on the pollutant concentration, either gravimetric or spectral methods can be used. Accurate analytical methods have to be adopted, as the effluent consists of a mixture of organic compounds. Since these compounds are subjected to biochemical degradation or chemical transformation, steps should be taken to prevent further chemical reaction immediately after sample collection. For spectral analysis they should be available in concentrated form. Therefore they can be concentrated by evaporation/distillation or isolated by processes such as solvent extraction, selective precipitation, fractional crystallization and selective adsorption.

2.6.3 Measurement of Metal Pollutants

The major sources of heavy metals are alloy industries, electroplating lathes, refineries, fertilizer plants, turnery industries and electronic industries. Out of several heavy metals, a few are generally encountered in industrial effluents. They include copper, cadmium, chromium, lead, nickel and zinc and unfortunately these metals are associated with dreadful effects in exposed organisms. Few metals may be safe under controlled conditions but when influenced by certain environmental conditions they may become toxic. And therefore it becomes essential to check out the amount of heavy metals in the effluent prior to their discharge. Several methods are adopted to evaluate the metal concentration and they are

1. Gas chromatography
2. High performance liquid chromatography (HPLC)
3. Fluorimetry
4. Emission spectroscopy
5. Atomic absorption spectrophotometer (AAS)
6. Polarography

2.6.4 Measurement of Anions and Dissolved Gases

The most common anions present in the effluent are SiO_2, S^{2-}, SO_4^{2-}, SO_3^{2-}, F^- and CN^-. These anions after concentration by anion exchange chromatography are analysed by suitable methods. Fluorides, sulphide and sulphite are measured spectrophotometrically whereas chlorides are estimated by gravimetric method. The sources of the common anions in the effluent are listed in Table 2.6.

Table 2.6 Sources of common anions in the effluent

Anions	Sources
Cyanide	Electroplating industries
Sulphide	Tannery waste, septic tanks, oil refinery, tannery and viscose rayon waste
Sulphite	Pulp and paper industry wastes
Chlorides	Sewage

Among dissolved gases, O_2 is estimated by Winkler's method. In the case of nitrate interferences, sodium azide can be used. Dissolved chlorine is analysed by spectrophotometry.

2.7 INDUSTRIAL WASTES

The beginning of industrialization was very humble in the seventeenth century. However in the middle of the eighteenth century it geared up its pace as "Industrial revolution" and still continues to set its footmark in every aspect of life. At one stage, the nation's economic growth, raise in living standards, employment, increase in consuming power altogether have made industrial progress inevitable thus resulting in the development of new industries and refining and reforming old ones. It should be remembered that all these were achieved at the cost of our precious natural resources. In the beginning, it was mere pressure on the environment but later on, with a huge increase in "non-biological" industrial wastes, the incompetence of nature to assimilate these substances began to reflect in the form of pollution everywhere.

Based on the nature of the industrial wastes, pollutants have been compartmentalized in the biosphere components such as hydrosphere, lithosphere and atmosphere. There they settle, accumulate and spoil the total environment. Taking these things into consideration, industrialization cannot be put to an abrupt end as it directly reflects a country's economic progress. Therefore as an alternative, strict rules have been implemented to prevent untreated waste discharge. In this regard, the Department of Environment under the Ministry of Environment and Forests submitted a report on the guidelines for siting of Industries in August 1985. The report was concerned with the preservation of ecological balance and improvement in the living conditions of urban people. As per the report, starting up of an industry requires the permission from the state government. Moreover the entrepreneur should assure the state and the central governments that he will instal environmentally safe equipment and follow the prescribed measures for the prevention and control of pollution. However present conditions like scarcity of pure water, smoke and dust-filled air and sterile barren land reveal loopholes in pollution control laws and lack of awareness among the public. Rectifying this problem will only be a partial solution. Precautionary measures taken in favour of environmental protection before starting an industry will minimize the possible adverse effects on the environmental resources and quality of life. Accordingly, this matter can be reviewed considering three aspects

    i.   siting of industries

    ii.   area selection

    iii. waste discharge

Siting of industries   Industrialists should be aware of the severe damage that an industry can cause to the air, water, land, flora, fauna and public health. Therefore while setting up an industry, it should be remembered that the quality of life shoud be maintained and the surrounding ecosystem should be preserved. This is accomplished by keeping economic progress and environmental consideration in pace with each other. However it is ironical to note that entrepreneurs are more concerned about the industrial growth thus paying very little attention to its pollution effects. They are not permitted to start up industries in commercial and residential areas. This is due to the fact that installation of industries in far away places minimize their impact on the public. Taking these as an advantage industrialists convert majority of agricultural lands into industrial sites. Moreover, industries near the forest cover tend to extend their land area without legal permission. This results in the depletion of forest sources for the sustenance of industry. As an initial step, they should get the clearance of the central and state Air and Water Pollution Control Board. Certain industries may either produce or involve hazardous chemicals and the installation of these industries require

    i.   an approval from state director of industries,

    ii.   submission of report to the state and central government that the industry meets the effluent standards and

    iii. assurance from state pollution control board that the proposal submitted for siting up an industry will not in any way damage the environment.

Based on the nature of the project proposal, the Environment Impact Assessment (EIA) studies will be carried out. Its function is to identify and evaluate the possible impact of the industry on the environment. Generally the EIA study takes into consideration the physiochemical nature and biological composition of soil, air and water in the selected area and also in the nearby areas.

Area selection   In spite of techno-economic gain from industrial outputs, industries should not in any way affect their surroundings. Hence to be on the safer side their installation in certain areas should be restricted. These areas include religious and historical places, areas around monuments, hilly regions, the seashore, coastal areas, natural reserves, natural wetlands, parks, sanctuaries, human settlements, seismic zones, forest areas and agricultural areas.

Waste discharge   Industries discharge wastes in the form of solids, liquids and gases, and majority of these are discharged without proper treatment. Liquid wastes are called effluents and solid wastes are called sludge. These wastes contain complex chemical compounds that affect the organisms and the ecosystem where they are discharged. Therefore they should be properly treated. Discharge of wastes below the permissible limits can be achieved when an industry possesses the following features.

    i.   Regular servicing of processing equipments.

    ii.   Alternative processing methods that demand less water and chemicals.

    iii.   Adoption of advanced treatment technologies that emanate completely treated effluent.

    iv.   Recycling and reuse methods to reduce the volume of effluent production.

    v.   Developing an R & D unit within the industry which evaluates suitable treatment methods and standard of effluent coming out of the treatment plant.

    vi.   Extraction of valuable chemicals from the waste.

    vii.   Skilled labours who can overcome problems in operation.

    viii. Developing a green belt of about ½ km around the industry.

Disposal options   After treatment, effluents can be disposed into the nearby land area or aquatic bodies. In the case of aquatic environment, running water is preferable as it can repair itself by self purification process.

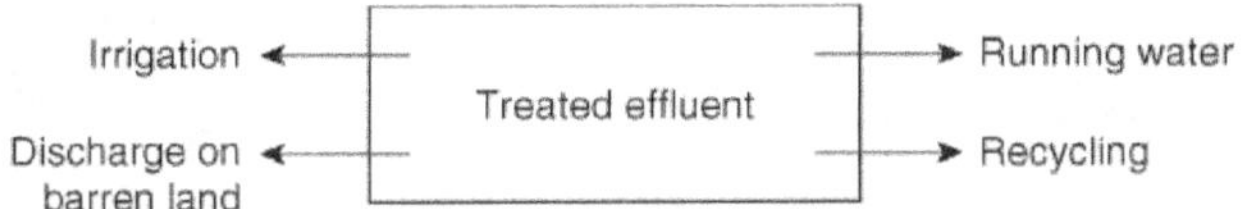

However frequent monitoring of water quality is essential. The composition of industrial waste cannot be generalized and the treatment methods differ accordingly. Therefore to get a general view, few industries are discussed in the following sections.

2.7.1 Tannery Industry

Leather industry in India is one of the largest export industries. There are 660 registered leather industries under the Factories Act. Tamilnadu leads in the export of finished leather goods accounting for 80% of the leather goods. Tannery is the basic unit of the leather industry which requires large amount of water and a number of chemicals which when discharged as effluent are known to pollute the environment. India produces about 100 millions of hides and skin which form about 12% of the output of the whole world. There are more than 2500 tanneries in our country. Leather industry is mainly concentrated in Tamilnadu, West Bengal, Maharashtra, Punjab, Karnataka, Andhra Pradesh, Bihar and Kanpur. In Tamilnadu, tanneries are located in Ranipet, Amlur, Periamet, Vaniambadi, Melpudupet, Chettithangal and Visharam.

Manufacturing process Tanning industries in India have made tremendous strides in developing newer and quicker processing techniques. Vegetable tanning process is the oldest method which poses intense pollution problems. Hence a majority (80%) of the tanning industries have switched over to chrome tanning.

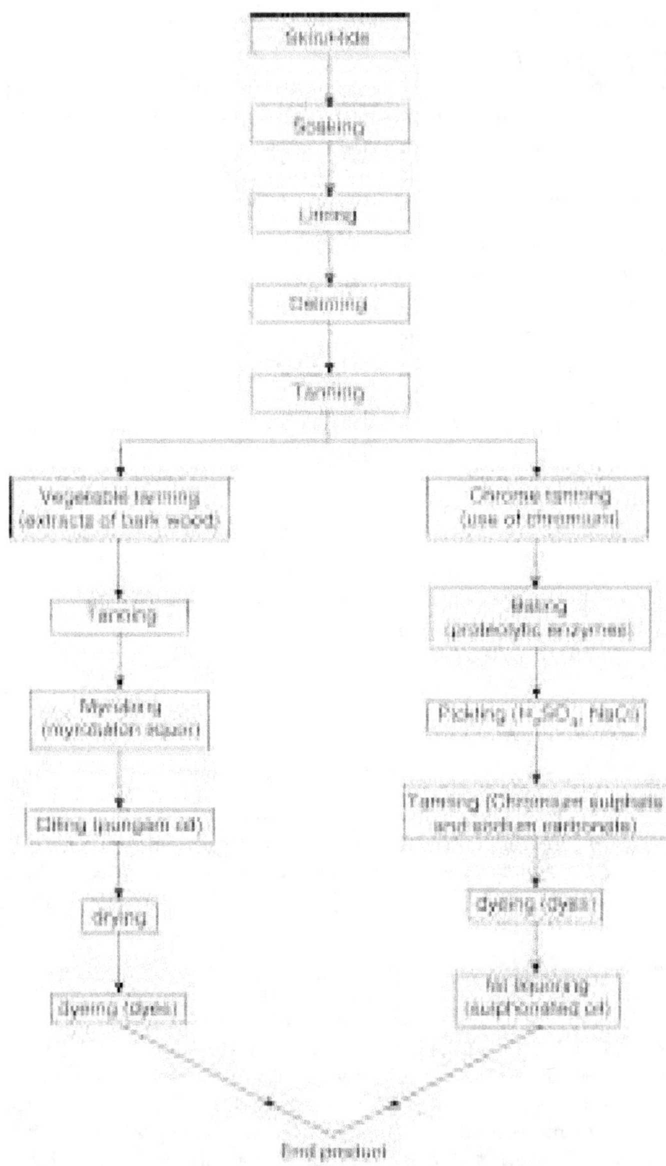

Characteristics of tannery effluent

Parameter	Concentration
colour	dark brownish
odour	highly disagreeable foul odour
temperature	30°C
pH	4.07
total solids	7800
BOD	1200
permanganate value	320
chlorides	15000
sulphates	380
ammoniacal nitrogen	213
oil and grease	16
tannin and lignin	50
total suspended solids	5500
total hardness	10500
total chromium	3590

Tannery effluent in general is dark brown in colour with foul odour, saline sodiac containing large amount of organic and inorganic substances, chlorides, tannin, protein, chromium and high BOD and COD values. The amount of dissolved and suspended solids are considerably more. Sulphides and tannin content impart colour and odour to the effluent. Several reports have confirmed high levels of copper, chromium, cadmium, iron, manganese and lead in the tannery effluent.

Steps involved in processing of skin	Possible outcome
Soaking	odour, suspended solids, sodium chloride, dirt, albumin
Liming	sulphide, lime, calcium, carbonates, suspended solids, colloidal proteins, ammoniacal and total organic nitrogen
Dehairing and defleshing	suspended solids, hair, flesh, sulphides, BOD
Tanning and myrobing	acidic pH, suspended solids, BOD, odour
Bating	ammoniacal and organic nitrogen
Pickling	acidic pH, sodium chloride, BOD
Dyeing	coloured chemicals
Fat liquoring	oil and grease

Effect of tannery effluent on environment   Tannery effluent is a potential pollutant of considerable importance. When discharged into the land or aquatic bodies, it spoils the total ecosystem. Tanneries release partially treated effluent into low-level areas in streams and lands which finally accumulates as large ponds and affects both the surface and ground water quality through seepage. In soil it is known to alter the physical and chemical characteristics of soil, affect water intake by soil and influence the metabolic processes of soil microorganisms. The lethality of tannery effluent in soil is due to high concentration of salts, proteinaceous and fatty substance that render soil unfit for cultivation. Also, the effluents influence the pH, moisture content, temperature and organic content of the soil which in turn affects the development of different groups of microorganisms.

Treatment methods   Except for the presence of chromium, the effluent is rich in organic context. As an initial step, the effluent should be subjected to preliminary treatment such as screening (filtering large-sized suspended particles), sedimentation (removal of colloidal particles) and neutralization (adjusting the pH). Both anaerobic (oxidation pond and activated sludge process) and anaerobic (USAB, anaerobic lagoon and anaerobic filters) treatment methods are suitable for this type of effluent.

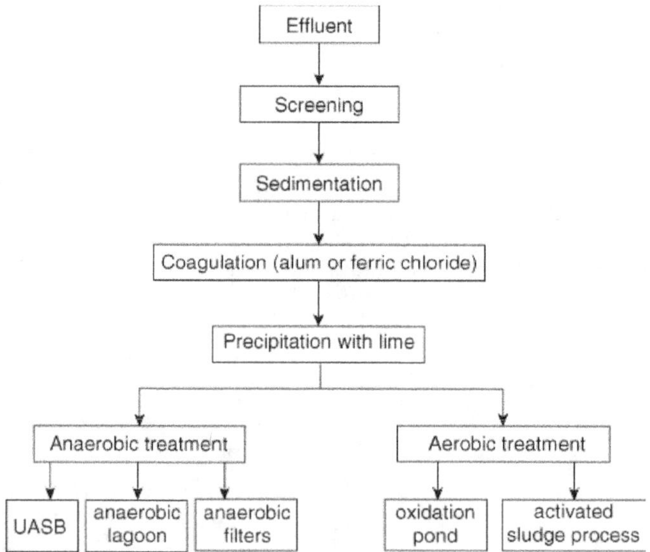

2.7.2 Sugar Mill

Recently, India has emerged as the largest sugar producer in the world. Sugar is one of the significant agricultural products and the industries processing sugar are vital for India's economy. The sugar industry is the third largest industry in the country and releases huge amount of waste water containing high concentration of a number of chemical pollutants such as carbonates, bicarbonates, nitrates, phosphates, oil and grease, total suspended solids, total dissolved solids, volatile solids and other toxicants.

Manufacturing process   Sugarcane is crushed with a roll to get the juice and bagasse. After heating, it is filtered and then concentrated by evaporation. Condensed extract is crystallized and centrifuged to form sugar.

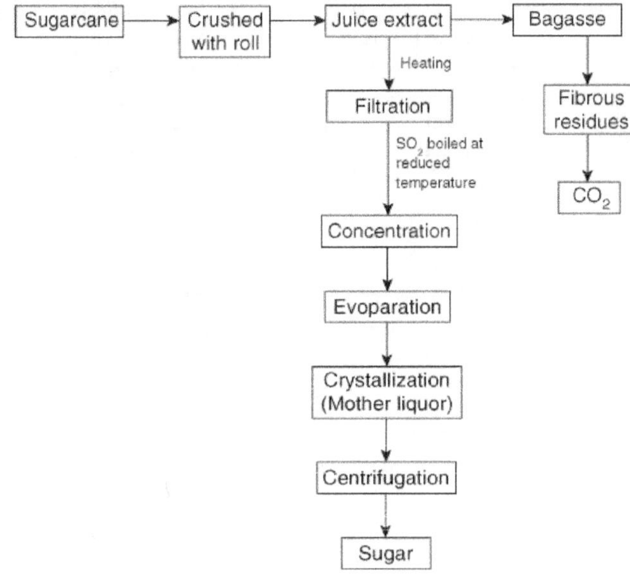

2.7.3 Distilleries

Distillery industries which are an ancillary unit of sugar industries are one of the most polluting industries and distillery effluent in the form of spent wash is one among the worst pollutants produced by these industries both in magnitude and strength. Distillery waste water generally known as stillage, stom, vinnage or durn wastes is disposed into the land or water without treatment. Each distillery produces approximately 30 billion litres of spent wash per day which is discharged into the environment. India produces around 6500 million litres of alcohol in about 200 distilleries every year, with an output of 130 billion litres of waste water.

Manufacturing process

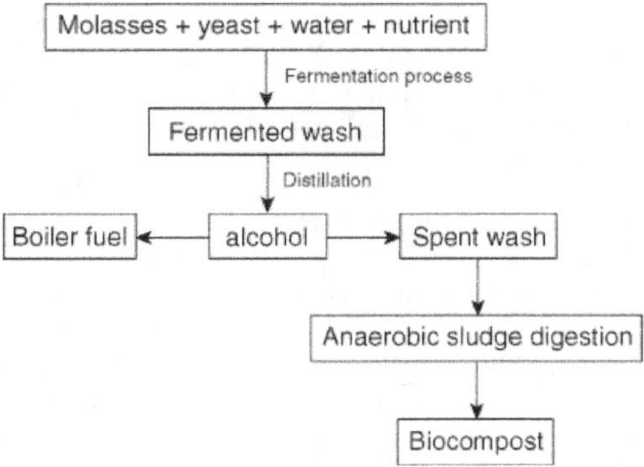

Here the molasses is fermented which is subsequently distilled to produce alcohol. The liquid waste produced is called as spent wash and the alcohol is used as fuel.

Characteristics of distillery effluent   Distillery effluent is a highly coloured effluent with objectionable odour. It is rich in BOD(>10000). The organic content is in both the suspended and dissolved form. Also it has considerable amount of nitrogen, chlorides, sulphate and potassium.

Characteristics	Concentration
colour	dark brown
pH	3.9
total solids	109600
total dissolved solids	63000
total suspended solids	35600
BOD	40000
chlorides	4240
sulphides	4400
ammoniacal nitrogen	977
potassium	10200

Effect of distillery effluent on the environment   When the raw effluent or partially treated effluent is released into the waters, the effect is equal to that of toxic effluents. This is because the effluent with its irresistible odour and a very high BOD destroys the whole aquatic ecosystem. The effluent-induced changes in the aquatic environment are

i.   increase in BOD

ii.   decrease on DO

iii.   appearance of colour

iv.   death of aquatic plants

v.   anoxic condition

vi. H_2S production

Treatment options   Treatment of distillery effluent should bring the level of pollutants within the standard limits. Since the effluent is very hot, storing them in lagoons for a specific period of time is mandatory. Then preliminary methods like screening and neutralization can be done to favour the pH adjustment and removal of large-sized particles. When the favourable condition for the microbial growth is ensured, the effluent is subjected to anaerobic treatment followed by aerobic treatment. Anaerobic treatment involves anaerobic digestion and anaerobic lagoon.

Anaerobic digestion   Anaerobes in general are very sensitive and therefore feeding the raw effluent with high organic loading rate would collapse them. To overcome this problem, the digester should be started up with a low organic loading rate (OLR) and hydraulic loading rate (HLR). This can be accomplished by appropriate dilution. However it demands more water. OLR and HLR of the effluent can then be progressively increased to achieve the maximum OLR, i.e. the COD of the raw effluent. Methane is obtained as a by-product. Production of H_2S should be checked at regular intervals as it is toxic to the anaerobes prevailing inside the reactor. To prevent this, iron salts can be added which readily combine with sulphide to form ferric sulphide.

Anaerobic lagoon Even though methane is not produced by this method, it remains as a best choice for majority of distilleries in India. BOD reduction rate is more than 90% and the drawbacks encountered in this system are long retention time and odour nuisance.

Aerobic treatment methods It includes trickling filters, activated sludge process, aerated lagoons and oxidation ditches. The retention time required for these treatments differs.

2.7.4 Pulp and Paper Mill Industry

Pulp and paper mill is an important one among the highly polluting industries in India. During the last two decades, there has been a tremendous expansion in pulp industry which ranks third in the world in terms of freshwater withdrawal. There are about 305 paper mills in India with an installed capacity of 3014 lakh tons. It is one of the largest consumers of water. An integrated pulp and paper mill generates 220 to 320 m^3 of effluent portion of paper manufactured. Wood is the major raw material which produces approximately 80% of the pulp capacity and rest comes from non-wood materials like baniyan cloth, straw, bamboo and grasses.

Manufacturing process The process of paper making from pulp involves stock preparation, fibre suspension, formation of paper well, couching, pressing section and drying section, cooling stage and calendaring. The sources of waste water generated from pulp and

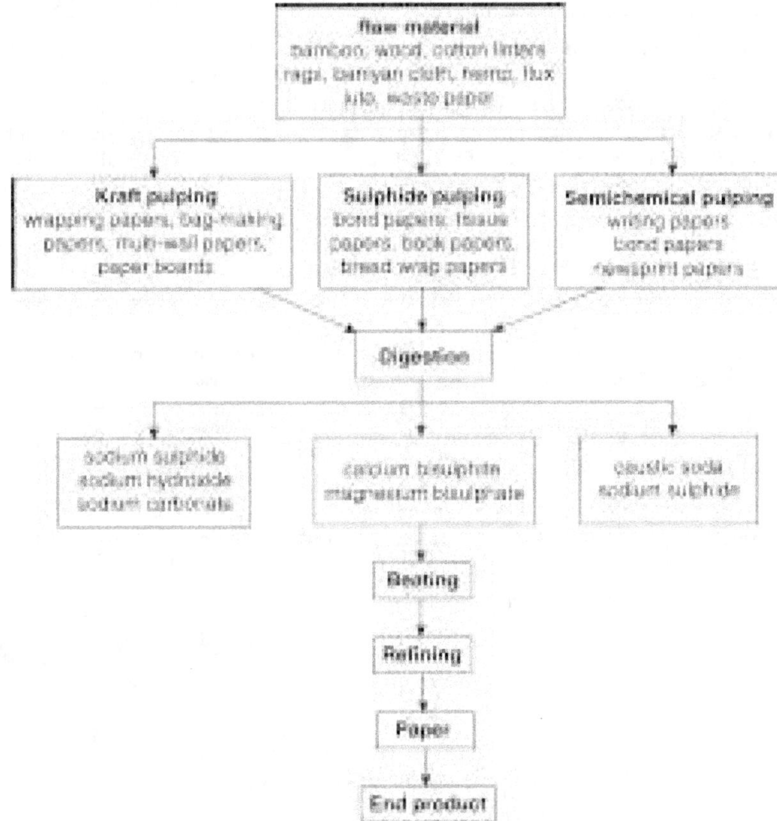

paper mill is from digester, chipper house, paper machine washing, floor washing and equipment washing from boiler godown. Wood and allied materials are digested with chemicals to convert them into pulp and the pulp is refined and passed over a fine screen to form paper. Pulping is the process in which wood or other cellulose raw materials are digested with chemicals under high temperature and pressure so that cellulose fibres of wood are detached from its binders such as lignin, resin, etc. All the pulping processes include washing and bleaching as final steps which contribute to the bulk of effluents. During bleaching, the pulp is treated with chlorine, caustic alkali and calcium hypochloride solution.

Steps involved in pulp and paper mill processing	Characteristics of effluent
kraft pulping	alkaline pH, BOD, COD, SS, Sodium, dimethyl sulphide, methyl mercaptan, caustic soda
sulphite pulping	acidic pH, BOD, lingo sulphonic acid, reducing sugars
semichemical pulping	alkaline pH, BOD, COD, SS, sodium, dimethyl sulphide, methyl mercaptan, caustic soda
soda pulping	alkaline pH, COD, suspended solids, sodium, dimethyl sulphide, methyl mercaptan, caustic soda

Characteristics of pulp and paper mill effluent Pulp and paper mill effluents consist of organic and inorganic matter, sodium, potassium, phosphorus, calcium, magnesium, sulphur, chlorides, fatty acids, resin acids, hydrogen sulphide, organic sulphides, methanol, ethanol, acetone, terpene, phenolics and heavy metals. The concentration of heavy metals in the paper mill effluent is in the order of zinc > copper > lead > nickel > cobalt > manganese. The colouring body present in the waste water is organic in nature as it is comprised of wood extractives, lignins and its degradation products formed by the activity of chlorine and lignin. It contains solids of both suspended and dissolved forms. Chlorides, sulphates, nitrates, bicarbonates, phosphates, sodium, potassium, calcium and manganese constitute dissolved solids and strong wood chips, bits of barks and cellulose fibres constitute suspended solids. Effluent at different units in a pulp and paper mill vary in BOD and it is high in spent liquor and very low in pulp mill effluent. The alkali extraction of bleach plant increases the chlorine content of the effluent to a considerable level.

Characteristics	Concentration
colour	light brown
pH	5.21
total solids	1100
total suspended solids	138
total dissolved solids	962
BOD	1590
COD	2080
sulphides	--
residual chlorine	--
tannin and lignin compounds	62
oil and grease	1.6

Effect of effluent on environment   The effluent consists of considerable amounts of non-biodegradable lignin that imparts dark colour to the receiving streams. The colour is persistent and affects the aesthetic value of the aquatic environment. Also, the presence of phenol, wood pieces, baniyan cloths, bark layers, cellulose fibres, sulphite and free chlorine in the effluent reach the bottom and influence the reproduction of fishes and invertebrates, i.e., in excess amount, they occupy considerable area including the spawning grounds of aquatic organisms and in case these organisms have limited habitat range, their population will soon be eliminated. Moreover the effluent cannot be discharged by removing its organic content alone as it consists of hazardous chemicals such as methyl mercaptan, pentachlorophenol and sodium pentachlorophenate. These components with their toxic properties destroy the ecosystem in which they are discharged. Individual components such as resin, fatty acid and chlorophenolic compounds are said to bring acute toxicity in fish. Depletion of dissolved oxygen is the indirect effect, the reason being the presence of oxygen-demanding wastes.

Treatment options   Primary treatment of the effluent involves screening, skimming, coagulation, sedimentation and floatation. Secondary treatment helps to oxidize the organic content of the effluent. Of the various aerobic treatment methods, activated sludge processes prove to be the best with 80% COD reduction rate. Anaerobic treatment methods prove to be equally satisfactory and upward anaerobic sludge blanket reactor gives the maximum reduction rate.

2.7.5 Textile Industry

Textile mills in India are one of the oldest sectors operated at large scale and household level. Among the large numbers, over 700 are organized and they are present in cities like Ahmedabad, Bombay, Coimbatore, Kanpur and Delhi.

Manufacturing process (Wet process)

Slashing   The thread is sized with starch to make it suitable for weaving. Sizing gives strength to the thread. Since addition of starch enhances the BOD of the effluent, chemicals like carboxymethyl cellulose/polyvinyl alcohol are used.

Desizing   In this process removal of sizing materials is accomplished by treating the cloth either with dilute sulphuric acid or with enzymes. After this treatment the clothes are washed well.

Scouring   The cloth is exposed in a steam produced by boiling an alkaline solution (a mixture of soda ash, sodium silicate, sodium peroxide and detergents). This process completely removes the impurities such as grease, wax and fats in the cloth.

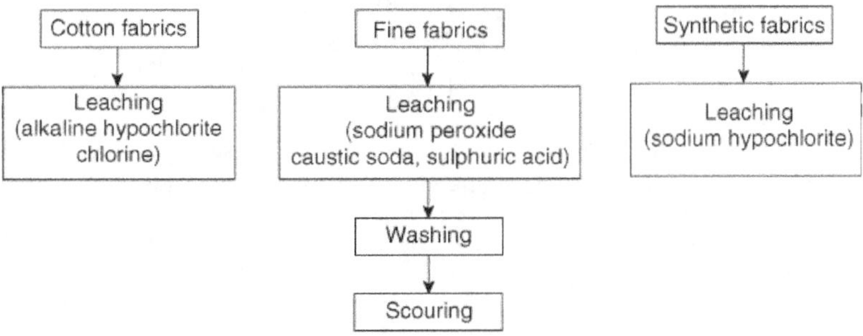

Bleaching   Bleaching process accomplishes the removal of colour materials from the cloth and makes it white and the chemicals used for bleaching cotton fabric are alkaline hypochlorite or chlorine. Then to remove traces of alkali and chlorine, the cloth is subjected to treatment with sodium sulphide (scouring) and the process differs with the type of fabrics and based on the fabric used the effluent consists of free chlorine, hypochlorite, caustic soda and sodium peroxide.

Mercerizing   Mercerization is a process by which the clothes are given lustre, strength and dye affinity. Processing of rough clothes does not involve this process and they can be readily dyed following leaching.

Dyeing   Dyeing process involves several types of dyes including a) naphthol dyes, b) sulphur dyes, c) developed dyes, d) direct dyes, e) vat dyes and basic dyes. Based on the dye type used, effluent characteristics differ.

Printing and finishing   Printing can be done as manual printing or machine printing. Finishing imparts smoothness and resistance to the clothes and the effluents of these processes contain starch, dextrin, natural and synthetic waxes and synthetic resins.

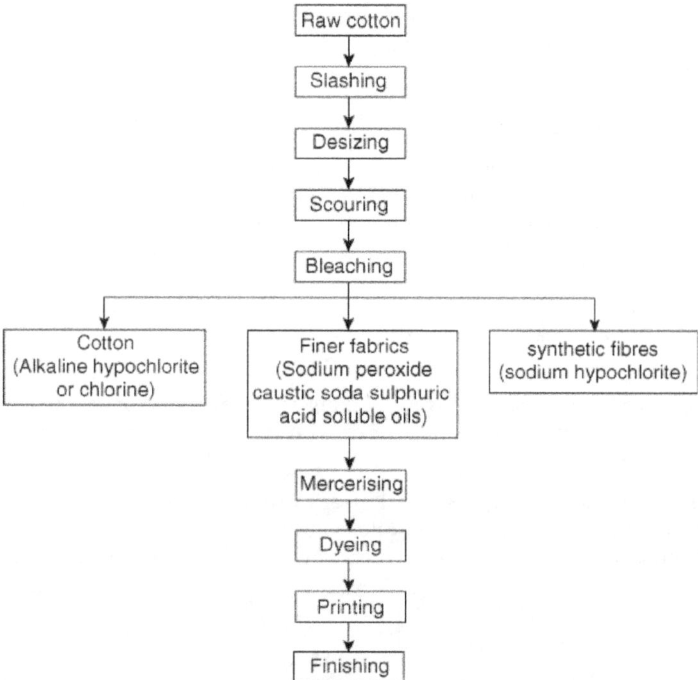

Effluent characteristics   The effluent is alkaline in nature with pH greater than 9. Total solids exist in other forms, i.e., suspended and dissolved forms. COD of the effluent ranges between 300 and 400. Also it contains heavy metals like chromium and zinc. Considerable amounts of phenols, oil and grease are also present.

Characteristics	Concentration
pH	9.5
total solids	3040
total dissolved solids	2010
total suspended soils	1030
BOD	250
COD	390
chlorides	420
total chromium	4.5
zinc	0.8
phenolic substances	0.06
oil and grease	3.6

Steps involved in Textile processing	Characteristics
slashing	BOD
desizing	acidic pH, BOD, SS and DS
scouring	alkaline pH, soda ash, sodium silicate, sodium peroxide, detergents, suspended and dissolved solids
leaching	free chlorine, hypochlorite, alkaline pH
mercerizing	alkaline pH
dyeing	dyes
printing	dyes, BOD, oil, resin, soap gums
finishing	synthetic waxes, synthetic resins, BOD

Effect of textile effluent on environment   The colours and dyes of the effluent persist in aquatic bodies and affect the aesthetic value of the environment. In addition, they form a thin film on the surface and inhibit the photosynthetic activity. Alkaline pH of the effluent affects the aquatic biota. Colloidal organic content accompanied by the oil scum prevents the diffusion of atmospheric oxygen into the water. Continuous accumulation of organic matter depletes the dissolved oxygen content. Apart from this, effect of compounds like sulphide, free residual chlorine, aniline dyes and heavy metals like chromium in aquatic organisms are apparent, i.e. they are toxic. The presence of bioaccumulating substances such as biphenyl 1-2-4 trichlorobenzene and O-phenyl phenol has made the treatment of this effluent inevitable.

Treatment methods   Preliminary treatment methods include screening, equalization, neutralization and chemical coagulation. Secondary treatment includes aerobic treatment methods. Tertiary treatment involves reverse osmosis and electrolysis.

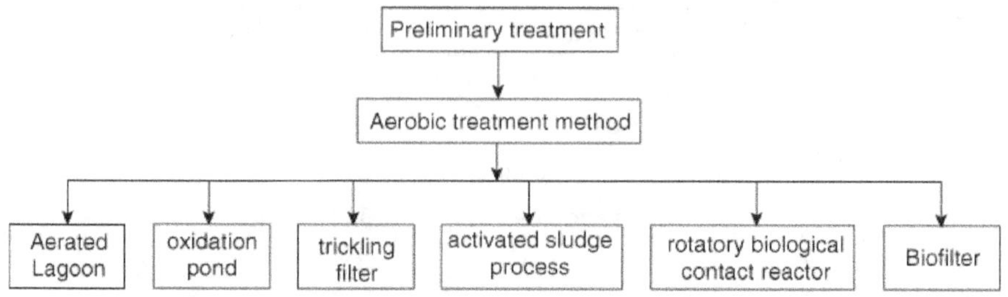

2.7.6 Dairy Industry

India has the largest number of milk producers compared to all other countries of the world. Basically the dairy industry is food-based and therefore has spread in each and every part of our country both at large-scale and at household level.

Manufacturing process   Milk is collected from different places and then stored in clean milk cans. Based on the product requirement, it is subjected to different processes. For the preparation of bottled milk, it is pasteurized. For other milk products, the cream is separated out. Cream-removed milk is used for cheese production and whey is obtained as a by-product of this process. Cream when churned gives butter and butter milk is the by-product of this process.

To acquire other milk products, whey, skimmed milk and butter milk are condensed in a vacuum pan. Skimmed milk is dried by this process and forms skimmed milk powder whereas whey gives rise to milk sugar and hog feed. Butter milk in powdered form is used as a main ingredient in poultry feed.

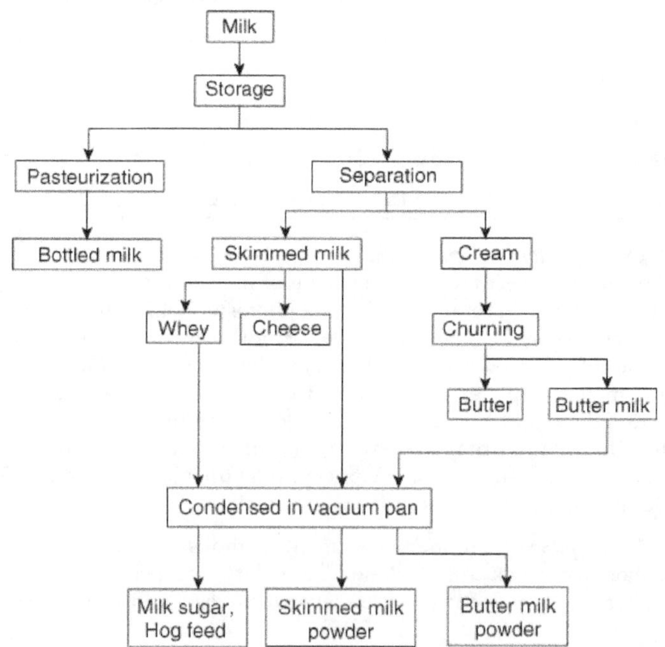

Effluent characteristics Characteristics of the effluent vary with the processing methods involved. However in general it is highly putrescible in nature with poor sludge setting property. Except for the usage of detergents or (acid, alkali) cleaning agents, no specific chemicals are involved in the manufacturing process. Therefore, effluents consist of milk, milk products and enormous quantity of water. The pH of the effluent is alkaline and the organic content is considerably high. Also the effluent is rich in oil and grease. Another major problem encountered in this effluent is the presence of solids (settleable and dissolved).

It has an obnoxious odour and the colour of the effluent is milky white. Evaluation of effluent characteristics of the sample should be immediately done otherwise the chemical composition of the effluent will be altered thus misleading the analysis.

Characteristics	Concentration
pH	9.0
Colour	milky white
Odour	putrid milk odour
TSS	360
TDS	910
DFS	286
DO	–
BOD	650
COD	1130
ammoniacal nitrogen	19.2
total kjeldahl nitrogen	39.4
Sulphide	0.3
oil and grease	82
surfactants	4.1

Inorganic substances like calcium, magnesium, sodium chloride, sulphate and bicarbonates are present in high concentration. Storage of raw dairy wastes tends to putrefy and therefore shifts its pH towards the acidic side. The concentration of dairy waste water varies considerably on a day-to-day basis. In weekly composite samples, the total COD concentrations vary from 1400 to 4700 gm^3 and the filtered COD concentrations vary from 650 to 2300 gm^3. Waste water temperature is between 10 and 30°C.

Effect of effluent on the environment While colour of the effluent affects the aesthetic value of the receiving waters, its alkaline pH causes damage to aquatic life. Its high BOD depletes the dissolved oxygen content of the aquatic system and in due course of time creates anaerobic conditions. Also the fouling smell of the effluent is a nuisance to the public, and in large volume it becomes the breeding ground for flies and mosquitoes. Moreover, high concentration of dairy effluent has been found to have toxic effects on certain fish and algal species.

Treatment methods Primary treatment involves screening (removal of unwanted large-sized substances), skimming (removal of oil), coagulation (removal of colloidal substances), and sedimentation (removal of settleable solids). It is followed by the secondary treatment of the two biological methods, where aerobic methods are found to give better results. However adjustment of pH is done (neutralization) prior to treatment. Among the several aerobic treatment methods, activated sludge process, trickling filters, oxidation ditch, aerated lagoon and waste stabilization ponds are reported to give maximum BOD reduction rate. Anaerobic methods if adopted should be followed by aerobic processes to give the effluent a quality well below the permissible limits.

Source of effluents	Constituents
milk spillage, drippings, can washing tankers, bottles, utensils and floors	alkaline chemicals
wash water from butter processing unit.	buttermilk, cream
waste water from powder milk plants, yogurt manufacturing unit, cheese manufacturing unit	spilled milk, cream, cheese, alkaline chemicals

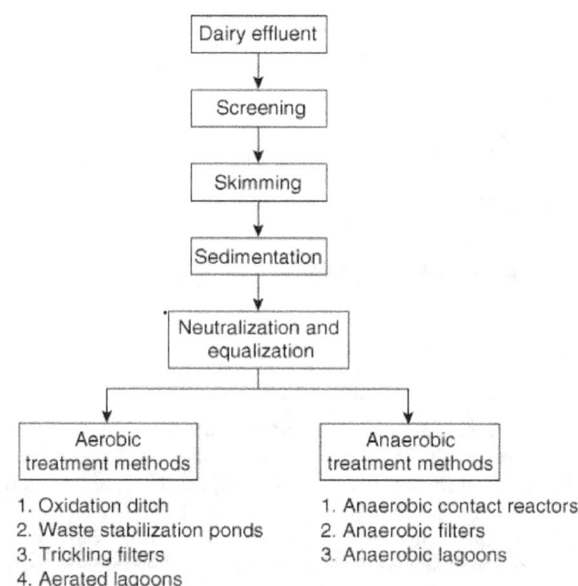

2.7.7 Sago Industry

Sago industry is an agro-based seasonal industry using tapioca root or tubers as basic raw material. Tapioca is one of the richest sources of starch. In India, the starch production has been reported to be 14 lakh tonnes out of which tapioca contributes about one-third. This excludes tapioca starch converted to sago which is estimated to be around 2.5 lakh tonne per annum. Sago procured from edible starch is a very popular food as it is easily digestible and is widely used in several parts of India for feeding invalids and infants. Also it can be used to prepare a variety of dishes. A number of processes are in practice to fortify sago and starch products.

Manufacturing process Washing and peeling of the tuber is done to remove and separate all adhering soil and the protective epidermis. Rasping or disintegration destroys the cellular structure and ruptures the cell walls to release the starch as discrete, undamaged granules from other insoluble matter. Screening or extraction separates comminuted pulp into waste of fibrous material and starch milk. Purification or dewatering separates solid starch granules from their suspension in water by sedimentation. Drying removes sufficient moisture from damp starch cake which favours its effective storage. Pulverization, shifting and bagging are the finishing processes.

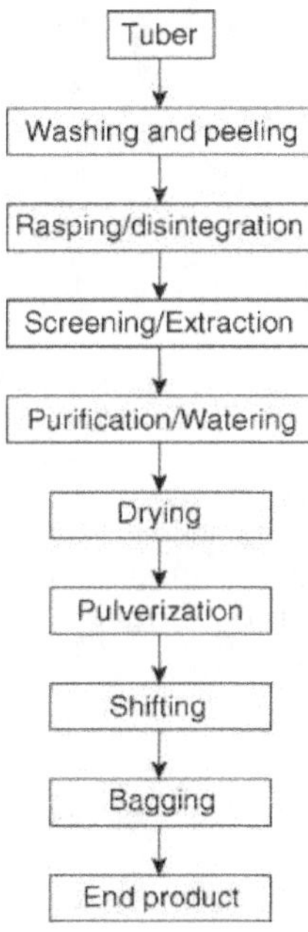

Characteristics of Sago effluent Sago effluent has acidic pH with high concentration of organic content. Settleable and dissolved solids constitute the total solids whose concentration is more than 4000 mg/l. Also it consists of total nitrogen and phosphorus. Presence of high BOD in the effluent alters its composition if kept unpreserved.

Sago effluent	Concentration
pH	3.8
BOD	1100
COD	3256
total solids	5738
total dissolved solids	2020
total suspended solids	3718
total nitrogen	5.4
phosphorus	1.2

Effect of effluent on the environment   These effluents pose a serious threat to the quality of life and environment where these factories are situated. In this industry, effluents from the washing tank and sedimentation tank form the major source of pollution. These effluents when discharged into the water bodies deplete the oxygen content and create anaerobic conditions unsuitable for aquatic life.

Treatment methods   In the treatment of sago effluent, anaerobic treatments followed by aerobic treatment methods give successful results.

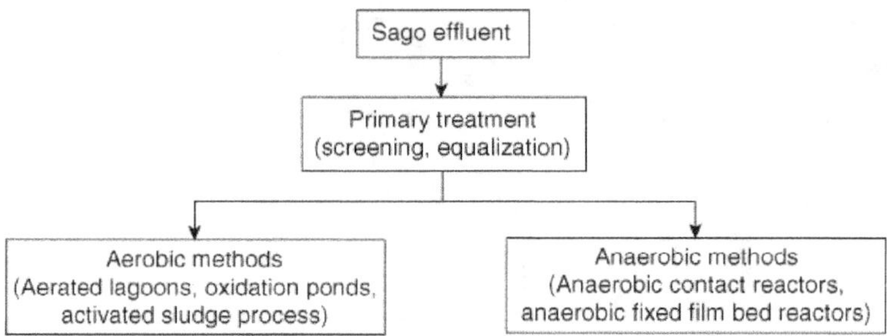

2.7.8 Fertilizer Industry

Food demand for the growing population has been met by intensive agricultural practices which in turn require the regular application of fertilizers. Therefore for the past two decades, the number of fertilizer industries are on the rise. Basically they are generated in both solid and liquid form.

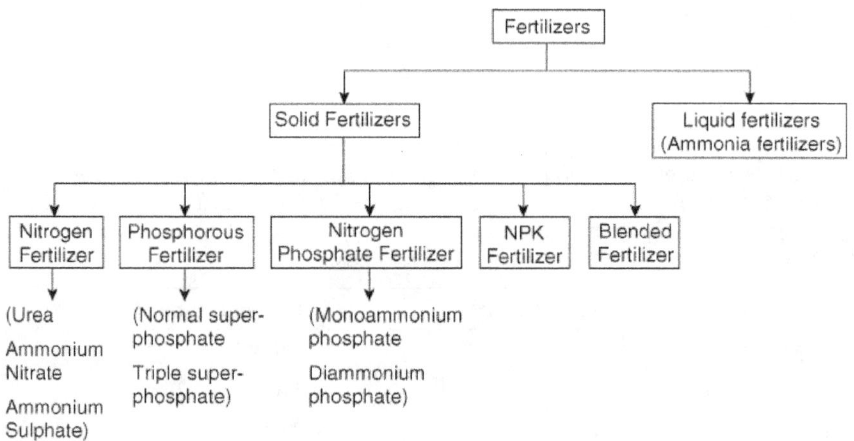

Manufacturing process   The processing unit of nitrogen fertilizers generates effluent from the boiling and cooling units. The major constituent of these effluents is little amount of fertilizers, conditioning agents and biocides. The concentration of fertilizer may increase in the case of accidental spillage.

Phosphate fertilizer production involves the following steps.

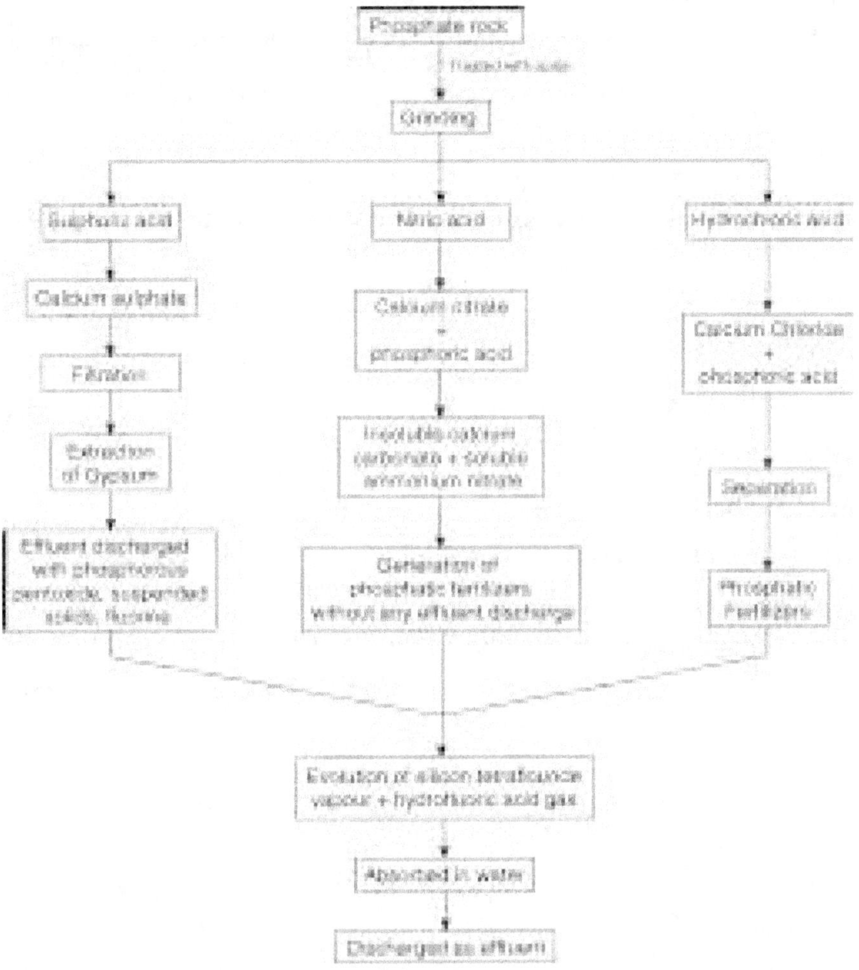

Nitrogen phosphate fertilizers include the following:

Mono ammonium phosphate   Slurry of ammonia and phosphoric acid applied over dried raw material produce monoammonium phosphate.

Diammonium phosphate   Monoammonium phosphate in combination with ammonia forms diammonium phosphate.

Manufacture of NPK fertilizers is accomplished by

i.   blending of dried fertilizer and

ii. mixing of diammonium phosphate with potash

Blended fertilizers are produced by just mixing different fertilizers in required amounts. The combination to be applied depends on the nature of soil quality.

Effluent characteristics　Effluent discharged from nitrogen fertilizer production process consists of suspended solids, fluorides, acid, water condensate and phosphorous pentoxide. Similar is the case of effluents from superphosphate and triple superphosphate processing units. Diammonium phosphate production releases effluents containing ammonia fluoride, phosphate and dust. Volume and characteristics of the effluent depends on the raw material used and the process adopted.

Characteristics of fertilizer industry

Characteristics	Ammonium nitrate	Superphosphate	Nitrogenous fertilizer	Phosphate fertilizer
pH	85	-	-	-
TSS	325	375	1100	-
TDS	2090	812	1750	9660
BOD	280	-	-	-
COD	535	105	-	-
Ammoniacal nitrogen	720	2040	210	-
Organic nitrogen	102	-	-	-
Nitrate nitrogen	65	-	-	-
Sulphate	980	188	35	2865
Phosphate	4.2	0.2	20	-
Fluoride	-	2046	-	-
Chloride	-	138	13	-
Free ammonia	-	-	20	-
Calcium	-	59	160	-
Chromium	-	-	7	-
Zinc	-	-	Trace amount	5
Oil	-	-	505	-
Alkalinity	-	-	-	2500
Silica	-	-	-	40

Effect of effluent on the environment　The upgradation of the nutrient status of soil by fertilizers is done whenever the soil exhibits its inability to supply certain nutrients to the growing crop when other nutrients and environmental factors are adequate to promote growth. Generally, the fertilizers are made up of organic and inorganic materials, a common term applicable to natural and synthetic products. Common fertilizers practically in use are nitrogenous fertilizers, phosphate fertilizers, potassium fertilizers and compound fertilizers, enriched with few eventual elements in specific concentrations. They very well supplement nutrients supplied from organic sources and those deplete in the soil. Since they exhibit acidity or alkalinity in their residual form, they could as well promote soil reactions. Altogether they boost the productivity and the close relation between the upheaval of green revolution and fertilizer cannot be as such denied. On the other hand, the impact of nitrogenous fertilizers on the soil quality is beyond the limit. They build up either acidity or alkalinity in the soil by inducing the plants to release the respective ions (H^+, OH^-). The mechanism would be even more clear by the fact that it is the form of nitrogen, either an anion (NO_3^-) or cation (N_4^+) that induce the plant to balance the charge by releasing H^+/OH^- ions. Excessive input of nitrogen fertilizers such as urea and ammonium sulphate leads to acidification of soil. The adverse effects of nitrates on human health are methaemoglobinemia in children, respiratory illness and tumour formation with N-nitroso compounds. Carcinogenic nitrosamines formed as a reaction product of ingested amines with nitrite in the human stomach pose a severe health hazard in humans.

Pollution of ground water by nitrate occurs in areas with high rainfall, intensive irrigation and land with organic soils. Hardly surprising is the fact that these fertilizers have affected a few locations in Punjab, Haryana, and Tamilnadu. Even a level of nitrate beyond the acceptance limit of 45 mg/l was observed in ground water samples of Bihar, Orissa, Maharashtra, Madhya Pradesh, Karnataka, Andhra Pradesh, Gujarat and Rajasthan.

Treatment methods   Based on the contstituents of the effluent, suitable treatment methods can be opted. However general steps involved in treatment are

Preliminary treatment	→	Neutralization, sedimentation
Secondary treatment	→	aerobic/anaerobic treatment methods that involve specific microbes to favour nitrogen conversion process.
Tertiary treatment	→	ammonia stripping, ion exchange, treatment with lime and chalk.

2.7.9 Electroplating Industry

To protect the basic metals from corrosion, they are coated with heavy metals like chromium, cadmium, nickel, zinc, copper, silver and gold and the process is termed as electroplating. It operates at large scale and therefore discharges considerable amount of effluent which is potentially toxic.

Manufacturing process   The steps involved in the manufacturing process are

1. *Degreasing*   removal of oil or grease from the base metal surface.
2. *Stripping*   metal to be plated is treated with acid to remove rust and scales from its surface.
3. *Plating*   plating of metals.
4. *Bathing and polishing*   plates are washed in running water and then polished.

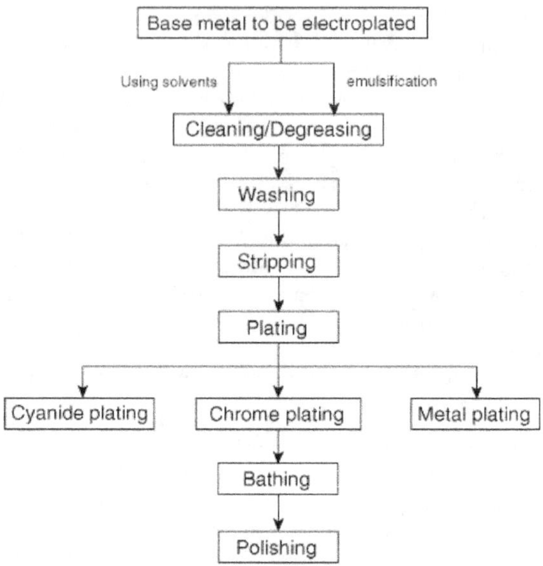

Effluent characteristics The volume and characteristics of the effluent depends on the type of plating methods involved. Even the pH of the effluent cannot be generalized as it is based on the use of acids or alkali in the cleaning and stripping process. However combined effluent has sodium hydroxide, phosphate, silicate, carbonates, organic emulsifiers and wetting agents. Acids such as sulphuric acid, nitric acid and hydrochloric acid are also present.

Characteristics	Average Concentration
pH	8.2
basicity	415
total solids	4280
total suspended solids	840
chromium (UI)	0.8
cyanide	3.1
copper	-
zinc	0.2
nickel	0.1
silver	-

Effect of effluent on the environment Presence of heavy metals and cyanide make this effluent toxic. Therefore they are carefully treated prior to the discharge in aquatic bodies. If left untreated into sewers, they combine with hydrogen ions to form hydrogen cyanide and affect the water system. In some cases, solvents of the effluents explode to cause irreversible damage to the sewer treatment plant. Acids of the effluent corrode the sewage pipes. In aquatic bodies they affect the flora and fauna even in trace amounts. Iron compounds render colour to the streams.

Treatment methods   The preliminary treatment methods include skimming and neutralization. Biological treatment involves biosorption of heavy metals using

i.   aquatic plants
ii. algae
iii. immobilized microbial cells

2.7.10 Pesticide Industry

All the chemicals that are used to kill pests collectively come under the composite term pesticides. In agriculture this includes herbicides (weeds), insecticides (insects), fungicides (fungi), nematocides (nematodes) and rodenticides (vertebrate poison). Based on their composition they are classified under three major divisions.

Division I   Organochlorine compounds
Division II   Organophosphate compounds
Division III Carbonate compounds

Since the production of pesticides does not produce any effluent, the manufacturing process of the pesticides are not discussed here. However the purification process gives rise to acidic wash waters with considerable amount of pesticides.

Characteristics of effluent (BHC)   The effluent has acidic pH. Both the methyl orange and total acidity are considerably more. Among the different types of solids, dissolved solids contribute to about 80%. The effluent contains moderate amount of COD. Presence of BHC complicates its treatment.

Characteristics	Concentration
pH	3.5
methyl orange acidity	360
total acidity	440
total suspended solids	10
total dissolved solids	880
dissolved fixed solids	250
COD	70
chlorides	530
benzene hexachloride	56.3

Effect of effluent on environment   Residual pesticides move from the area of application to different ecosystems under the influence of physical and biological processes. Its possible drift towards the non-target surface is made possible by run-off, stream flow, rain water, and deposition from air and spray on water surface to kill pests. In the water, they exist either in the dissolved or suspended form. In the case of suspended solids they get adsorbed to silt and organic particulates and at more concentration they settle at the bottom. Active ingredients in pesticide formulation, impurities in ingredients, additives that are mixed with active ingredients like wetting agents, diluents or solvent extenders, adhesive buffers, preservatives and emulsifiers and the intermediate metabolic products formed during the chemical, microbial or photochemical degradation of active ingredient find their way to the nearby water bodies and their drift outside the targeted site is worrisome. This is because of their ability to persist in the environment for a very long time. Different pesticides have markedly different effects on aquatic life which makes generalization very difficult. Many of these effects are chronic and they include cancer, tumour and lesions on fish and animals, reproductive inhibition or failure, suppression of immune system, disruption of ecosystem, cellular and DNA damage, teratogenic effects, poor fish health marked by low red to white blood cell ratio, excessive slime on fish scales and gills, intergenerational effects and other physiological effects such as egg shell thinning. If these chemicals get deposited in sediment or get concentrated in targeted organisms the impact is considered as localized (bioaccumulation). However if they get transmitted through trophic chain the effect is multifarious. This clearly shows that the pesticide plays an important role in suppression of targeted organisms. After a particular period of time, the suppressed organisms may get eliminated.

Treatment options

> *Preliminary treatment*   Neutralization, sedimentation
>
> *Biological treatment*   to remove the organic content if present
>
> *Tertiary treatment*   adsorption, activated carbon and alkaline chlorination

2.7.11 Petrochemical Industry

Petrochemical industries are involved in the manufacture of chemicals used in the processing of solvents, resins, fibres, fertilizers, synthetic rubber and plastics.

Manufacturing process   The source of raw material is petroleum refineries (gaseous, liquid and solid hydrocarbons). By suitable separation processes individual components of the mixture are segregated and subjected to chemical conversion process. This results in the production of intermediate products which are used in the processing of a wide range of chemicals.

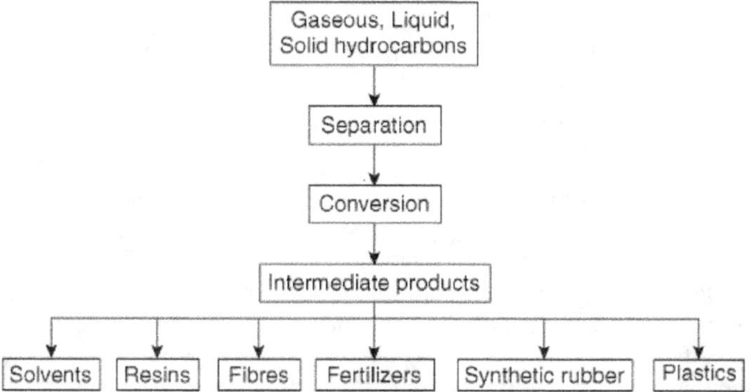

Effluent characteristics   Water used in the processing is discharged as effluent with high concentration of impurities, treated raw materials (hydrocarbons), alcohols, aldehydes, ketones, acid, ester, salt, phenol and oil. The presence of hydrocarbons increases its organic content with a high COD : BOD ratio. Another important component of this effluent is the presence of heavy metals and other substances. Besides, it possesses considerable amount of total dissolved solids.

Characteristics	Concentration
pH	5.5–8.7
total solids	4388
biochemical oxygen demand	2150
chemical oxygen demand	6380
ammoniacal nitrogen	562
phosphates	12
oil and grease	40

Effect of effluent on environment   Effects of effluent on the aquatic bodies are as follows:

  i.   damage the aesthetic value
  ii.   affect the flora and fauna
  iii. obstruct the photosynthetic activity
  iv.   alter the physiochemical characteristics of the aquatic system
  v.   inhibit the self-purification ability

Surface active agents of the effluent when discharged into the water bodies cause foam and damage the scenic value of the environment. Oil spread on the water surface, prevents the light penetration and affects the primary productivity. Acids or alkalis in the effluent shift the pH and collapse the total environment. Since the organic content of the effluent is resistant to biological breakdown, it cannot be reduced by indigenous microorganisms.

Treatment methods

Preliminary treatment	equalization, pH adjustment, coagulation, flocculation, skimming.
Biological treatment	aerobic, aerated lagoons, activated sludge process, oxidation ponds.

Aerobic–anaerobic digestion

2.8 TREATMENT OF INDUSTRIAL WASTES

The self-purification ability of receiving waters are highly limited and they cannot withstand increasing loads of organic and inorganic pollutants (King and Bull, 1964). Environmental deterioration is the consequence which induced the government to enforce strict laws as a pollution prevention strategy. However, development of any technical advancement in industrialization is accompanied by the competing demands on the limited finances available for development (Droste and Sanchez, 1983). Added to this, industrial effluent treatment methods, though considered to be the prime need from public health point of view, are seldom given priority (Murugesan, 1988). Moreover the treatment process is a non-productive activity and the recommended conventional methods demand high investment and hence a general solution to convince both the sides, environmental regulation and convenience of industrialists, is lagging. Any treatment process adopted to treat an effluent should satisfy the industrialist in terms of results, durability and expenses, final mode of disposal and labour availability. Above all the effluent that comes out of the treatment plant should not degrade the environment, soil, air and water. Since most of the methods adopted at present do not fulfil the above requirements, the effluents are released as such into the environment. To eliminate the problem of investment, low-cost conventional methods can be tried. These methods tend to reduce the complex pollutants in the effluent by simple means.

Composition of waste water depends on the type of community generating it, with its components varying from complex organic substances to simple dissolved inorganic salts. Reckless release of these wastes leads to environmental deterioration but when handled properly the vast treasure of resources can be recycled leaving back the ecologically safe treated water into the environment. This can be attributed by the multitude of treatment methods which purify industrial effluents to a degree equivalent to drinking water quality.

Industrial wastes are treated in three steps.

1. Preliminary or primary treatment
2. Biological or secondary treatment
3. Tertiary treatment

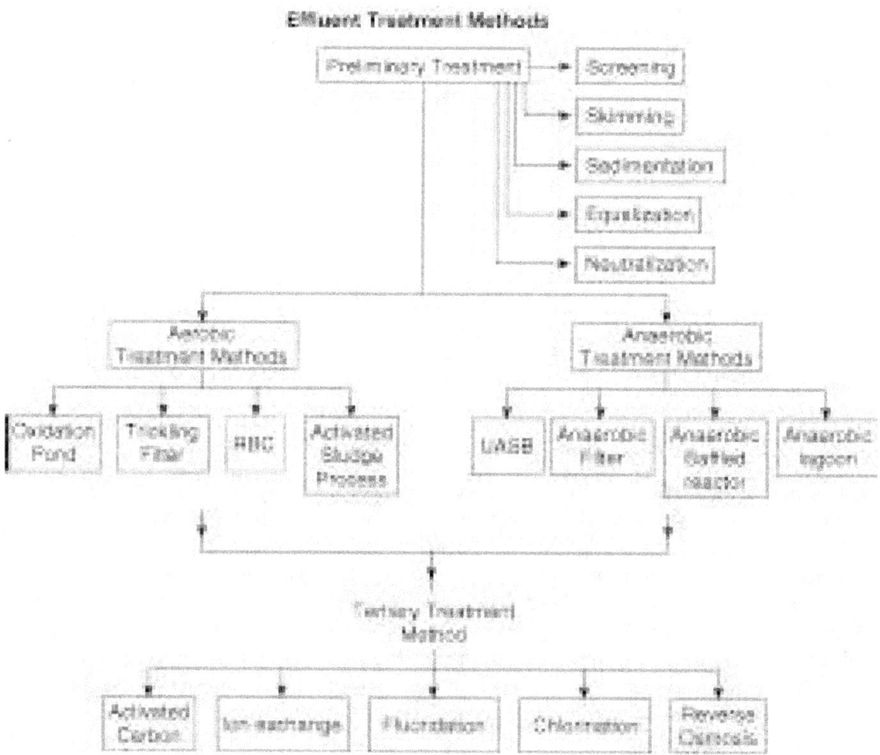

Figure 2.8   Schematic diagram showing general treatment pattern of industrial waste waters

2.8.1 Primary Treatment

Here the physiochemical characteristics of the effluent are adjusted in such a way that the effluent is made suitable for biological treatment. This is because the optimum conditions that are necessary for ecological treatment plant execution involves certain preliminary methods such as screening, settling, coagulation, flocculation and floatation. All these processes ensure removal of suspended solids, colloidal particles, oil and grease from the effluent thus favouring its suitability for biological treatment. The collection of effluents from various processing units to a collection tank and optimization of its pH are the equalization and neutralization processes. It is to be noted that the effluents of the same industry but from different processing units may vary in their structural composition and should be thoroughly mixed so that the pollution load of the total effluent can be evaluated. This gives a proper idea about the system design to be implemented in the secondary treatment.

Primary treatment involves:

1. screening
2. skimming
3. sedimentation involving physical and chemical processes
4. neutralization
5. equalization

Screening Screens of various sizes are used to remove large particles such as grit, gravel, pebbles, leaves and other wastes. Three types of screens are used a) horizontal b) vertical and c) wire mesh. Based on their size they are classified as coarse, medium and fine screens.

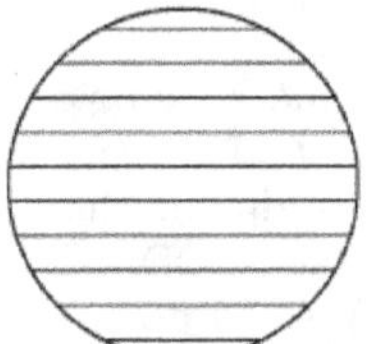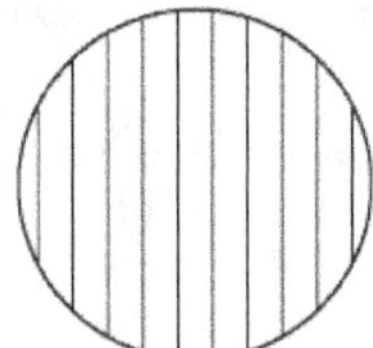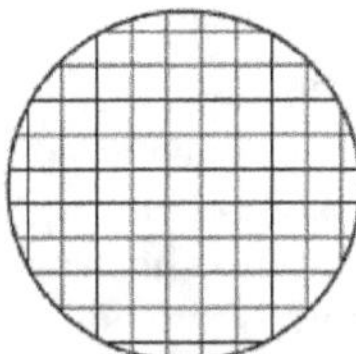

Figure 2.9   Different types of screens used in screening process

Skimming This process ensures the removal of oil from the wastes. The effluent tank is subjected to high pressure at its bottom so that bubble formation occurs which collects the oil and scum as a layer on the surface. This layer can then be separated efficiently.

Sedimentation Settleable solids in the effluent can be successfully removed in this process and this can be accomplished by a) mechanical sedimentation and b) chemical sedimentation. In the mechanical process, a central rod supported by a shaft is allowed to rotate inside the effluent reactor. Sludge is collected at the bottom and the cleared effluent at the top. Chemical sedimentation is done by using alum or ferric chloride. It favours the coagulation of settleable solids which is then removed.

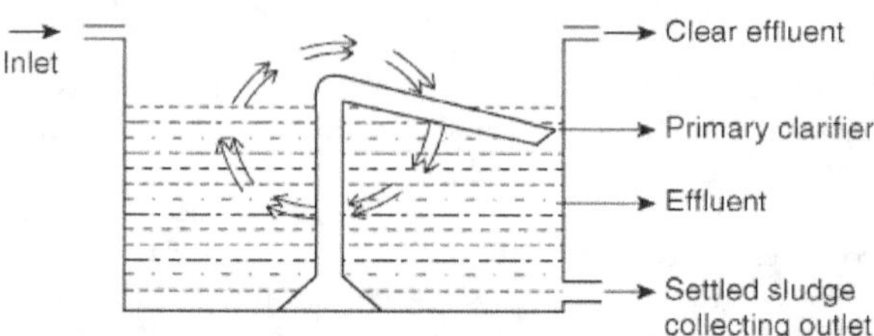

Figure 2.10   Sedimentation tank

Equalization and neutralization   Effluents from different processing units possess different characteristics and adoption of specific treatment methods for each processing unit is highly expensive. So they are collected in a common collection tank, a process called equalization. Adjustment of effluent pH in the collection tank is called neutralization. This is because only at neutral pH can the effluent support a wide range of microorganisms needed in the secondary treatment.

2.8.2 Secondary Treatment

In biological systems, the rate of substrate supply should be continuous so that F/M (Food/Microorganism) ratio is always maintained high. This keeps the microorganisms in active log phase and assures rapid metabolic rate. Therefore the treatment design is based on the F/M ratio of the substrate. Also, the settleability of sludge should be encouraged whereby the capsules of individual bacterial cells merge to form a slimy layer on pebbles such as in trickling filters or in flocs as in activated sludge formation. This becomes necessary because, more the concentration of bacteria in aggregates, maximum will be the rate of reaction. For effective biological treatment, the suggested optimum range of nutrients in any waste is BOD (100) : N (5) : P(1).

Biological treatment methods that include both aerobic and anaerobic methods are relatively neglected and less properly understood. The effective performance lies in the choosing of the right method that suits the composition of effluent. Mere addition of sewage to the waste water will not bring about the expected BOD/COD reduction. Generally the pollutants can be classified into two types. a) Readily degradable organics (food industry) and b) Complex organics (organochemical industry).

Selection of the appropriate method based on the above classification would however be effective and the following criteria should be considered while executing a bio-treatment plant. They are a) cost of investment b) adaptability of process to various organic and hydraulic loading rates c) less energy-demanding d) effective treatment performance e) limited land requirement f) and less sludge production. The two major categories of ecological treatment namely, aerobic and anaerobic process are well-known. But the treatment design, regulating factors, operational flexibility, process stability and control and treatment pattern of various methods that come under these two categories are hardly studied. Evaluation of carrier materials, mode of operation, and the organic and hydraulic loading rate of waste applied however makes the process successful.

Growth and metabolism of microbes involved in treatment process   In aerobic biological treatment process, the conversion of dissolved organic material to new cellular structure is accomplished by a variety of heterotrophic microorganisms. Several researches evaluated the presence of bacteria, fungi, protozoa, rotifers and other higher forms of life. Among these, bacteria form the majority of the mixed population group. A better understanding of the processes like adsorption, absorption, microbial enzymatic potency to bring out transformation and its growth kinetics provides us more information regarding the selection of appropriate method (Lawrence et al., 1970). The two distinct metabolic processes by which microorganisms stabilize the organic substances in an aquatic environment are respiration and synthesis of new cells. Through respiration, O_2 is consumed and utilized for the oxidation of organic or inorganic substances to liberate energy. With the energy acquired, they synthesize new microbial cells and the end products in the metabolic pathway are CO_2, water and ammonia. While observing the growth pattern of microorganisms, four different stages are observed. They are initial lag phase, exuberant growth-exhibiting log phase, growth-retarding stationary phase and death phase.

Food/microorganisms ratio   The acclimatization of microorganisms to the environment differs with the waste water applied for treatment. Its switch-over from lag phase to active growth phase depends on the amount of food present in the substrate and this linear relationship is expressed as food/microorganism ratio. However, the substrate phase proceeds with the demand of food. Increased O_2 consumption along with the new cell formation slows down the stationary phase, which finally leads to the death phase.

Aerobic treatment technology

In this treatment method, the aerobic microorganisms oxidize the organic matter to give carbon dioxide and water. Some of the aerobic reactors are activated sludge process, rotatory biological contact reactor, aerated lagoons and trickling filters

Activated sludge process   Here the oxidation of organic matter is accompanied by the biosorption and flocculation process and therefore gives a maximum efficiency (85–90%). Oxidation process is aided by the continuous aeration of culture biomass whereas the settleable solids are collected in the sedimentation tank as flocs. As macro-invertebrates cannot inhabit activated sludge reactor, problem of houseflies is not encountered. To facilitate microbial growth, the BOD : nitrogen : phosphorus ratio is maintained as 100 : 5 : 1 and the wide pH range (6–9) can withstand a wide variety of microorganisms. Another important advantage is that the settled sludge can be recycled, aerated and added with mixed liquor suspended solids (MLSS) to act upon the organic content.

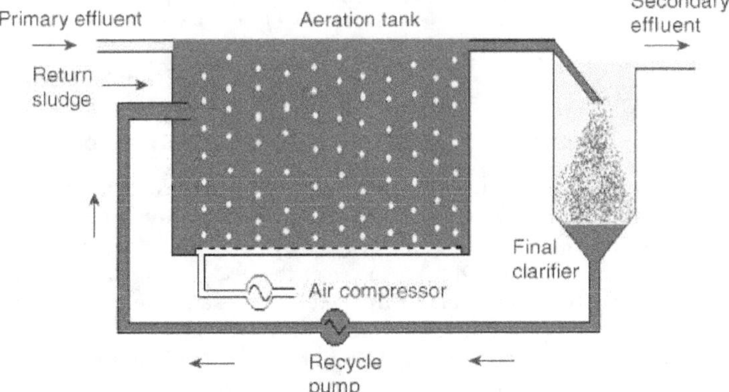

Figure 2.11   Activated sludge process

Rotatory biological contact reactor (RBC reactor)   This effluent reactor consists of a horizontal shaft across its surface. Numerous discs (2–3 mm diameter; 10–20 mm wide) are attached to the horizontal shaft and during the operation they are rotated so that the diffusion of atmospheric oxygen into the effluent is facilitated. Microbial growth on the surface of the disc does the oxidation process. Retention time is about 10–15 days and this method is highly efficient in terms of BOD/COD reduction rate. Minimum space requirement, easy operation, tolerance to shock loading, infrequent biomass washout, foam, aerosol and air stripping control are some of the advantages of the RBC reactor. However limited knowledge about its operation, excessive sludge production, fouling odour and limited aeration possibility restrict its commercial application.

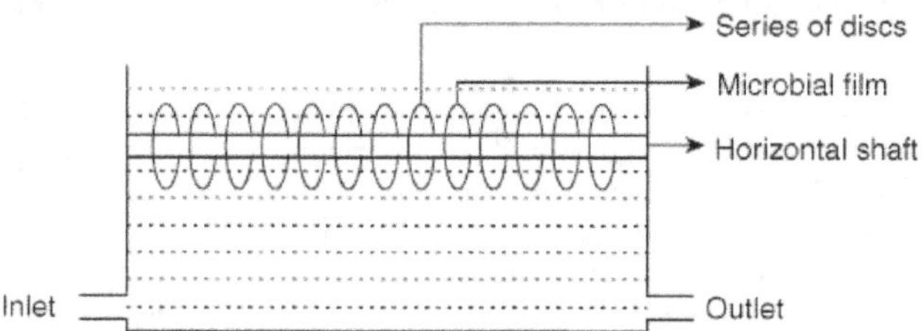

Figure 2.12   Rotatory biological contact reactor

Aerated lagoons/oxidation ponds   These consist of cemented large tanks with 5 m depth in which the effluent to be treated is collected. By suitable mechanical devices the tank is aerated. This facilitates the growth and proliferation of microbial biomass which in turn utilizes the organic content of the waste. It is reported that a retention time of about a week is sufficient enough to accomplish 90% COD reduction. In spite of its simple operation and easy maintenance, this system is popular due to large space requirement, environmental nuisance and bacterial contamination of the lagoon which further needs biological purification.

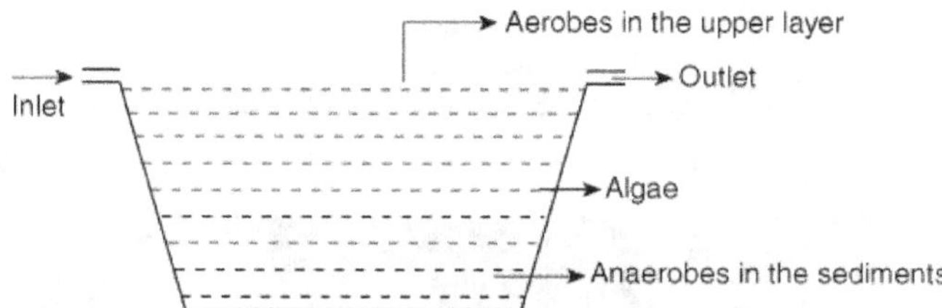

Figure 2.13   Aerated lagoons/oxidation ponds

Trickling filters   It is a circular or rectangular tank (3 m depth) made up of a tightly packed bed made up of several substances like sand, gravel, pebbles, coal, coke and synthetic resins. Usually the larger particles such as gravel or pebbles constitute the topmost layer. A fine wire mesh is spread over the bed which allows the direct contact of waste water with the slime layers of the bed. At the same time it prevents the biomass tearing off from the media due to high flow rates. The effluent to be treated is collected in the tank, and the microorganisms bring about the oxidation process by utilizing its nutrients. A rotatory wheel attached to a central shaft is kept over the mesh in such a way that it is partly submerged in the waste water. With the aid of a motor, the rotatory wheel is rotated and the water is sprinkled through its orifices or nozzles thus facilitating surface aeration. This arrangement makes the air diffuse through the media and even oxidizes the lower layer. The efficiency of this process depends on the pH, temperature, waste water loading, oxygen level and the presence of toxic compounds. The main disadvantage of this process is the high cost of construction.

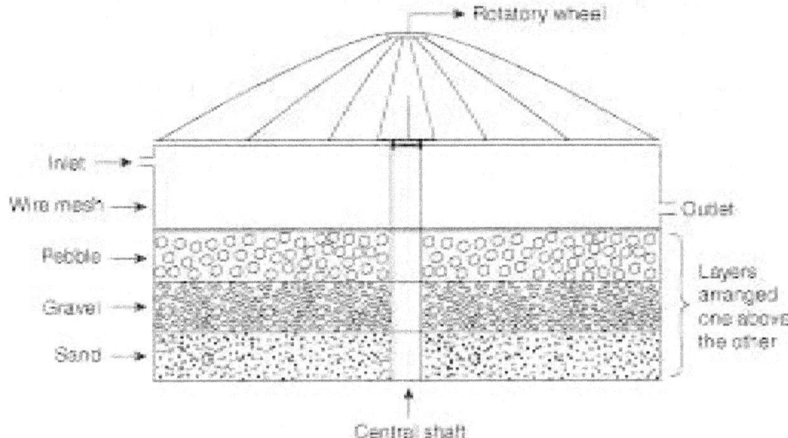

Figure 2.14   Trickling filters

Anaerobic treatment technology

Anaerobic digestion is a fermentation process by which the anaerobic microorganisms convert organic matter into CO_2 and methane using nitrates and sulphates as final electron acceptors. For many years this technology has been considered as a best option for sewage treatment. In recent years, advanced anaerobic reactors have been developed to overcome the drawbacks of the older models like sensitivity to high-strength waste waters and poor performance against toxic compounds. Also it is far more advantageous than the aerobic treatment due to the low quantity of sludge production, lower energy consumption, biogas production, better efficiency at high organic loading rates, no environmental nuisance (odour, aesthetic value), rejuvenation of microorganisms when required, short retention time and limited space.

During anaerobic degradation, the substrate undergoes a three-stage process. In the first stage, a group of anaerobic microorganisms, primarily cellulolytic bacteria act upon organic polymers. The reaction is an enzyme hydrolysis of polymers to the individual monomers. These monomers are fermented to various intermediates primarily acetate, propionate and butyrate. Additional acetate is produced by a second group of organisms termed as acetogenic bacteria. The acetate production is probably accompanied by a co-reaction of CO_2 reduction with hydrogen gas. Acetic acid becomes the substrate for a group of strictly anaerobic bacteria. These bacteria convert acetic acid to methane and carbon dioxide. This methane along with the methane formed by the bacteria that reduces CO_2 utilizing hydrogen gas or formate produced by other species accounts for the methane produced in this process. There are different methanogenic bacteria responsible for the production of methane. Methane formed during digestion, being insoluble in water, is lost to the gaseous phase. It can be collected and used as full.

However it is relatively soluble in water and readily combines with the hydroxide ions (OH) in the system and produces HCO_3 ions. However the reaction depends on several factors such as pH, bicarbonate concentration, temperature and substrate composition. Merits of the anaerobic treatment include successful treatment performance, less sludge production, low operating cost, anoxic condition, biogas production and low nutrient requirement. Since the performance of this treatment technology is dependant on the active growth of microorganisms and its utilization of substrates, certain factors are to be considered for screening the waste water to be treated. They are the origin and nature of waste water, temperature, concentration of suspended solids, presence of toxic compounds and their persistence and targeted reduction (Hickey and Owens, 1981). The prime advantages in adopting anaerobic technologies are the minimum energy requirement and biogas production steps involved in anaerobic microbial conversion.

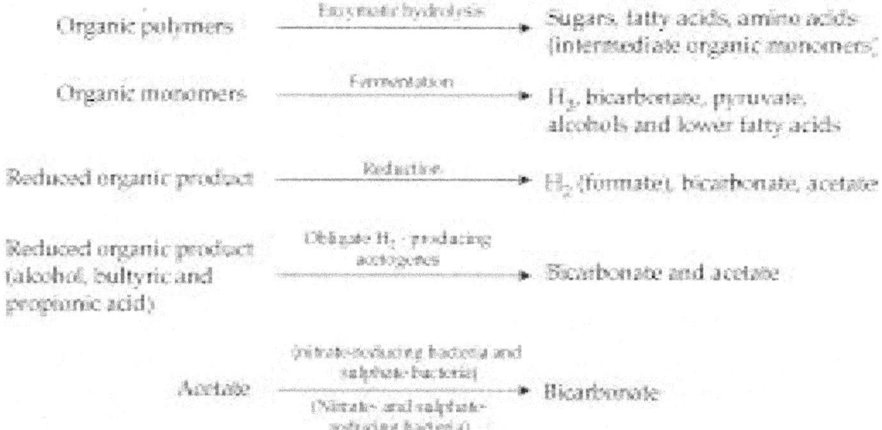

Despite the key factors, availability of substrate and viable microbial population, anaerobic treatment technology are reported to be more sensitive to pH, temperature, ionic strength, salinity, nutrients and toxic or inhibitory substances. Optimum pH range for anaerobic activity is near neutral and therefore excessive acid production and its accumulation are said to prevent methane production. (Chen et al., 1985). Higher salinities and temperature above 35°C (mesophilic) and above 60°C (thermophilic) are inhibitory to methanogenesis. Besides, total ionic strength and sulphates exert a significant control over the viability of the microbial population. Nitrogen, phosphorus and trace elements are the primary nutrient requirements for microorganisms and their appropriate level in the waste water should be evaluated so as to get an effective removal. For successful results, COD : N ratio should be 400 : 1 at high organic substrate loading and 1000 : 1 for low organic substrate loading respectively. The amount of N and P needed for anaerobic growth is calculated as 7 : 1 while the trace elements include iron, Ni, Mg, Ca, Na, Ba, tungstate, molybdate, selenium and cobalt.

Some of the anaerobic reactors are discussed below.

a.   Conventional anaerobic digester

b.   Anaerobic contact reactors

c.   Packed bed reactors

d.   Membrane bioreactors
e.   UASB
f.   Fixed film reactors
g.   Anaerobic baffled reactors
h.   Anaerobic filters

Conventional anaerobic digester   Conventional anaerobic digester tanks are designed in such a way as to have a continuous flow through a channel where the effluent fed is mixed either continuously or intermittently. Mixing ensures uniform temperature throughout the digester, which in turn ensures physical and chemical uniformity, intimate contact between microbial biomass and waste applied, rapid dispersion of metabolic end products and toxic products if present. Also it ensures prevention of surface scum formation and deposition of silt, grit and other heavy inert solids. Decontamination of wastes with high level of suspended solids and soluble organics, uniformity in environmental conditions such as pH, temperature and ionic strength, easy monitoring, deduction of toxicity below the accepted limits and good internal mixing that leads to equal distribution are some of the advantages of these systems.

Anaerobic contact reactor   Anaerobic contact digesters consist of a continuously stirred tank and sludge separator and provide a simple-to-operate anaerobic treatment system. It favours excellent contact between the biomass and the effluent. Activated biomass is attached along the tilted plates at the bottom of the reactor and the effluent is allowed to flow in a horizontal pathway perpendicular to the tilted plates. External sludge separator splits the effluent from the biomass and sends the clear fluid through the outlet. This prevents the mixing of biomass with the effluent and also maintains a high biomass concentration. The advantage of this system is excellent solid removal combined with the minor resuspension. Since the gas bubbles produced due to continuous stirring get attached to the solid particles which cause turbulence and prevent the settleability of solids, effluents with high loading rates cannot be treated.

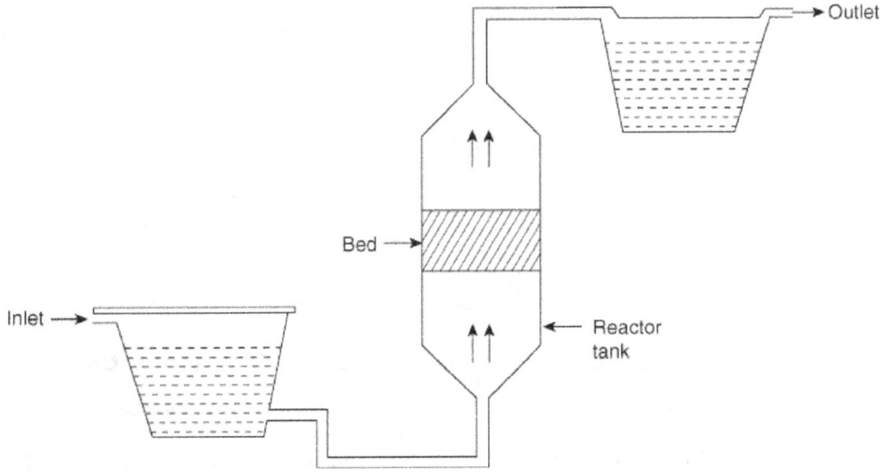

Figure 2.15   Packed bed reactors

Packed bed reactors   The problem caused by the clogging of the filters and inefficiency in treating high-strength wastes in anaerobic contact reactors have resulted in the construction of packed bed reactors. This reactor is found to be 30 times more efficient than the contact digesters. Here the microbial biomass is maintained in flocculated form and is packed in an enclosed vessel. The effluent is passed in an upward direction. This prevents the sloughing of suspended solids into the effluent. An interesting thing in this reactor is that it ensures better performance without the use of carrier media.

Membrane bioreactors   These reactors are used to treat complicated wastes such as effluents with a wide range of toxic compounds. Here the microbial biomass is attached to a silicon rubber membrane. Based on the composition of effluents, heterogeneous biofilms characterized by the complex assemblage of cell types and gradients of physicochemical parameters are used. The problem encountered in this reactor is that specialized cultures inoculated to treat targeted toxic compounds have to be replaced every time.

Upward anaerobic sludge blanket reactor (UASB)   Upward anaerobic sludge blanket reactor is a complete and proven technology for the treatment of a wide variety of waste waters. With respect to its design, sufficient information is available for the installation of the appropriate systems including the pre- and post-treatment steps. It consists of a circular reactor with a provision for gas collection at its top. Sampling ports provided along the sides of the reactor are used to study the sludge profile. Two feed inlet points to ensure complete mixing with a common feed distributor are connected to a peristaltic pump that allows the effluent to flow inside. In order to achieve high sludge hold up, the reactor is equipped with a gas solid separator (GSS) device, the function of which is to separate and discharge biogas from the reactor; to prevent the washout of viable bacterial matter; to enable the sludge to slide back into the digester compartment; to serve as a kind of barrier for rapid excessive expansions of a sludge blanket; to provide a polishing effect and to prevent the washout of floating granular sludge.

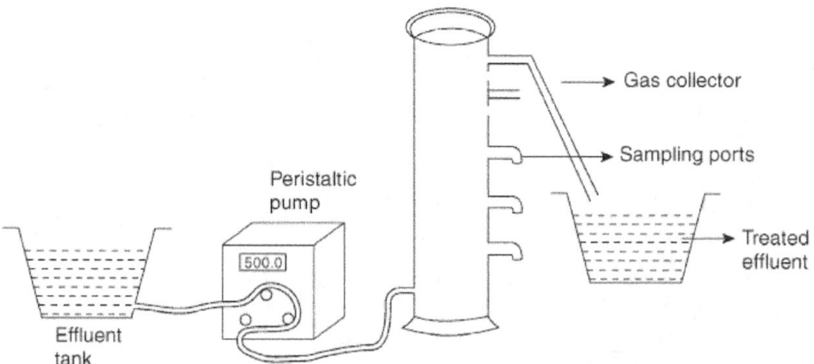

Figure 2.16   Upward anaerobic sludge blanket reactor

During the reaction process, the sludge blanket as a whole acts as a dynamic unity. At low loading rates, the sludge bed expands as a dynamic unity. At low loading rates, the sludge bed expands as a result of occluded gas produced by anaerobic digestion. Thus the liberation of the entrapped gas causes a local turn over of the sludge bed and the gas eruptions succeed each other rapidly, giving the surface of the sludge blanket the appearance of a boiling fluid. The half-freed particles formed by the tearing forces of fast upward moving separate gas bubbles are kept dispersed in the liquid bulk above the sludge blanket. When the waste water strength is increased, reflocculation occurs to form larger, more easily settling flocs and a balance between sludge ejection and entrapment in the sludge blanket is obtained.

Fixed film reactors   In the case of fixed film expanded bed and fluidized bed reactors, microbial biomass colonized on a solid material is used. The supporting material may be a series of parallel channels or a tightly packed bed formed of sand/pebbles. The effluent flow can be operated in an up-flow or down-flow mode. In the case of hybrid reactors, the lower part of the reactor operates like UASB and the upper part is filled with floatable materials like polyurethane foam or plastic pieces to act as fixed film reactor.

Anaerobic baffled reactor   With the walls (baffles) built from top to bottom across the reactor tank, complete mixing and equal distribution of both the effluent and microorganisms is ensured. At high hydraulic loading rates, the only disadvantage that withhold this type of treatment system is the frequent washout of mother liquor suspended solids (MLSS). However under normal conditions, the baffles help the biomass to be retained in the tank itself.

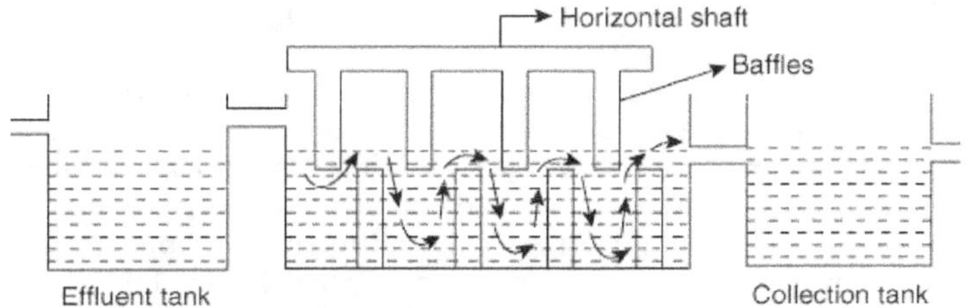

Figure 2.17   Anaerobic baffled reactor

Anaerobic filters   They do not need a separation device to accumulate the biomass in the reactor. Retention is done through the adsorption and adhesion on an inert support

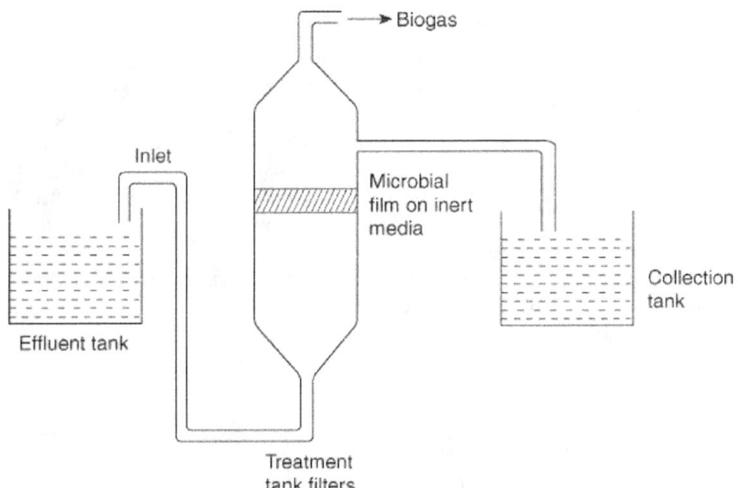

Figure 2.18   Anaerobic filters

material by means of slime films with a very strong binding. The sludge is captured and firmly retained within the macroporous structure of the carrier material. This greatly prevents the biomass washout. High cost of carrier material and potential risk of clogging are the two factors that inhibit its usage in small scale treatment plants. Therefore the choice of the carrier material in terms of specific surface, total porosity and shape is very important.

Expanded and fluidized bed process   Expanded bed systems account for increased settled bed volume by 15–30% while fluidized bed process produces 25–300 % bed expansion. The most commonly used media are spherical silica sand of 0.2–0.5 mm diameter and 2.65 specific gravity. However, granulate active carbon and basalt can also be used. In anaerobic filters instead of a separable device that accumulate biomass in the reactor, the process facilitates retention through adsorption and adhesion on an inert support material by means of slime growth. This prevents frequent washout of biomass. The two major disadvantages that restrict its utility in large scale treatment plants are high cost of clogging material and the potential risk of clogging. The major problem encountered in anaerobic filter, the risk of clogging due to accumulation of inert solids is tackled in his process.

Fixed film anaerobic reactor (FFAR)   FFAR is a typical example for cost-effective design. Here, the higher the biological floc concentration, the maximum will be the system performance. In this rear, FFAR utilizes the fixed film that supports the development of active biomass thus achieving prolonged MCRT (mean cell residence time) without the danger of cell washout. The relative stability of the system helps to recover from shock loads, resist temperature fluctuations and obtain better gas production rate. Classified as up-flow fixed and down-flow fixed film lead reactors in terms of effluent inlet, the system provides a stable and operable form of anaerobic treatment technology.

2.8.3 Tertiary Treatment

It involves processes like coagulation, sedimentation, filtration, disinfection, ion exchange fluoridation, softening, demineralization, colour and odour removal and removal of iron and manganese.

Coagulation process   The process of coagulation involves the agglomeration of suspended particulate matter. For the process to be accomplished, the charge on the surface of the suspended flocs should be opposite to the charge of the coagulants used. Usually the charge on the suspended particulates is negative and is expressed in terms of millivolts. Therefore synthetic coagulants or polyelectrolytes with positive charge are used. Also anionic coagulants with negative charge, non-ionic polyelectrolytes with both positive and negative charge are available. Therefore the effectiveness of coagulation depends on the magnitude of zeta potential and the magnitude of electric charge on the particle surface. The principal coagulants that are used in waste water treatment are aluminium sulphate, ferric chloride, ferrous sulphate, activated silica, ozone and the reactions involved are as follows.

Aluminium sulphate

$$Al_2(SO_4)_3 + 3Ca(HCO_3)_2 \rightarrow 3CaSO_4 + 2Al(OH)_3 + 6CO_2$$
$$Al_2(SO_4)_3 + 3Ca(OH)_2 \rightarrow 3CaSO_4 + 2Al(OH)_3$$
$$Al_2(SO_4)_3 + 3Na_2CO_3 + 3H_2O \rightarrow 2Al(OH)_3 + 3Na_2SO_2 + 3CO_2$$
$$Al_2(SO_4)_3 + 6NaOH \rightarrow 2Al(OH)_3 + 3Na_2SO_4$$

Ferric chloride

$$2FeCl_3 + 3Ca(HCO_3)_2 \rightarrow 3CaCl_2 + 2Fe(OH)_3 + 6CO_2$$
$$2FeCl_3 + 3Ca(OH)_2 \rightarrow 3CaCl_2 + 2Fe(OH)_3$$

Ferrous sulphate

$$FeSO_4 + Ca(HCO_3)_2 \rightarrow Fe(OH)_2 + CaSO_4 + 2CO_2$$
$$FeSO_4 + Ca(OH)_2 \rightarrow Fe(OH)_2 + CaSO_4$$
$$2Fe(OH)_2 + \tfrac{1}{2}O_2 + H_2O \rightarrow Fe(OH)_3$$

Copper sulphate

$$CuSO_4 + Ca(HCO_3)_2 \rightarrow Cu(OH)_2 + CaSO_4 + 2CO_2$$
$$CuSO_4 + Ca(OH)_2 \rightarrow Cu(OH)_2 + CaSO_4$$

Polymerisation reaction

$$2[Fe(H_2O)_5(OH)]^{2+} \rightarrow [Fe(H_2O)_8(OH)_2]^{4+} + 2H_2O$$

Sedimentation   It is a physical process ensuring the settling of suspended particulates. Baffles provided inside the tank favours continuous stirring of the tank, however with variation in its speed. As soon as the rotation of the baffle slows down, high density sludge particles settle at the bottom and the process is accomplished within 15–60 min. depending on the suspended solids. To reduce the continuous working of baffles, the effluents can be allowed to follow a circular or horizontal flow pattern. The sludge clarifier at the bottom prevents the mixing of settled sludge with the effluent.

Filtration   The main purpose of filtration is to separate the suspended particulates from the water by means of filters. Filtration is of two types—surface filtration and in-depth filtration

With the aid of supporting media, surface filtration can be performed whereas in-depth filtration requires a filter bed. The total height of filter bed is about 80 cm in which anthracite coal, sand and gravel are arranged in layers. The percolation of water along the bed retains the solid particles and it is interesting to note that the real filtration takes place even in the first 3–4 cm and the cleaning can be done by back washing. Care should be taken while washing the bed as it may expand it. Simultaneous spray of air and water from the bottom to top can prevent the expansion of the bed. Moreover, based on the water quality, filtration speed can be varied. For instance, slow sand filters are used for water with high content of impurities whereas pretreated water is subjected to rapid sand filtration. This ensures a smoothening effect.

Disinfection   The main aim of disinfection is to kill the microorganisms, the causative agents of contagious diseases. Also it eliminates the smell and taste of the aquatic system. Commonly used disinfectants are chlorine gas or hypochlorates of calcium and sodium. Chlorine reacts with water to give hypochlorous acid (HOCl). However at low pH, HOCl is converted to OCl^- which is ineffective. HOCl kills the microorganisms by penetration and oxidation of cell matter. But its action on microorganisms is prevented by several interfering compounds such as Fe, Mn, Ca, Mg, NO_2^-, CO_3^{2-}, HCO_3^- and NH_3. Therefore it is essential to add chlorine in excess. Also, the products acquired by the reaction between ammonia and hypochlorate vary along the pH range. i.e., at pH greater than 8.0, monochloramines are produced; between 4.3 and 5.0, dichloramines are produced; and at pH < 4.3, trichloramines are produced.

$NH_3 + HOCl \rightarrow NH_2Cl + H_2O$ monochloramines

$NH_3 + 2HOCl \rightarrow NHCl_2 + 2H_2O$ dichloramines

$NH_3 + 3HOCl \rightarrow NCl_3 + 3H_2O$ trichloramines

These chloramines, though germicidal in their action, cannot be utilized for disinfection purposes because they act very slowly.

Removal of iron and manganese   The presence of iron (Fe) and manganese (Mn) in water makes it unfit for domestic and industrial purposes as well. The more their concentration in water the maximum will be the colour change and metallic taste. This deliberately makes it unpalatable. On the other hand, they lead to corrosion of pipes and cause laundering problems. Also they cannot be used for industrial processing. Therefore their removal is the essential criteria in water and waste water treatment. It involves two steps.

1.   Peptization/oxidation
2.   Aeration or treatment with Cl_2, ClO_2 and $KMnO_4$

The reactions are as follows.

$Fe^{2+} + O_2 + 8OH^- + 2H_2O \rightarrow 4Fe\,(OH)_3$

$Mn^{2+} + 2ClO_2 + 2H_2O \rightarrow MnO_2 + 2O_2 + 2Cl^- + 4H^+$

$Mn^{2+} + O_3 + H_2O \rightarrow MnO_2 + O_2 + 2H^+$

$3Mn^{2+} + 2MnO_4^- + 2H_2O \rightarrow 5MnO_2 + 4H^+$

Other options are ion exchange method and zeolite treatment.

Ion exchange   The reactions involved in this method are as follows.

$H_2R + Fe^{3+} \rightarrow Fe - R + 2H^+$

$H_2R + Mn^{3+} \rightarrow Mn - R + 2H^+$

The ion exchange method favours the removal of small quantities of Fe and Mn. However it cannot be adopted in small-scale plants as it is highly expensive. Zeolites are complex aluminum silicates of alkali metals and their usage is better than the ion exchange method because of better performance and low cost.

$Na_2Z + Fe^{3+} \rightarrow Fe - Z + 2Na^+$

$Na_2Z + Mn^{3+} \rightarrow Mn - Z + 2Na^+$

Softening of water   The presence of chlorides and sulphates of calcium and magnesium render hardness to water. The form in which these substances exist determines the strength of hardness. For example $Ca(HCO_3)_2$ and $Mg\,(CHCO_3)_2$ give temporary hardness to water while $CaSO_4$, $MgSO_4$, $CaCl_2$ and $MgCl_2$ render it permanently hard. Soap test can evaluate hardness of water. Hard water does not give lather with soap. Thus after its confirmation, steps should be taken to soften the water. Three methods are practiced to soften hard water. They are

a.   Lime soda ash process
b.   Demineralization method
c.   Calgon process

Lime soda ash process   Addition of calcium hydroxide $Ca(OH)_2$ and sodium carbonate (Na_2CO_3) to water is the first step in the softening process. The quality of these substances to be added is determined based on the hardness of the water. Precipitated calcium and magnesium salts can be easily removed and the reactions are as follows.

Calcium hydroxide

$$Ca(HCO_3)_2 + Ca(OH)_2 \rightarrow 2CaCO_3 + H_2O + CO_2$$
$$Mg(HCO_3)_2 + Ca(OH)_2 \rightarrow MgCO_3 + CaCO_3 + H_2O + CO_2$$

Sodium carbonate

$$CaCl_2 + Na_2CO_3 \rightarrow CaCo_3 + 2NaCl$$
$$MgCl_2 + Na_2CO_3 \rightarrow MgCO_3 + 2NaCl$$
$$CaSO_4 + Na_2CO_3 \rightarrow CaCO_3 + Na_2SO_4$$
$$MgSO_4 + Na_2CO_3 \rightarrow MgCO_3 + Na_2SO_4$$

Demineralization method   In this process, hydrogen form (cation exchange resin (CH^+)) and hydroxy form (anion exchange resin (OH)) are used to eliminate respective ions from water. Vinyl styrene copolymer with sulphonation, acidic cationic exchange resin, vinyl styrene base with labile OH group and basic anionic exchange resin are commercially available and the reactions are as follows.

$$H_2R + CaCl_2 \rightarrow CaR + 2HCl$$
$$H_2R + MgCl_2 \rightarrow Mg–R + 2HCl$$
$$H_2R + CaSO_4 \rightarrow CaR + H_2SO_4$$
$$H_2R + MgSO_4 \rightarrow Mg–R + H_2SO_4$$
$$RNH_3 – OH + HCl \rightarrow RNH_3Cl + H_2O$$
$$RNH_3 – OH + H_2SO_4 \rightarrow RNH_3SO_4 + H_2O$$
$$RNH_3 – OH + H_2O \rightarrow RNH_3OH + H_2O$$

Nowadays, natural ion exchange resins such as sodium zeolite $(Na_2OAl_2O_3 \cdot SiO_2)$ have gained more interest.

$$Ca(HCO_3)_2 + Na_2Z \rightarrow CaZ + 2NaHCO_3$$
$$CaSO_4 + Na_2Z \rightarrow CaZ + Na_2SO_4$$
$$Mg(NO_3)_2 + Na_2Z \rightarrow MgZ + NaNO_3$$

After use, these resins can be regenerated for repeated use. Regeneration of resins to regain their original form H^+ and OH can be done with HCl, NH_4OH (synthetic resins) and NaCl (natural resins). However during the process, the chambers containing cation and anion exchange resins must be disconnected.

Synthetic resins

$$CaR + 2HCl \rightarrow H_2R + CaCl_2$$
$$MgR + 2HCl \rightarrow H_2R + MgCl_2$$
$$RNH_3Cl + NH_4OH \rightarrow NH_4Cl + RNH_3 – OH$$
$$RNH_3 SO_4 + 2NH_4OH \rightarrow (NH_4)_2SO_4 + RNH_3 – OH$$

Natural resins

$$CaZ + 2NaCl \rightarrow Na_2Z + CaCl_2$$
$$MgZ + 2NaCl \rightarrow Na_2Z + MgCl_2$$

Calgon process   Insoluble phosphate salts of calcium and magnesium present in the water can be removed by the addition of sodium hexametaphosphate. The precipitated salts can then be removed.

$$MgSO_4 + 2NaH_2PO_4 \rightarrow Mg(H_2PO_4)_2 + Na_2SO_4$$
$$3CaSO_4 + 3Na_3PO_4 \rightarrow Ca_3(PO_4)_2 + 3Na_2SO_4$$
$$3Ca(HCO_3)_2 + 2Na_3PO_4 \rightarrow Ca_3(PO_4)_2 + 6NaHCO_3$$

Taste and odour removal   Chemical based odour in water is due to the presence of phenol, H_2S, chloramines and chlororganics. Decay of biological components mainly algae produces foul smell. This can be rectified by aeration. But the chemical based odour problem needs special treatment such as aeration, coagulation, break point chlorination and application of suitable chemicals. If the water has more colour, activated or granular carbon can be used for better adsorption of impurities, the former is more effective than the latter. However interfering compounds like dyes and detergents should be eliminated. Alternatively chemicals such as ClO_2, O_3, $KMnO_4$ and $CoSO_4$ can also be used.

Fluoridation   Fluoride at optimum level (0.4–1mg/l) promotes the formation of dental enamel and protects the teeth against caries. When this level goes beyond the limit (1.5 mg/l) it causes destruction of enamel, dental disorder, decalcification, mineralization of tendons, digestive and nervous disorders. Under these conditions, the water has to be treated with $Ca_3(PO_4)_2$, $Al_2O_3SiO_3$ or $Ca(OH)_2$.

EXERCISES

1. Concentration of heavy metals in the paper mill effluent are in the order of ______________.

2. Nitrogen fertilizer plant generates effluent from ______________ and ______________ unit.

3. ______________ weight of the average individual faeces is nothing but coliforms.

4. Classify water based on its utility.

5. What were the different treatment methods adopted to treat water in ancient times.

6. Describe the nutshell of standards published in 1914.

7. In what way were the standards published in 1925 different from the previous standards.

8. What were the main objectives of the 1962 standards?

9. Name the two major nutrients implicated in eutrophication process.

10. Correlate loss of species diversity with eutrophication and explain.

11. Discuss the changes that take place in a water body during the transformation from aerobic to anaerobic condition.

12. Why are stagnant waters more prone to 'pollution effect' than running waters?

13. Explain the toxic effect of heavy metals on aquatic organisms on the basis of their reactive nature.
14. Define LC_{50}.
15. Organisms that can withstand lethal concentration cannot be expected to have nil effect—justify.
16. What is more dangerous—bioaccumulation or biomagnification?
17. What would be the pH of rivers embedded with rocks made of limestone?
18. Name the acidifying species of nitrogen and sulphur and explain the process by which they reach the earth.
19. List out the harmful effects of thermal pollutants.
20. Discuss the process of self-purification in rivers.
21. Why should the analyst visit the sampling site?
22. Discuss the importance of time and frequency of sampling in the sampling programme.
23. How will you take samples in a wide water source?
24. What are the parameters that come under conservative material sampling?
25. Illustrate with diagrams the different points of a water body for the sampling of nonconservative materials.
26. Based on the objective of the assessment study, the parameters to be analysed vary. Discuss.
27. Explain the advantages of composite sampling methods over grate sampling method.
28. What are the different types of samples?
29. What is the maximum holding time for ammonia, nitrite, nitrate and organic carbon?
30. What is a spiked sample?
31. In what way does regular servicing of processing equipment reduce pollution load.
32. Out of the two tanning methods, which is more preferable—vegetable or chrome tanning. Why?
33. What are the steps involved in the processing of skin?
34. What are the common pollutants in sugar mill industry.
35. Which step in the pulp and paper processing generates effluent in bulk?
36. Explain the steps, mercerising, dyeing, printing and finishing in textile processing.
37. In textile industry, which chemical is used to bleach cotton fabrics?
38. Write down the characteristics of the dairy effluent.
39. Which is the raw material for the sago industry?
40. What are the two processes in the manufacture of NPK fertilizers.
41. Write down the manufacturing process in electroplating industry.
42. What are the three major divisions of pesticides?
43. List out the chronic effects of pesticide effluent on aquatic organisms.
44. How will you design a treatment method for a particular effluent?
45. Conventional biological methods cannot be applied to waste water containing hazardous substances. Explain.
46. Compare the characteristics of dairy effluent with that of pesticide effluent.

47. Design a treatment plant for both an organic and inorganic effluent.

48. Other than wood, what are the raw materials used for paper production?

49. What are the chemicals used in the bleachng process of pulp and paper industry?

50. Which is the substance in the pulp and paper mill effluent that imparts colour?

51. What are the treatment options of petrochemical effluent?

52. What is the composition of alkaline solution in scouring process of textile production?

53. Name the predominant heavy metals in textile mill effluent and what is their source?

54. What are the inorganic constituents in dairy effluent?

55. List out few aerobic treatment methods to treat dairy effluent.

56. Discuss the adverse effects of nitrates on human health.

57. What are the heavy metals used in the electroplating industry?

58. Explain the bathing and polishing process of electroplating industry.

59. Name the compound of electroplating industrial effluent which renders colour to the streams.

60. Give an account of manufacturing process in petrochemical industry.

61. Explain how O_2 depletion in aquatic system affects life directly and indirectly?

62. Discuss the effect of pesticide effluent on the environment.

63. Describe the manufacturing process of electroplating industry.

64. What is the role of phosphate fertilizer in soil.

65. Write down the different aerobic methods by which sago effluent can be treated.

③

EFFLUENT AND SEWAGE ANALYSES

3.1 INTRODUCTION

It is well known that the analytical procedures can be carried out in a precise manner only when the concept is clearly understood. It is for this reason that this chapter attempts to provide background information, ecological significance and application for every parameter in concern. The procedures given here are widely accepted standard methods. Slight modification in the lab techniques and chemicals present are the outcome of our own experience. However, accuracy in analytical results depends on several factors such as instruments and reagents used, method and type of sampling, etc. Analyses carried out using regularly serviced instruments give reliable results. Certain high quality chemicals may be expensive and in such cases, low-grade chemicals can be purchased and then purified by simple methods. It should be noted that all the chemicals cannot be treated as mentioned above and hence the analyst should decide the need depending on the type of method followed, nature of parameter and significance of analysis. Regarding sampling, much has been discussed in chapter 2. Even then, frequency and time of sampling, sampling procedures, sampling location and method of sampling cannot be generalized as they are different for different effluents. For example, processing of raw materials to a final product involves a series of steps and in each step, a variety of chemicals are used. Therefore waste water coming out from different processing units differ in quantity and quality.

The sampling location varies with the purpose of study. To analyse the concentration of individual chemicals, it is better to take samples from the outlet of different processing units than to take the samples from combined effluents, whereas to assess the effect of effluent on water or soil, grab samples at regular intervals are preferred. While taking the combined effluent it should be noted that the effluent is not mixed with sewage or canteen wastes. Samples taken at waste water treatment plants should be labelled as follows:

Name of the treatment	: _______________________________
Sampling date	: _______________________________
Inlet/outlet	: _______________________________
Detention date	: _______________________________
Characteristics to be analysed	: _______________________________
Storage time	: _______________________________

3.2 SAMPLE PREPARATION

Sample preparation is an important step in quantitative analysis. This is because all the parameters cannot be evaluated by simple analysis. Some constituents in the sample may be in minute quantities and the chemicals used to evaluate their amount are too sensitive and are prone to chemical transformations by interfacing compounds. Therefore as a precautionary measure these compounds have to be eliminated so that the constituents to be analysed are left behind. Based on the parameter to be evaluated the sample preparation procedure varies. They are

1. precipitation
2. filtration
3. ignition
4. acid digestion

3.2.1 Precipitation

1. Add suitable chemicals to the sample and precipitate the solids.
2. Transfer the precipitated solids into a suitable solvent/reagent and then proceed with the analysis.

3.2.2 Filtration

1. Filter the sample with membrane filter or glass fibre filter paper.
2. Select the pore size of the paper according to the need.
3. In the case of ash filter paper, take the initial weight and then filter the sample.
4. During filtration, see to it that the samples along with the settled solids are completely transferred to the filter paper.
5. To prevent the dissolved solids sticking onto the filter paper, wash the filter paper with distilled water thrice.
6. Ignite the filter paper at 550°C in a muffle furnace. This temperature ignites the paper leaving behind the filtered substance.
7. Dissolve the substance in suitable solvent and proceed with the analysis.

3.2.3 Ignition

The drying process removes the available free water and therefore wastes with considerable amount of organic matter can be evaluated by drying the sample at 103°C. In certain cases, wastes may contain more amount of inorganic constituents and organic matter in limited quantities. To eliminate the organic content, dry the sample at 180°C. It is at this temperature that the oxidation of organic matter occurs so that the inorganic content is maintained stable for analysis.

3.2.4 Acid Digestion

Evaluation of certain inorganic constituents and heavy metals demand the complete elimination of suspended solids, colour and organic matter from the sample. These samples should be digested with acids. The commonly used digestion methods are given below.

1. *Nitric acid digestion*   Calcium, magnesium, sodium, potassium, lithium and strontium.

2.   *Nitric acid and sulphuric acid digestion*   Aluminium, barium, beryllium, cadmium, chromium, copper, iron, lead, manganese, nickel, selenium, silver, vanadium and zinc.

3.   *Nitric and perchloric acid digestion*   Aluminium, barium, beryllium, cadmium, chromium, copper, iron, lead, manganese, nickel, selenium, silver, vanadium and zinc.

The effluent characteristics discussed in this chapter are as follows.

Physical constituents	colour, turbidity, odour and temperature
Chemical constituents	pH, acidity, alkalinity, CO_2, total solids, suspended solids, dissolved solids
Organic constituents	DO, BOD, COD, nitrogen
Inorganic constituents	chlorides, nitrates, nitrites, sulphate, sulphide and fluoride
Metallic constituents	nickel, copper, mercury, chromium, zinc, lead, manganese, iron, calcium, magnesium, sodium and potassium
Accessory constituents	arsenic, phenols, tannin and lignin, surfactants, boron, cyanide.

PHYSICAL CHARACTERISTICS

3.3 COLOUR

Water flooding over the swampy areas or coloured soil layers are often coloured and they cannot be used for consumption or even for industrial purposes without proper treatment. When the water comes in contact with plant material like leaves, needles of conifers and wood, colouring material are formed. These colouring materials are plant extracts formed by the plant material at different stages of decomposition. Among the different plant extracts, tannin, humic acid and humates (ferric) are the principal components which impart colour. Colour formation in water is of two types.

1. True colour
2. Apparent colour

3.3.1 True Colour

Since natural colour is due to the presence of negatively charged colloidal particles, by mere addition of coagulants they can be removed. In general, trivalent metallic ions such as aluminium or iron are used. However colouring substances from artificial sources such as effluents of dyeing units in textile industry and pulp and paper mill impart strong colour to the water to the extent that they cannot be removed easily. Pulp and paper mill waste water has dissolved lignin derivatives and chemicals added for processing of paper complexed with the natural products which intensify the colour and make its removal complicated. Highly coloured substances are quite resistant to biodegradation. Therefore water containing colouring substance has harmful or toxic properties. The colour of the water is usually light yellow and the disinfection process performed will lead to the production of still more toxic compounds. Addition of chlorine results in the formation of chloroform, trihalomethanes and chlorinate organics. Basically, in drinking water, the presence of colour and odour is more bothering than its chemical safety because coloured water is not aesthetically acceptable. Therefore standards for colour are established by the United States Environmental Protection Agency and according to it, potable water should not have colour exceeding 15 units. Since colour is caused by a variety of substances both natural and artificial, an arbitrary standard is maintained for visual comparison.

3.3.2 Apparent Colour

The presence of coloured suspended matter imparts colour to water and the colour can be easily removed by separating the suspended solids from the water. To make accurate estimation of true colour, apparent-colour-producing suspended solids should be removed. To facilitate this removal, the sample should be subjected to centrifugation. Filtering the sample should not be done as the filter medium may adsorb the colour.

Potassium chloroplatinate (K_2PtCl_6) added with small amount of cobalt chloride gives colour which exactly resembles natural colours. Therefore it is used as standard colour and different shades of standards can be produced by adjusting the volume of cobalt chloride. Usually 1 mg/l of platinum is kept as standard unit of colour and the stock solution is prepared by adding 500 mg/l of platinum with little amount of cobalt chloride. From this, we can prepare a series of working solution—500 colour units can be prepared by dilution ranging from 0 to 70 colour units. Samples with colour units less than 70 can be measured by direct comparison, while colour units greater than 70 units should be diluted in such a way that the colour unit is brought down to 70 units. During calculation the results obtained can be multiplied by the dilution factor.

Since the standard solution is required to be prepared often, instruments with a variety of coloured glass discs have been developed. The devices can be operated even by untrained personnel in laboratories and are useful in field measurements where the preparation of standard colour solutions is not possible. However this method is not recognized due to the colour variation in the glass discs. Moreover, the glass discs may show different colour characteristics due to external influences such as fingerprints, dust, etc. For colouring substances that do not come under the colour range of standards 0–70 colour units, spectrophotometric method can be employed.

3.3.3 Significance

1. Measurement of colour is very essential for an environmental engineer to find out whether a chemical treatment to eliminate trihalomethane is required.
2. It is also used to select the suitable chemicals and their required dosage in chemical treatment.
3. To find the aesthetic quality of the potable water and to check whether the water used for industrial processes is colour-free.
4. To go in for a more suitable treatment method, as colour removal is an expensive process with high investment and operation cost.
5. To ensure the effectiveness of treatment in treated water and to check whether water discharged into the receiving streams or rivers are within maximum permissible limits.

Expt. 1   *To measure the colour of the effluent sample*

MATERIALS REQUIRED

Nessler's tube

Colour Standards

Dissolve 1.245 g of potassium chloroplatinate and 1.0 g of crystalline cobaltous chloride in a little distilled water. To this add 100 ml of conc. hydrochloric acid. Make up to 1 l. Colour value of this solution is 500 colour units. Prepare several dilutions of this standard solution ranging from 0.5 to 7 ml at 0.5 ml interval. Make up each dilution to 50 ml. Transfer the dilutions into standard Nessler's tubes. The colour value of these test tubes is as follows: 5, 10, 15, 20, 25, 30, 35, 40, 45, 50, 55, 60, 65 and 70. Mark the colour value in the respective tubes, close them, and store for further use.

PROCEDURE

1. Take 50 ml of the sample.
2. Remove the suspended particulates in the sample by centrifugation.
3. Carefully transfer free supernatant to the Nessler's tube.
4. Place the standard tubes against a white background.
5. Compare the colour of the sample with that of the standard.
6. Note down the colour value.
7. If the colour of the sample does not match with the standard, dilute the   sample and repeat the procedure as described above.

CALCULATION

Colour (unit) = Estimated colour × Dilution factor

3.4 TURBIDITY

3.4.1 Principle

The presence of suspended particulates in water inhibiting the passage of light through it is termed as turbidity. To make this definition more simple, it can be expressed as "visual interference". The suspended particles may be colloidal or coarse dispersions and their size decides the degree of turbulence. In stagnant waters colloidal and fine dispersions are the two main constituents which induce turbidity whereas in rivers, coarse particles contribute more to turbidity, and include the topsoil washed out from the fields, coarse particles from the mountain areas, domestic waste water, and industrial waste water. They contain clay, silt, and organic and inorganic substances. Organic matter induces the proliferation of microbial groups, which in turn enhances the turbidity. Similarly inorganic constituents like nitrogen and phosphorus induce the algal bloom that spreads out to increase the turbidity. Therefore both the organic and inorganic substances turn the water turbid. Turbidity can be identified even by visual appearance and hence turbid water is not preferred as potable water. Moreover turbid water is always associated with water pollution especially waterborne diseases. Filtration even though sounds to be the cheap and best option for turbid waters, it is very difficult to adopt this method in drinking water supplies because in water with high degree of turbulence, operation and maintenance cost of filters is high as it requires special construction. Moreover high turbidity shortens the filter runs and needs frequent cleaning. To destroy the harmful pathogens, several disinfecting agents like chlorine, chlorine dioxide and ozone are applied. However the success of disinfectant application depends on the direct contact between the microorganisms and disinfectant. But in turbid water, the organisms encase themselves in the suspended particulates (organic and inorganic) and escape from the disinfectant. Therefore water should be made clear before disinfection.

Initially Jackson Candle Nephelometry method was used for turbidity measurement. For calibration, silicon dioxide was kept as arbitrary standard, 1 mg SiO_2 /l = 1 unit turbidity. Then it was replaced by Nephelometry method. Instead of silica, formazin polymer is used for standard preparations and the results are expressed in nephelometric turbidity units (NTU). It measures the scattering of light from the turbidity-causing particles and therefore considered to be a more sensitive and reliable method. Nowadays standards like styrene divinyl benzene are available.

3.4.2 Method of Determination

The main components of the instrument include

1. Light source
2. Photoelectric detectors
3. Monitor/readout device

The instrument can measure samples with a range of turbidity (0.02 NTU–40 NTU). Formazin polymer is used as standard and while using it, it should be noted that 40 NTU = 10 JTU. Samples having the turbidity values greater than 40 NTU should be sufficiently diluted to bring the level below 40 NTU. The turbidity is then measured by multiplying the results with the correction factor.

3.4.3 Significance

Measurement of turbidity is very important in domestic water supplies to determine

a. whether a primary treatment such as coagulation and filtration process is required
b. coagulant dosage and the effectiveness of treatment adopted. This prevents the blocking of sand filters by excessive coagulants.
c. turbidity at a rapid rate than measuring the suspended solid content of the sample by gravimetric method.

Expt. 2	*To measure the turbidity of the effluent sample*

MATERIALS REQUIRED

Nephelometer

Colourless and clean glass tubes for sample loading

Standard solution

Solution (a) 1 g of hydrazine sulphate is dissolved in 10 ml of distilled water.

Solution (b) 1 g of lexam ethylene tetramine in 10 ml of distilled water.

Transfer 5 ml each from solution (a) and (b) in a 100-ml volumetric flask. Leave it undisturbed for one full day at 25°C. Make up to 100 ml. The NTU value of this solution is 400. This solution can be used for one month from the stock solution. Pipette out 1 ml and make up to 100 ml using distilled water (40 NTU).

PROCEDURE

1. Adjust the nephelometer at 100 using 40 NTU.

2. Then transfer the sample into the nephelometer tube.

3. See to it that the sample does not contain any air bubble during measurement. Read out the value on the scale.

4. If the sample possesses an NTU greater than 40, dilute it using distilled water and repeat the procedure as described above.

3.5 ELECTRICAL CONDUCTIVITY

3.5.1 Principle

Solids present in the sample conduct current through the solution which can be measured using EC meter.

Expt. 3   *To measure the electrical conductivity of the effluent sample*

MATERIALS REQUIRED

Conductivity meter
Conductivity cell
Thermometer

PROCEDURE

1. Read the temperature of the sample and adjust the temperature knob of the conductivity meter accordingly.

2. Shift the selector switch to 1000 and adjust to CAC mark.

3. Take the sample in a beaker and immerse the conductivity cell in it.

4. After connecting the cell terminals to the sockets, note for the deflection in dial.

5. In the case of negligible deflection, shift the selector switch to X 100 and repeat the step described above.

6. Record the deflection at this level.

7. Shift the selector switch to X 10 and perform the steps described above.

8. Disconnect the cell terminals, wash with distilled water and put off the instrument.

CALCULATION

$$\text{Electrical conductivity} = \text{Deflection value in the dial} \times \text{Value of the selector switch}$$

3.6 TEMPERATURE

Temperature of the effluent sample is an important parameter that has to be recorded during effluent analysis.

Expt. 4   *To determine the temperature of the effluent sample*

MATERIALS REQUIRED

Laboratory glassware

Thermometer

PROCEDURE

1. Transfer 50 ml of the sample in a clean beaker
2. wipe and clean the thermometer with blotting paper and immerse into the beaker
3. Stir the sample well before noting down the temperature.
4. allow sufficient time to achieve a constant temperature
5. Note down the temperature, take the thermometer outside the beaker, wipe it clean and recap it for further use.

CHEMICAL CHARACTERISTICS/CONSTITUENTS

3.7 pH

pH is a measure of hydrogen ion concentration in water. In other words, it is the acid or alkaline condition of water.

3.7.1 Principle

It is a known fact that water is a mixture of hydrogen and hydroxyl ions and the neutralization reaction of acids and bases results in water with equal amount of hydrogen and hydroxyl ions. When it is in pure form, about 10^{-14} g molecules of water gets equally dissociated into H^+ and OH^- ions which represent a value of 10^{-7} for both the ions. If the hydrogen ions are in excess, it leads to acidity and excess of hydroxyl ions leads to alkalinity of water. For convenience sake, the pH is expressed as the negative logarithm of hydrogen ion concentration and the pH value of pure water is therefore 7. However in natural water bodies, this is not the case as photosynthetic activity by aquatic plants results in more amount of CO_2 consumption thus favouring alkaline pH. The condition will be different at times, the reason being the lack of photosynthesis and respiration. However this lowers the pH only to some extent as carbonates in the form of $CaCO_3$ tend to shift the pH towards the alkaline side. Several factors like temperature, aeration and input from external source also interfere with the pH. pH of the effluent determines the type of treatment to be given. In waste water treatment, pH is an important criterion for coagulation, disinfection, water softening and corrosion control. Also in biological treatment methods, adjustment of pH favourable for the microbial activity is essential. Above all, pH of the ecosystem determines the condition suitable for life. It is a parameter of great significance in the checking of water quality for consumption in domestic water supplies. In biological treatment methods, it is used to find out other parameters like acidity and alkalinity.

3.7.2 Measurement of Hydrogen Ion Activity

Pure water, on dissociation, yields equal amount of hydrogen and hydroxyl ions equal to about 10^{-7} mg/l.

$$H_2O \rightarrow H^+ + OH^- \tag{1}$$

When the equation is substituted in the equilibrium equation

$$\frac{[H^+][OH^-]}{[H_2O]} = K \tag{2}$$

However low degree of ionization cannot diminish the large concentration of water and hence the constant K is assumed to equal the value 1.

$$[H^+][OH^-] = K$$

Ionization constant for pure water is $= [H^+][OH^-] = 10^{-7} A \times 10^{-7} = 10^{-14}$. Addition of any acid to water increases the hydrogen concentration in water. This is due to the fact that its ionization brings down the hydroxyl activity so that the combination of both the activities equals the ionization constant 10^{-14}. Similar is the case of any base added to the water. However it should be remembered that addition of acids or bases can in no way bring down the (H^+) or (OH^-) ion concentration to zero.

The pH term is represented as $pH = -\log(H^+)$, and the scale as mentioned above ranges from 1 to 14. Water possessing pH of 7 at 25°C is said to be neutral.

3.7.3 Measurement of pH

Initially measurement of pH using suitable devices was not possible. Even though hydrogen electrode is considered as an absolute standard for the measurement of pH, it does not provide accurate results in field condition and also for solutions containing materials absorbed on platinum black. To make it suitable for a wide variety of wastes, several indicators that are able to produce colour at different pH range are evaluated. Later, with the development of glass electrodes, measurement of hydrogen ion activity in waters with less interference of other ions became possible and undoubtedly the results obtained were reliable. Based on the pH range of the solution, different types of electrodes are used. For instance, water with alkaline pH (>10) and high temperature can be measured using glass electrode whereas spear type electrode is best-suited for semi-solid substances. For standardization of instrument, buffer solution with a pH range between 1 and 2 is used.

Since pH should be expressed in terms of hydrogen ion activity, pH is represented as (H^+) and therefore pH 5 is represented as $(H^+) = 10^{-5}$. Also it should be noted that it does not evaluate the total acidity or total alkalinity. This can be clearly understood by taking two acid solutions of similar normality (N/5O) and neutralization value. When the pH of these solutions is measured, the value obtained is highly variable. i.e.

$$(H^+) \text{ of } CH_3COOH = 10^{-3}$$
$$(H^+) \text{ of } H_2SO_4 = 10^{-1}$$

This shows that the pH value depends on the degree of ionization. From this, hydroxyl activities can also be interpreted

$$pOH = 14 - pH$$

In the case of CH_3COOH it is 11, obtained by subtracting 3 from 14.

$$\text{i.e. } pOH = 14 - 3 = 11$$

For sulphuric acid it is 13 and this can be explained as follows

$$pOH = 14 - pH$$
$$= 14 - 1$$
$$= 13$$

Evaluation of hydroxyl ion activity gives an idea about the time requirement for the water-softening process and iron and aluminum salts in calculation process.

Expt. 5   ***To measure the pH of the effluent sample***

MATERIALS REQUIRED

pH meter

Buffer solutions 4 and 9.

Standard pH tablets are commercially available. Dissolve the appropriate pH tablet in 100 ml distilled water.

PROCEDURE

1. Warm up the instrument for 15 minutes before use.
2. Adjust the pH knob to read 7.
3. Carefully connect the electrode with the pH meter.
4. Wash the electrode in distilled water, wipe it dry and then immerse it in the buffer solution (4.0).
5. Adjust the temperature knob to the temperature range of buffer.
6. Similarly adjust the buffer knob to the pH of the buffer solution, i.e. 4.
7. Now turn the selector switch to zero position.
8. Wash the electrode with distilled water, wipe it dry and immerse it in the buffer solution of pH 9.2 and dip the electrode in the sample.
9. According to the pH range of the sample, adjust the selector switch to either 0–7 or 7–14.
10. Read the pH of the sample.
11. Remove the sample; shift the selector switch again to zero position.
12. Put off the instrument.
13. Always keep the electrodes dipped in distilled water.

3.8 ALKALINITY

3.8.1 Principle

Volumetric titration of alkalinity by sulphuric acid is expressed in terms of $CaCO_3$. Based on the pH of the sample, the indicator added responds to the titrant. For example, if the pH of the sample is > 8.3, the phenolphthalein indicator turns from pink to colourless at 8.3.

$$CO_3^{2-} + H^+ \rightarrow HCO_3^-$$

and when the titration is continued at pH 4.5, methyl orange indicator turns from orange to red.

$$HCO_3^- + H^+ \rightarrow H_2CO_3$$

For the samples with pH < 8.3, phenolphthalein does not show any change and the end point is indicated by the colour change of methyl orange. This is due to the fact that the sample lacks carbonates. Three constituents of alkalinity, hydroxides, carbonates and bicarbonates determine the titration curve. For example, beyond pH 10, hydroxides are completely neutralized and carbonates are converted to bicarbonates at pH 8.3. Therefore the end point shown at pH 8.3 is termed as phenolphthalein alkalinity whereas the end point at pH 4.5 is termed as total alkalinity as it denotes the amount of carbonates, bicarbonates and hydroxides. However the quantity of each form of alkalinity cannot be determined by the method mentioned above. Especially for treatment of effluents, it is customary to calculate hydroxide, carbonate and bicarbonate alkalinities individually. To proceed with the calculation, let us assume five different possibilities.

1.   Hydroxide only
2.   Carbonate only
3.   Hydroxide + carbonate
4.   Carbonate + bicarbonate
5.   Bicarbonate only

For samples with hydroxides alone, the amount of acid added to reach pH 8.3 is enough for its complete neutralization. Therefore hydroxide alkalinity equals phenolphthalein alkalinity. For samples with carbonates alone, half of the total carbonates will be converted to bicarbonates at the phenolphthalein end point which is exactly the half of the methyl orange end point. This gives a clear picture about the estimation of carbonates, i.e. carbonate alkalinity equals total alkalinity. To measure the samples with hydroxide and carbonates, the amount of acid added from the phenolphthalein end point to the methyl orange end point is considered. This amount represents half the amount of original carbonate in the sample and therefore

Carbonate alkalinity = 2 (amount of acid utilized from phenolphthalein end point to
methyl orange end point) × 1000 ml of sample

Hydroxide alkalinity = Total alkalinity – Carbonate alkalinity

Sample with carbonate and bicarbonate can be estimated as follows. Since carbonate alkalinity is exactly twice the amount of phenolphthalein end point, it is expressed as

Carbonate alkalinity = 2 (phenolphthalein end point) × 1000 ml of sample

Bicarbonate alkalinity = Total alkalinity – Carbonate alkalinity

Bicarbonate alkalinity represents the total alkalinity. This clearly shows that all the three forms of alkalinity exist in equilibrium and even a minute fluctuation in any one form will ultimately influence the other two and this conversion depends on the pH of the aquatic system.

3.8.2 Significance

In waste water treatment, coagulants are used to precipitate the colloidal and suspended particulates as hydroxides. This will release the hydrogen ions which ultimately act on alkalinity of water. The optimum pH at which the coagulant is effective is obtained by the buffering action of the alkalinity and therefore the alkalinity value should be in excess of the amount that can be neutralized by the coagulant hydrogen ions released.

The estimation of alkalinity is very essential for the evaluation of lime and soda ash dosage in water-softening process. It is also determined to prevent corrosion in water pipelines and to evaluate the buffering capacity of waste waters and sludges. Among the different forms of alkalinity, water with hydroxide alkalinity is strictly restricted and therefore along with the determination of total alkalinity, the identification of different forms is needed. It is essential to find out the suitability of water in biological treatment.

Expt. 6   *To determine the total alkalinity of the effluent sample*

MATERIALS REQUIRED

Laboratory glassware

Titration assembly

REAGENTS

a.   Sulphuric acid (0.02 N)

b.   Sulphuric acid (0.1 N).

2.8 ml of conc. H_2SO_4 diluted to 1 l using distilled water. Take 200 ml from this stock solution and dilute to 1 l using distilled water.

c.   Phenolphthalein indicator

1 g of phenolphthalein is dissolved in 100 ml of ethyl alcohol. After complete dissolution, add 100 ml distilled water. Add NaOH reagent (0.2272 N) in drops till a faint pink colour appears.

d.   Methyl orange indicator

0.1 g of methyl orange is dissolved in 200 ml of distilled water.

PROCEDURE

1.   Estimation of alkalinity should be done immediately after the collection of the sample.

2.   In 50 ml of sample, add 2–3 drops of phenolphthalein indicator.

3.   If the solution shows pink colour, titrate it against sulphuric acid.

4.   Appearance of slight pink colour indicates the presence of hydroxides or carbonates whereas colourless sample confirms the presence of free CO_2.

5.   Note down the colourless end point as p.

6.   Now add 2–3 drops of methyl orange indicator to the same flask and proceed with the titration till the solution turns from yellow to orange.

7.   Record the value as t and calculate the alkalinity using the formula.

CALCULATION

$$\text{Phenolphthalein alkalinity} \quad p = \frac{p \times 1000}{v}$$

$$(\text{as } CaCO_3 \text{ mg/l})$$

$$\text{Total alkalinity} \quad t = \frac{t \times 1000}{v}$$

$$(\text{as } CaCO_3 \text{ mg/l})$$

where,

p = ml of titrant used for phenolphthalein alkalinity

t = ml of titrant used for both titrations

v = volume of sample

3.9 ACIDITY

3.9.1 Principle

The presence of excess carbon dioxide in water depends on its concentration in water and atmosphere, i.e. it tends to be in equilibrium between atmosphere and water. If the level exceeds in atmosphere, a part of it enters into the surface waters by absorption and therefore it is a normal component in the aquatic system. Another source of acidity is the organic content in polluted waters. This may be the case of ground water and hypolimnion portion of stratified water where the oxidation of organic matter leads to the production of CO_2. Respiration of aquatic organisms contribute significantly to the CO_2 concentration in water and if the level goes up in water, the excess CO_2 escapes from the water to the air. Dissolved oxygen content in the water cannot be correlated with the carbon dioxide level as it is the final product of both aerobes and anaerobes. If water contains sufficient amount of calcium and magnesium carbonate, CO_2 concentration is controlled by the formation of carbonates.

$$CO_2 + CaCO_3 + H_2O \rightarrow Ca^{2+} + 2HCO_3$$

Industrial wastes such as wastes from metallurgical processes and manufacture of synthetic organic products contribute to mineral acidity. Similarly abandoned mines and ore dumps are rich in sulphate and iron pyrites which on entering surface waters due to run-off, acidify them.

$$2S + 3O_2 + 2H_2O \rightarrow 4H^+ + 2SO_4^{2-}$$

$$FeS_2 + 3\tfrac{1}{2}O_2 + H_2O \rightarrow Fe^{2+} + 2H^+ + 2SO_4^{2-}$$

Both these materials under the action of sulphur-oxidizing bacteria will be converted to sulphates or sulphuric acid and effluent-containing heavy metals such as Fe (III) (and Al (III)) under hydrolysis is converted to ferric hydroxide.

$$FeCl_3 + 3H_2O \rightarrow Fe(OH)_3 + 3H^+ + 3Cl^-$$

Carbonate-bicarbonate exchange takes the responsibility of buffering aquatic system. Based on the range of CO_2 concentration in waters, it is considered that water having pH less than 8.5 contains acidity. However at neutral pH (7), excess amount of this is neutralized. Between pH 4 and 8, CO_2 dominates, and below pH 4 strong mineral acids contribute more to the acidity. Below pH 8.0, carbon dioxide and strong mineral acids are said to contribute to the acidity.

Acidic effluents are of two types namely mineral and organic acids. Mineral acids are those effluents released from chemical, fertilizer, pesticide, battery manufacture, electroplating, viscose rayon manufacture, mining, iron and copper picking industries, etc. Organic acids are those effluents released from rayon, fermentation plants, distilleries and dyeing industries. Till now no deleterious effect associated with excessive CO_2 has been reported and the source of acidity in water is the effluent containing malt and carbonated beverages. On the other hand acidity is also formed by the hydrolysis of oxides of nitrogen and sulphur, a combustion product of fossil fuels in power plants and automobiles with rain. This results in the formation of acid products, sulphuric and nitric acid. This lowers the pH of poorly buffered lakes which in turn victimize the aquatic life. Also, the low pH induces the mobilization of aluminum, iron and manganese from the soil out of which aluminum precipitates in the gills of fish and leads to their death.

Acidity is caused by CO_2 in water with pH >4 or strong mineral acids in water with pH <4. The main source of CO_2 in water is atmospheric diffusion even though a small amount is produced by the biological oxidation of organic matter. In the case of industrial wastes, hydrolysis of trivalent metals imparts mineral acidity. The titration of the sample to pH 4.5 is methyl orange acidity and to end point of 8.3 is called total acidity. Water with pH >8.5 contains acidity and are evaluated by the phenolphthalein pH. The neutralization of acids is completed at pH 4. Great care should be exercised to protect the sample from being contaminated by CO_2. Since water rich in CO_2 has partial pressures greater than atmospheric CO_2 it is very difficult to carry the water samples with the original CO_2 concentration to the laboratory. Therefore estimation at the sampling site is preferable. For samples that have to be transferred to the laboratory, use submerged tubes.

3.9.2 Phenolphthalein Acidity

At times, the estimation of total acidity formed by both mineral acids and weak acids may become necessary. In the presence of heavy metals, boiling of the samples should be done prior to titration. This is because high temperature increases the rate of metal salts which in turn speeds up the titration. Sodium hydroxide is used as titrant and the weak acids get neutralized only at pH 8.3. Therefore only indicators like phenolphthalein or metacresol purple can detect the end point.

3.9.3 Methyl Orange Acidity

Waters with acidic pH (<4) have methyl orange acidity and the mineral acids are the sole contributors of this type of acidity. They are neutralized only when the sample reaches a pH of 3.7, which can be readily identified by using methyl orange indicator. However now, bromophenol blue is more widely is used as titrating agent (N/50) since at this concentration 1 ml is equivalent to 1 mg of acidity.

3.9.4 Indicators for Measurement

The same alkaline reagent can be used to estimate both carbon dioxide and mineral activity as the difference lies only in their neutralization to varied pH. To represent this they are represented by the name of their respective indicators i.e. mineral acidity is called as methyl orange acidity and both the mineral acid and CO_2 is called as phenolphthalein acidity. While sampling, to avoid gas bubbles, allow the sample to overflow the container and close it with an air-tight stopper.

While performing the titration, two things have to be considered. Firstly that the estimation procedure in any case should not expose the sample to the CO_2. Secondly, unless the sample is stirred well during titration, the end point may be erroneous i.e., the sample may show the colour formation well before the end point. Even though both the factors sound to be entirely contradictory to an analyst, steps/measures have to be taken to satisfy both the needs. Accordingly, take the sample in a graduated beaker in such a way that it overflows out and pipette out the excess sample with great care. Now reduce the volume of the sample to the desirable level. Then titrate it against the NaOH and do not bother to stir well. Note down the end point and then take the sample and perform the titration without stirring. This will minimize the exposure of the sample to CO_2. When the titration reaches the end point noted previously, do it slowly and carefully. The final end point is the appearance of pink colour which persists for about 30 seconds. NaOH prepared as the titrant should be free of sodium carbonate.

$$2NaOH + CO_2 \rightarrow Na_2CO_3 + H_2O$$

$$Na_2CO_3 + CO_2 + H_2O \rightarrow 2NaHCO_3$$

On the other hand, sodium carbonate can be directly used as titrant. Estimation of CO_2 can be done indirectly from pH and alkalinity and total dissolved solids. The problem that is encountered in this method is that even a minor variation of 0.1 in the result of above mentioned parameter can cause an error of 25% and mislead the analyst. Therefore for samples having carbon dioxide >2 mg/l, titration method can be employed and if the method is conducted with care, accurate result is guaranteed. However samples with < 2 mg/l CO_2, excessive error may establish and the value of CO_2 can be obtained by calculation.

3.9.5 Significance

It is an important parameter and its evaluation is very much essential

    i.   in waste water treatment to choose the suitable treatment method.

    ii.   in ground water use to choose the suitable methods to overcome corrosive characteristics formed due to carbon dioxide.

    iii.   to determine the dosage of chemicals such as lime required to remove the excess CO_2 and to evaluate a suitable treatment method.

Expt. 7   *To determine the total acidity of the effluent sample*

MATERIALS REQUIRED

Laboratory glassware

Titration assembly

a.   Sodium hydroxide (0.1N)

Dissolve 4 g of NaOH in little distilled water and make up to one litre.

b.   Sodium hydroxide (0.05N)

Take 5 ml of 0.1 N sodium hydroxide and make up to one litre.

c.   Phenolphthalein indicator

Dissolve 0.1 g of phenolphthalein in 100 ml ethyl alcohol. To this, add 100 ml distilled water. Mix well and store for future use.

PROCEDURE

1.   Transfer 100 ml of the sample in a 250-ml volumetric flask.

2.   To the flask content, add 3 drops of methyl orange indicator.

3.   Observe for the colour change from orange to yellow.

4.   If the solution changes to pink colour, titrate it with 0.05N sodium hydroxide.

5.   The end point is sharp with a colour change from pink to yellow.

6.   Note the reading. Let it be m.

7.   Now continue to titrate the sample after adding phenolphthalein indicator.

8.   Observe for the colour change from yellow to pink. Note the reading. Let it be p.

CALCULATION

$$\text{Methyl orange acidity} = \frac{m \times 0.05\text{N of NaOH} \times 50{,}000}{\text{Volume of sample}}$$

$$\text{Phenolphth alein acidity} = \frac{p \times 0.05\text{N of NaOH} \times 50{,}000}{\text{Volume of sample}}$$

$$\text{Total acidity} = \frac{(m + p) \times 0.05\text{N of NaOH} \times 50{,}000}{\text{Volume of sample}}$$

3.10 TOTAL HARDNESS

3.10.1 Principle

Presence of alkaline earth metals such as calcium, magnesium, iron, manganese and strontium in water determines its hardness. Out of these, Ca and Mg play an important role. These cations when combined with chlorides and sulphates tend to remain permanent. In some cases, excess of carbonates and bicarbonates impart temporary hardness but they can be removed by boiling the water. Hardness of water can be ascertained by soap test. Soap gives beautiful lather with soft water. Evaluation of hardness is done by EDTA titration.

$$CaCl_2 + 2EDTA \rightarrow Ca(EDTA)_2 + 2NaCl$$

Hard water demands more soap to produce lather and also produces scales in hot water pipes, heaters, and boilers at high temperatures. Even though synthetic detergents replace soap, in hot water, scaling problem cannot be controlled. Generally surface water is softer than ground water until it is polluted by chlorides of calcium and magnesium salts. Principal hardness-causing agents are divalent cations (Ca^{2+}, Mg^{2+}, Sr^{2+}, Fe^{2+}, Mn^{2+}) and anions (HCO_3^-, SO_4^{2+}, Cl^-, NO_3^- and SiO_3^{2-}). When these ions react with soap, they form precipitates and with some anions they form scales. As soon as the rain water reaches the ground, a portion of it percolates to form soil water. This water is rich in CO_2, the source being bacterial action. To exhibit equilibrium with atmospheric acid, excess CO_2 is transformed to carbonic acid shifting the pH of the medium to the acid range. The carbonic acid formed is sufficient enough to convert the insoluble calcium carbonates into soluble bicarbonates. Since limestone, in addition to carbonates, possesses sulphates, chlorides and silicates, dissolution of bicarbonates favour their entry too in the water.

$$CaCO_3 + H_2CO_3 \rightarrow Ca(HCO_3)_2$$

$$MgCO_3 + H_2CO_3 \rightarrow Mg(HCO_3)_2$$

The presence of these substances imparts hardness to the water. Based on their level, water is classified as soft (0–75 mgl^{-1}); moderately hard water (75–150 mgl^{-1}) hard water (150–300 mg^{-1}) and very hard water (7300 mg1^{-1}) and hardness is expressed as $CaCO_3$. It is estimated titrimetrically using ethylene diamine tetra-acetic acid (EDTA) or its sodium salt as the titrating agent. Being a chelating agent, EDTA forms a stable complex with the available free calcium and magnesium ions to form a wine red complex.

In the titration method, addition of blue colour Erichrome Black 'T' indicator combines with free hardness ions [Ca^{2+}, Mg^{2+}) to form weak complex of wine red colour. During the titration, EDTA combines with the hardness ions to form a more stable complex than the Erichrome black "T" indicator. This leaves the indicator free, causing the solution to turn blue.

In general, two types of hardness are recognized

- calcium and magnesium hardness
- carbonate and non-carbonate hardness

3.10.2 Calcium and Magnesium Hardness

Calcium and magnesium contribute to the maximum portion of hardness in water. Calcium hardness can be analysed separately while magnesium hardness is obtained by subtracting the calcium hardness from the total hardness. In some cases, like the lime requirement for lime soda ash softening, knowledge of magnesium hardness in water gains more importance.

3.10.3 Carbonate and non-carbonate Hardness

When the amount of total alkalinity equals the part of total hardness, the value of total alkalinity is taken as carbonate hardness.

Carbonate hardness = total alkalinity > total hardness equals or is greater than the total hardness, the value of total hardness itself is taken as carbonate hardness.

Carbonate hardness = total hardness < alkalinity

The principle sources of carbonate hardness are carbonates and bicarbonates forming scales on boilers and they are formed by the action of carbonic acid on limestone. The analysis of carbonate hardness is necessary because it is these components that get precipitated into calcium carbonate at high temperature. However continuous boiling can remove them from water and therefore the hardness rendered by them is said to be temporary. Non-carbonate hardness is said to be permanent as they can neither be precipitated nor removed by boiling. The principle sources of non-carbonate hardness are the cations associated with anions like chlorides, sulphates and nitrates.

3.10.4 Significance

Estimation of hardness is very essential

    i.   to determine whether a softening process is required

    ii.   to choose the type of softening process

    iii. to determine the suitability of water for consumption

---

Expt. 8   ***To estimate the total hardness of the effluent sample***

MATERIALS REQUIRED

    Laboratory glassware

    Titration assembly

REAGENTS

    a.   Ammonia buffer solution

        Dissolve 13.5 g of ammonium chloride in 114 ml of conc. ammonium hydroxide. Make up to 200 ml using distilled water.

    b.   Erichrome black 'T' indicator

        Dissolve 0.5 g of Erichrome black 'T' dye in 100 ml ethyl alcohol (80%).

    c.   EDTA solution (0.001M)

        3.723 g of sodium salt of EDTA in little distilled water and make up to 1000 ml. Use polyethylene bottle for storage.

PROCEDURE

    1.   In 50 ml sample, add 1 ml of ammonia buffer solution and 4 drops of Erichrome black 'T' indicator.

    2.   Now the solution turns wine red in colour.

    3.   Titrate it against EDTA solution till the colour changes from wine red to blue. (The colour change is very sharp).

    4.   Record the end point, repeat the procedure to attain constant value.

    5.   Calculate the total hardness using the formula.

CALCULATION

$$\text{Total hardness} = \frac{E \times 1000}{v}$$

$$(\text{as } CaCO_3 \ mg \, l^{-1})$$

where,

E = volume of titrant in ml

V = volume of sample

3.11 SOLIDS

3.11.1 Principle

Natural water bodies are found to be abundant in a variety of salts such as carbonates, bicarbonates, chlorides, sulphates, phosphates, and nitrates of calcium, magnesium, sodium, potassium, iron and manganese, etc. Thus salts are very essential for maintaining the biological productivity of water. However when the level goes beyond the limit, the physico-chemical characteristics of water get changed, i.e., the density of water will rise thus reducing the solubility of O_2 and CO_2. Added to this, the osmoregulatory processes of aquatic organisms will be severely affected. Their prescribed limit is between 10 and 20 mg/l. Industrial waste water may have the TDS (Total Dissolved Solids) level even up to 1000 ppm. The excess level gives salty taste to water and leads to health complications like stomach pain and fever. The term "solids" comprise a wide variety of organic and inorganic materials in water. Suspended and dissolved particles in water are termed as Total Suspended Solids (TSS) and Total Dissolved Solids (TDS). Specific tests have been designed to evaluate different kinds of solids in water.

3.11.2 Dissolved and Suspended Solids

The presence of dissolved and suspended solids varies between the waste water. Potable water consists of mostly inorganic salts, dissolved gases and organic matter in small amounts. Majority of these constituents are present in dissolved form. Maximum permissible limit of TDS in potable water is 20 to 1000 mgl^{-1} and in hard water its level is more. For the analysis of total suspended solids, the weight of the filtered portion of the sample is taken whereas the unfiltered portion is represented as total dissolved solids.

3.11.3 Volatile and Fixed Solids

Volatile solids are abundant in domestic waste, industrial wastes and sludges. The procedure involves the combustion of the sample at which all the organic matter is lost by vaporization. The temperature fixed, (550°C) should allow the complete volatilization of organic matter but should prevent the decomposition and volatilization of inorganic matter. The loss of weight after combustion is taken as the organic matter content in the sample. It is at this temperature, carbon residues from the pyrolysis of carbohydrates oxidized is obtained.

$$Ca(H_2O)\ b \rightarrow ac + bH_2O$$
$$C + O_2 \rightarrow CO_2$$

Except for the trace amount of ammonium compounds that has escaped drying, volatilization of inorganic salts and magnesium carbonate are prevented

$$MgCO_3 \xrightarrow{\ 350°C\ } MgO + CO_2$$

In the case of volatile fixed solids, filtration procedure subsequently removes the dissolved inorganic forms and at 550°C ammonium compounds present as ammonium bicarbonate get volatilized leaving behind the volatile fixed particulates alone.

$$NH_4HCO_3 \xrightarrow{\quad 105°C \quad} NH_3\uparrow + N_2O\uparrow + CO_2\uparrow$$

It is very important to perform the procedure at controlled temperature as it is the only regulating factor to give accurate inferences. For example, calcium carbonate, the major inorganic constituent volatilizes at 85°C and contributes significantly to the volatile dissolved solids. To avoid this, always carry out the volatilization in muffle furnace and to get better result, burn the sample in a bunsen burner for about 15 minutes. This will eliminate the flammable materials from the sample. In the industrial waste water, organic matter may be in the form of short chain fatty acids, alcohol, ketones, aldehydes, and hydrocarbons which when evaporated, get lost and therefore to get a real picture about the organic matter, COD and total organic estimations can be done.

3.11.4 Settleable Solids

Certain solids, due to gravity force, when left undisturbed tend to settle at the bottom to form sludge. Such solids called settleable solids and in general they are coarse particles with a specific gravity greater than water. The measurement of settleable solids is very essential to study the physical behaviour of water. Even though the permissible limit for total solid content is 1000 mgl^{-1}, a level that exceeds 500 mgl^{-1} has laxative property and may cause adverse effect allergic to skin. However most of the natural wastes have solid content more than 500 mgl^{-1} and effects induced by total solids have not been reported so far. Among the different types of solids, dissolved solids gain more importance due to their undesirable effects when present in large amounts. For determination of total solids, platinum dishes are highly preferred as they exhibit constant weight before use. Do not use porcelain dishes as the result acquired is not accurate. This is due to the fact that the weight of the dish tends to fluctuate. Dissolved solids can be evaluated by an indirect method i.e. by estimating their specific conductance. The reason behind is that majority of the dissolved substances are in the ionized form and therefore measurement of specific conductance reveals the approximate amount of total dissolved solids in the sample. Since this measurement is influenced by the relative concentration of the different types of ions and the ionic strength of water, practical estimation of total dissolved solids is done by multiplying the specific conductance with an empirical factor 6.23.

For the determination of suspended particulates, use glass fibre or filter paper whereas for settleable solids, adjust the sample to room temperature. Allow the sample to filter in an imhoff cone for about one hour. Do not perform the experiment under direct sunlight, and record the results in mg/l.

3.11.5 Significance

Estimation of the totals solids was initially devised to evaluate the pollution strength of waste waters. However the results obtained do not give high degree of accuracy and hence COD and BOD tests are done to know the exact amount of oxidizable and non-oxidizable organic content in the sample. Determination of suspended solids is helpful to determine the type of preliminary treatment to be adopted, (physical/chemical); to select the suitable chemicals for flocculation; to determine whether the preliminary treatment is required for treated water; to determine the efficiency of treatment method and to check the level of TSS prior to its discharge into the receiving waters. Determination of TS in the sample is used to determine the type of softening procedure as the precipitation methods decrease the TS content whereas the exchange method elevate it. In treatment, all the suspended particulates are considered as settleable solids. Since the final conclusion about the pollution strength of industrial waste waters is drawn out by the results of several parameters, estimation of total solids content is considered to be an essential parameter only in domestic waste water. However, in industrial wastes, undesirable amount of dissolved inorganic states can be easily detected by total dissolved solids test. Also only when their level is estimated they can be subjected to biological treatment especially anaerobic treatment as the nature and concentration of TDS interfere with the anaerobic microbial action.

Expt. 9   *To determine the amount of total solids in the effluent sample*

MATERIALS REQUIRED

> Laboratory glassware
> Weighing balance
> Desiccator

PROCEDURE

1. Accurately weigh a clean, dry 100-ml silica crucible and record its weight as (W_I).
2. Transfer 100 ml of unfiltered sample in the silica crucible.
3. Evaporate the sample by placing it in a hot-air oven at 105°C for one hour.
4. Cool it in a desiccator.
5. Then take the crucible out of the desiccator and record its weight. Let it be (W_F).
6. Now calculate the total solids of the given sample as follows.

CALCULATION

$$\text{Total solids (mg/l)} = \frac{W_I - W_F}{S} \times 10,000$$

where,

> W_I = Initial weight of the crucible
> W_F = Final weight of the crucible
> S = Weight of the sample

Expt. 10   *To determine the amount of total dissolved solids in the effluent sample*

MATERIALS REQUIRED

> Laboratory glassware
> Weighing balance
> Desiccator

PROCEDURE

1. Accurately weigh a clean dry 100-ml silica crucible and record its weight. Transfer 100 ml of the filtered sample in the silica crucible (W_I).

2. Evaporate the crucible in a hot-air oven for one hour. Cool it in a desiccator and record its weight. Let it be W_F.

3. Now calculate the total dissolved solids of the given sample.

CALCULATION

Total dissolved solids (mg/l) $= \dfrac{(W_I)-(W_F)}{S} \times 10,000$

where,

W_I = Initial weight of the crucible

W_F = Final weight of the crucible

S = Weight of the sample

Expt. 11   ***To determine the amount of total suspended solids in the effluent sample***

Total suspended solids can be obtained by subtracting the total dissolved solids from total solids.

Total suspended solids = Total solids − Total dissolved solids

---

Expt. 12   ***To determine the weight of solids in the sludge***

PROCEDURE

1. Take 50 g of sample in a pre-weighed small porcelain dish of about 7 cm in diameter.

2. Homogenize the sample before use

3. Dry the sample at 103°C for several hours. Make sure that all the water content in the sample has evaporated. Otherwise keep the sample overnight.

4. Cool it in a desiccator

5. Record the weight of the dish.

6. Repeat the procedure until a constant weight is achieved.

CALCULATION

Weight of the dish $= a$

Weight of the dish + sample $= a + b$

Weight of sample $= b$

Weight of dish + dried sample $= a + c$

Weight of sludge solids $= a + c$

$$a + b = c/b$$

Percent of solids in sludge $= \dfrac{c/b}{b} \times 100$

Percent of volatile matter $= \dfrac{d}{c/d} \times 100$

Weight of dish + ash $= d$

ORGANIC CONSTITUENTS

3.12 DISSOLVED OXYGEN

The dissolved oxygen in water is either due to atmospheric diffusion or due to photosynthetic activity of aquatic plants. Except for a specific group of microbes (anaerobes), O_2 is indispensable for living organisms without which they cannot survive and therefore their presence in the water body indicates the occurrence of physical and biological processes within the system. Physical processes contribute minimum when compared to the biological, i.e. atmospheric diffusion is absolutely a physical phenomenon and the solubility of O_2 depends on several factors such as temperature, salinity and water movements whereas the photosynthetic activity of aquatic plants is a biological one whose role in O_2 supply is dependent on suitable conditions (physical, chemical and biological) available for growth of autotrophs.

3.12.1 Principle

Dissolved oxygen in water gets depleted by organic and inorganic enrichment from external sources. Organic pollutants otherwise called as O_2 demanding wastes and oxidizing inorganic substances such as H_2S, NH_3, nitrite and ferrous iron cause the O_2 level to drop in due course of time. For all living organisms O_2 is an essential criterion for energy-giving metabolic processes. At saturation point, dissolved gases of water have 38% of O_2 and the dissolution increases with temperature whereas in polluted waters the dissolution rate decreases. It is an index of the aquatic life in water. Determination of dissolved oxygen content evaluates the atmospheric condition of water. All atmospheric gases are soluble in water. However the solubility of N_2 and O_2 depends on the partial pressure and temperature. Since O_2 is the basic requirement for biological oxidation it is continuously utilized by the microorganisms and may soon get depleted and bringforth anaerobic conditions. However to carry out the self-purification process, O_2 should be maintained at a level of about 8 mgl^{-1}.

At 1 atm. pressure, oxygen can dissolve in considerable amounts in water. However the rate of dissolution varies with temperature, i.e., it shows a wide range of fluctuation between 0°C (14.0 mg/l) and 35°C (7 mg/l). This may be due to its characteristic poor solubility. And it is this ability that helps to identify the DO amount in waters of higher altitudes, polluted areas and high temperature. Apart from the natural atmospheric condition, oxidation of biological matter may also induce the temperature to increase. This ultimately leads to O_2 depletion and prevents the self-purification process in water. Presence of sodium chloride prevents dissolution of atmospheric O_2. Unlike carbon dioxide, O_2 determines the water quality as its presence is a direct indication of aerobes. In general, both the aerobes and anaerobes can bring about biological changes in water. Aerobes function by oxidation whereas the anaerobes achieve the breakdown by the reduction of certain inorganic salts such as nitrates and sulphates. Only the favourable environmental conditions determine their domination. For example, dissolved oxygen in sufficient amounts is the main criterion for the aerobes and therefore its evaluation is the prime key for the analysts to determine the health of aquatic life and pollution strength of water.

3.12.2 Sampling Procedure

Carefully take the sample as exposure of sample to air may give erroneous results. Performing the estimation in the field is not possible many times. The sample may be subjected to biological activity as the O_2 and microorganisms required for the oxidation process are present in the sample itself. Therefore it becomes essential to fix the sample as soon as the sampling is made. The addition of reagents for fixation purpose depends on the iodine demand of the sample.

a. Samples with high iodine demand – conc. H_2SO_4 0.7 ml
                                       sodium azide   0.002 g
                                         alkali iodide reagent 3 ml

b. Samples with low iodine demand – manganous sulphate 0.7 g
                                         alkali iodide 2 ml

Always keep the sample in an icebox until it is analysed. After fixation, the titration should be performed within the next six hours. Basically determination of oxygen is oxidation and reduction whose completion is indicated by the sharp colour change of starch indicator from brilliant blue to colourless solution. Normality of the standard solutions is prepared in such a way that 1 ml of the solution consists of 1 mg of chemical dissolved. Since equivalent weight of O_2 is 8, the normality of sodium thiosulphate is N/8. However this concentration is in excess for a limited sample volume used for all titrations (50 ml). Therefore considerable amount of sample is necessary and several analyses have reported that 200 ml of sample gives reliable results. Accordingly the titrant is prepared.

1 mg of chemical should be present in 1 ml of solution.

200 ml of sample = 1/5 of a litre

                    = N/8 × 5 of the titrant

                    = N/40 of the titrant

Now the sample is equivalent to the titrant prepared. Sodium thiosulphate will not retain water at room temperature and low humidity. Therefore the solution cannot be prepared by taking the required concentration of the chemical which when dried will fluctuate in its composition. To be on the safer side, always prepare it too concentrated and then standardize it against the primary standards (potassium dichromate, potassium bi-iodate). Calculate the equivalent weight of sodium thiosulphate from the following equation.

$$2Na_2S_2O_3 . 5H_2O + I_2 \rightarrow Na_2S_4O_6 + 2NaI + NH_2O$$

From this equation it is understood that one molecule of NaS_2O_4 = one atom of I. Therefore similar to that of iodine, NaS_2O_4 gives one electron for its oxidation to tetrathionate. So it is clear that the equivalent weight = molecular weight. Now for N/40 of the solution, 0.205 g of the chemical is needed. As mentioned above, 6.5 g should be prepared

Potassium dichromate

$$Cr_2O_7 + 6I^- + 14H^+ \rightarrow 2Cr_3 + 3I_2 + 7H_2O$$

Potassium bi-iodate

$$2IO_3^- + 10I^- + 12H^+ \rightarrow 6I_2 + 6H_2$$

The standards when titrated against the iodine ion in acid solution combine with the iodine ions and liberate iodine which in turn is directly proportional to the oxidizing agent. The amount of oxidizing agent is the amount of titrant to be used. Of the two standards dichromate standard is less preferred as it may hinder the starch end point by its greenish-blue colour. While preserving sodium thiosulphate, two things have to be considered (a) bacterial attack and (b) pH drop due to CO_2 intrusion. Under oxic conditions, sulphur bacteria convert the thiosulphate to sulphate and entry of CO_2 into the solution brings down the pH which converts the thiosulphate ion into SO_3^{2-} and S. This can be prevented by adding 0.4 g of NaOH which maintains the pH of the solution in the alkaline range. However it should be remembered that the NaOH should not be added in excess.

The principle behind is Mn^{2+} when shifted to the higher state of valence can oxidize I^- to I_2 in the presence of an acid. The oxidation of Mn^{2+} to the described state of valence needs alkaline condition. The reagents added fulfil the need and evaluate the dissolved oxygen in the sample. Other than manganese, several agents can act upon iodine and give erroneous results.

Manganous sulphate and alkaline potassium iodide react with each other to form manganous hydroxide.

$$MnSO_4 + 2KHO \rightarrow Mn(OH)_2 + K_2SO_4$$
$$\text{Manganous}$$
$$\text{hydroxide}$$

Based on the oxygen content in the sample, manganous hydroxide is oxidized to brown coloured basic manganic oxide.

$$2Mn(OH)_2 + O_2 \rightarrow 2MnO(OH)_2$$
$$\text{Manganic oxide}$$

Under acidic condition, i.e., addition of sulphuric acid induces the basic manganic oxide to liberate iodine.

$$Mn(SO_4)_2 + 2KI \rightarrow MnSO_4 + K_2SO_4 + I_2$$

The liberated iodine is equivalent to that of dissolved oxygen in the sample and its titration against sodium thiosulphate with starch indicator reveals the DO content in the sample

$$2Na_2 S_2O_3 + I_2 \rightarrow Na_2 S_4O_6 + 2NaI$$

However it takes some time for the reaction to occur. Moreover, all the flocculated material of the sample should be allowed to come in contact with the reagent added. To ensure this, vigorous shaking is needed (for about 20 seconds). Soon after the floc formation, the floc is allowed to settle and then sulphuric acid is added. Addition of acid should be done when a clear liquid of about 5 cm is formed below the stopper.

$$MnO_2 + 2I^- + 4H^+ \rightarrow Mn^{2+} + I_2 + 2H_2O$$

Even though the liberated iodine is insoluble, its reaction with the excess iodide traps it in the solution to form tri-iodate. The solution is allowed to stand for another 10 seconds. Since the sample contains 203 ml of reagents (2 ml of $MnSO_4$ + 2 ml of NaOH + KI and 2 ml of H_2SO_4), the exact amount of 200 ml will not represent the sample alone.

3.12.3 Inhibiting Factors

Nitrite is the major inhibiting ion in dissolved oxygen estimation. While oxidizing I^- to I_2 it is reduced to N_2O_2 which can be oxidized by oxygen and raise the DO level above the original value.

$$2NO_2^- + 2I^- + 4H^+ \rightarrow I_2 + N_2O_2 + 2H_2O$$

$$N_2O_2 + \tfrac{1}{2}O_2 + H_2O \rightarrow 2NO_2^- + 2H^+$$

Also, the end point will not persist due to the reaction of I^- to produce I_2 giving the blue colour to the solution once again. Addition of sodium azide to the alkali iodide reagent will destroy the NO_2^- ion in acidic condition.

$$NaN_3 + H^+ \rightarrow NH_3 + Na^+$$

$$NH_3 + NO_2^- + H^+ \rightarrow N_2 + N_2O + H_2O$$

Dissolved oxygen membrane electrodes are useful for DO determination in field conditions. The membranes can be introduced into the water to take the DO at various depths. In aerobic waste water treatment plants, the DO determination is very essential to evaluate the aerator efficiency as an initial step. However they should be calibrated before use. Since they are sensitive to temperature, fluctuation in temperature should not be allowed. To prevent this, instruments equipped with thermostat are preferred.

Expt. 13   *To estimate the amount of dissolved oxygen in the effluent sample*

MATERIALS REQUIRED

Laboratory glassware

Titration assembly

REAGENTS

a. Sodium thiosulphate (0.1N)

   Dissolve 12.41 g of sodium thiosulphate in 500 ml distilled water. This is the stock solution.

b. Sodium thiosulphate (0.025N)

   Transfer 125 ml of stock solution and dilute to 500 ml using distilled water.

c. Alkaline azide solution

   i.   NaOH   25 g

   ii. NaI   6.75 g

   iii. NaN_2 0.5 g

   Dissolve (i) (ii) and (iii) in 500 ml distilled water.

d. Manganous Sulphate Solution

   Dissolve 40 g of manganous sulphate in 100 ml distilled water.

e. Starch Indicator

   Dissolve 1 g of starch in little distilled water. Make it to a paste and then dilute it to 100 ml. Heat the content gently until the solution appears clear.

PROCEDURE

1. Fill the DO bottle with the sample to the rim.

2. Carefully add 2 ml each of manganous sulphate and alkaline iodide azide solution.

3. While siphoning the reagents into the bottle, see to it that the pipette is immersed to at least three-fourth of the DO bottle.

4. Shake well. Observe for the brown coloured precipitate formation.

5. Add 2 to 3 drops of conc. H_2SO_4 so as to dissolve the precipitate.

6. Now pipette out 100 ml of the sample from the bottle.

7. Add 2 to 3 drops of starch indicator and the solution turns blue.

8. Titrate it against the 0.025N sodium thiosulphate till the blue colour disappears from the solution. Note the reading.

CALCULATION

Amount of dissolved oxygen in the sample

$$= \frac{\text{titrant value} \times 0.025N \times 8 \times 100}{\text{Volume of the sample}}$$

where,

0.025 = normality of titrant

8 = molecular weight of oxygen.

3.13 BIOCHEMICAL OXYGEN DEMAND (BOD)

3.13.1 Principle

BOD is the measure of amount of O_2 required by the microorganisms to oxidize the organic content in water. It aids in organic pollution assessment. Since the estimation of the amount of O_2 before and after incubation period is the only way to evaluate BOD, it is very important to provide conditions suitable for the proliferation of heterotrophs so as to enable them to utilize all the available O_2. This is because under aerobic conditions, in the presence of sufficient amount of ammonia, nitrification occurs where the ammonia is converted to nitrite and then to nitrate. Sharing of O_2 between heterotrophs and nitrifiers will give misinterpreted results. Therefore to suppress the growth of nitrifiers, the BOD samples are incubated at 20°C. At this temperature heterotrophs can flourish and undergo biodegradation. Similarly the physicochemical parameters for the sample should be altered to encourage the growth of heterotrophs. In this regard, pH of the sample should be adjusted between 6.5 and 8.3. Since the presence of chlorides inhibits microbial action, it should be removed by the addition of sodium sulphite. For industrial effluents with high proportion of metallic contaminants, conduct 4 hours permanganate test. Likewise effluents with high salt content (pollution load) can be determined by estimating the total organic carbon. All the effluents cannot meet the nutrient requirement of microorganisms and therefore dilution water with sufficient nutrients are added to the sample to enhance the microbial density. Also effluents with complex composition will not favour microbial growth. Under such conditions in addition to the dilution water, seeding of microorganisms is necessary. Limitations of highly polluted effluents are

- lack of O_2 due to high organic content
- metallic contaminants
- lack of nutrients

Thus BOD test is an index of water quality and can be used to evaluate the purifying capacity of water. While performing the test, care is taken to protect the samples from sunlight. Do not shake or agitate the sample bottles during incubation period. Two 300-ml samples are used and the DO of the first bottle is performed on the initial day and the second bottle in incubated under 20°C in the dark for 5 days and at the end of the incubation period it is estimated for its DO content.

The test is very important to determine the pollution strength of waste water and the purifying capacity of receiving waters. Since O_2 exhibits limited solubility at 20°C (9 mg/l), high strength wastes should be diluted in such a way that the dissolved oxygen is available for the biodegradation process throughout the incubation period. Also the effluent should not contain toxic substances that hinder the biological activity. Biodegradation is brought about by a variety of microorganisms and therefore industrial wastes should be supplemented with "microbial seed" and the microbial activity is directly proportional to the amount of organic matter and O_2 present in the waste water.

$$\text{Carbohydrate} + \text{oxygen} \rightarrow \text{carbon dioxide} + H_2O + NH_3$$

Other factors that govern the reaction are the microbial number and temperature. It should be noted that the BOD result obtained at the end of the 5-day incubation period is only a portion of the total BOD.

3.13.2 BOD Reaction

It is found that the rate of the reaction is directly proportional to the oxidizable organic matter, which in turn depends on the active microbial biomass. This is the condition until the microbial population achieves saturation. However after that, the reaction rate is based on the available food. This can be represented as

$$-dc = \frac{K'C}{dt}$$

where C is the concentration of nutrients (contaminants) at time t and K is the rate constant for the action. The equation clearly indicates that the speed of the reaction is directly proportional to the nutrients in the sample. Since C is not a stable parameter and fluctuates with time, it is represented as L denoting the ultimate demand and the equation is rewritten as

$$-dc = \frac{K'L}{dt}$$

Also, on the other side, amount of O_2 depleted is exactly equal to the amount of organic matter oxidized and therefore when a graph is drawn using O_2 depletion rate against time, a parabolic type of curve is obtained which is similar to that of a graph drawn using organic matter oxidation rate versus time.

Usually for bacterial seed, mixed cultures from soil or sewage are acquired as they contain heterotrophic bacteria that can utilize a variety of carbon wastes. Autotrophic bacteria, in particular nitrifying bacteria, utilize O_2 and non-carbon sources for energy production. Unless provision is made to eliminate them, it gives erroneous results. Fortunately their population size is minimum in polluted water and to increase in number, they need a suitable temperature (30–35°C) and this can be taken as an advantage for BOD determination. This is why the temperature for BOD determination is kept at 20°C at which nitrifiers take 8–10 days to proliferate. It is these nitrifiers that compete with heterotrophs for available oxygen and therefore absolutely interfere with the measurement of carbonaceous COD. Also the action of nitrifying population can be controlled by the addition of suitable inhibiting agents such as 2-chloro 6-trichloro methyl pyridine (TCMP). The reason behind the fluctuating 'K' value when, analysed, can be correlated to two factors (a) nature of the organic content present in the sample and (b) the type of microbial species and their efficiency to degrade the organic matter. The availability of the organic matter determines the ability of microorganisms to degrade them. For example organic matter in the form of coarse and colloidal suspensions should undergo hydrolysis before being taken up by microorganisms. 'K' value will be correspondingly high for simple substrates, whereas for complex substrates the rate of reaction slows down and may even stop when it comes in contact with synthetic materials. This is called as lag period and the lack of enzyme in the microorganisms is considered to be the sole reason. Indigenous microorganisms do not extend the lag period and very soon there will be the development of necessary enzymes. Domestic wastes encounter this problem when the bacteria specific to the wastes are not available. To overcome this, "seed water" can be taken from the location where the effluent/waste is being discharged or the seed can be prepared from the soil exposed to the waste. Also in BOD determination, the theoretical value does not go in agreement with the actual value. In general, the actual value is 85% of the theoretical value indicating the probability of organic matter to remain in the sample as residue. This portion is termed as humus which is highly resistant to biological attack.

3.13.3 Methods of Measuring BOD

Based on the pollution strength of waste water, measurement of BOD can be performed directly or by dilution method.

3.13.4 Direct Method

Direct method can be applied to the samples whose BOD concentration does not exceed 7 mg/l. The sample does not require dilution, however the DO level of the sample must be brought down to near saturation. Since the method does not involve any pretreatment of sample, calculation is direct.

3.13.5 Dilution Method

For polluted waters with considerable amount of BOD level, the sample has to be diluted. The purpose is to provide an O_2 level equal to the organic matter present, since it is well known that the O_2 utilized by the microbes is positively correlated with the oxidation process. Even though O_2 can be provided by aeration, excessive BOD amount would deplete them well before the 5th day. On the other hand, O_2 cannot be dissolved in water beyond the limit. Under these conditions, reduction of sample size is the only way. That is why dilution of high-strength waste waters is a prerequisite for BOD determination. It has been estimated that 10 percent dilution utilizes O_2 at one-tenth the rate of a 100 percent sample and above all the complete oxidation of organic matter depends on pH, temperature and osmotic conditions, amount of mixed group of microorganisms, amount of nutrients in the sample and absence of toxic materials.

3.13.6 Dilution Water

For the dilution water, there are two options—surface water and tap water. But they cannot be used due to a variety of reasons such as BOD of the water, variety of microorganisms including algae and nitrifiers, and mineral content. Apart from these, tap waters contain residual chlorine. It is for these reasons that synthetic dilution water is prepared using several chemical substances. To maintain a pH between 6.5 and 8.5, phosphate buffer solution is added.

The functional role of components in the synthetic dilution water is given below.

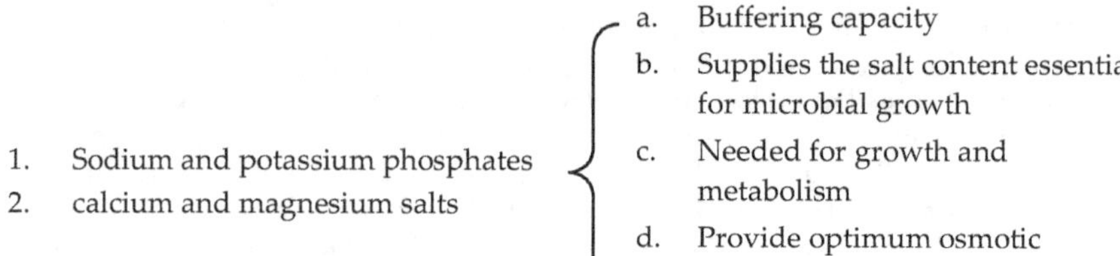

1. Sodium and potassium phosphates
2. calcium and magnesium salts

    a.   Buffering capacity
    b.   Supplies the salt content essential for microbial growth
    c.   Needed for growth and metabolism
    d.   Provide optimum osmotic conditions
    e.   Phosphate source

3. Ferric chloride, magnesium sulphate, ammonium chloride.   } Supply iron, sulphur and nitrogen

The dilution water should then be seeded. Since the amount of sewage added as seed may contain organic matter, a blank should be kept and DO content evaluated for the blank is considered as "0" day and the test sample is incubated. To minimize the result variation, at least run three blanks, for each set of BOD samples. Based on the strength of polluted waters, the dilution range should be determined. For waste water with known range, 2 dilutions are enough, whereas for unknown strength, four dilutions should be made.

3.13.7 Dilutions of the Waste

Dilutions should be made in such a way that the sample contains at least 2 mg/l of dissolved oxygen at the end of the incubation period and that the initial DO value should be above 8. The following table shows the range of BOD measurable by various dilutions. And based on this, the sample should be diluted. It is highly preferable to prepare 2 dilutions one higher and one lower than the range given in the table. The purpose of doing this dilution is to overcome the errors in the original estimate. Standard BOD bottles equipped with air-tight glass stoppers should be used. Since the experiment is to evaluate the organic content, care should be taken to keep the BOD bottles free from organic matter. Cleaning of bottles with detergents should be followed by hot-water rinsing. Otherwise the detergents may favour the growth of nitrifiers. The best option is to use chromic acid. However the bottle should be rinsed thoroughly at least four times with tap water and a final rinse with distilled water. For samples with low BOD i.e. lesser than 20%, blanks are not needed. But still, the samples should be aerated sufficient to raise this DO level.

Samples	Percent of sample to be added to the dilution mixture	Amount of sample (ml) to be added in 1 l solution of mixture
River water	No need for dilution	1000 ml sample
Polluted Water	50, 25, 10	500, 250, 100
Treated sewage	10–50	100–500
Raw sewage	1–5	10–50
Industrial effluents	0.05–20	0.5–200

3.13.8 Significance

i.   To evaluate the strength of domestic and industrial waste water in terms of oxygen needed to oxidize the organic matter.

ii. To determine whether the biologically oxidized organic matter is from domestic or industrial waste.

iii. To monitor the quality of receiving water bodies in terms of BOD.

iv. To evaluate the self-purifying efficiency of rivers.

v.   To select the suitable biological treatment type for the treatment of waste waters.

vi. To find out the efficiency of treatment processes by analysing the BOD level in treated waters.

Expt. 14   *To determine the BOD of the effluent sample*

MATERIALS REQUIRED

Laboratory glassware

BOD incubator

BOD bottle

REAGENTS

a.   Manganous sulphate

b.   Sodium azide (refer dissolved oxygen)

c.   Sodium thiosulphate

d.   Starch

e.   BOD-free water. Use double-distilled water

f.   Phosphate buffer solution

$Na_2HPO_4 \cdot 7H_2O$	33.4 g
K_2HPO_4	21.5 g
KH_2PO_4	8.5 g
NH_4Cl	1.7 g

Dissolve these contents in double-distilled water. Make up to 1 l. Set the pH to 7.2.

g.   Magnesium sulphate solution.

Dissolve 82.5 g of magnesium sulphate in double-distilled water. Make up to 1 l.

h.   Calcium chloride solution.

Dissolve 27.5 g of anhydrous calcium chloride in double-distilled water. Make up to 1 l.

i.   Ferric chloride solution

Dissolve 0.25 g of ferric chloride in double-distilled water. Make up to 1 l.

j.   Sulphuric acid (IN)

Dilute 2.8 ml of conc. sulphuric acid in 100 ml of double-distilled water.

k.   Sodium hydroxide solution (IN)

Dissolve 4 g of sodium hydroxide in 100 ml double-distilled water.

l.   Allylthiourea solution

Dissolve 500 mg of allylthiourea in double-distilled water. Make up to 1 l.

PROCEDURE

Determination of BOD involves the following steps.

1.   Preparation of dilution water

2.   Addition of microbial seed

3.   Dilution of the sample

4.   Determination of dissolved oxygen

Preparation of dilution water

1.   Transfer 2 litres of double-distilled water in a plastic can.

2.   Store it in the BOD incubator at 20°C for 2 hours.

3.   Aerate this water over night.

4.   At the end of aeration, the DO content of this sample should be above 8. If not, proceed with the aeration for another 2 hours.

5.   For every one litre of well-aerated water add 1 ml each of calcium chloride, magnesium sulphate, ferric chloride and phosphate buffer solution.

6.   Mix thoroughly and use for dilution of sample.

7.   Do not store this water and prepare fresh before use.

Addition of microbial seed

Since estimation of BOD content depends on the microbial ability to oxidize the organic matter, favourable conditions for the growth of microorganisms is very much necessary. However, this cannot be expected in all types of effluents especially for waste water with toxic substances. Under their circumstances, addition of microbes from external source is required. This can be done depending on the composition of wastes. Industrial wastes with low microbial load can be seeded with fresh sewage. Similarly industrial wastes with non-biodegradable substances and toxic metals can be seeded with microbial culture acclimatized with the industrial waste for about a week.

Dilution of the sample

BOD content of the sample well above its DO content will misinterpret the results. Therefore based on the expected BOD of the sample, it should be diluted with the dilution water. (For dilution refer table)

1. Take care not to entertain air bubbles while diluting the sample.
2. Now carefully transfer the diluted sample to two BOD bottles.
3. Again take utmost care to fill the BOD bottles without entertaining air bubbles.
4. Fill the sample up to the rim.
5. Check out the DO content in one bottle (initial DO) and keep the other bottle in BOD incubator at 20°C for 5 days.
6. At the end of the incubation period, determine its DO content (final DO).

Determination of DO

Refer DO

Calculation

$$BOD = \frac{(DO_I - DO_F) \times 100}{V}$$

where,

DO_I = DO of the sample at the 0 day

DO_F = DO of the sample on the 5th day

V = Volume of the sample

3.14 CHEMICAL OXYGEN DEMAND

BOD value of the given sample is limited to oxidizable organic matter leaving behind the inert biological and inorganic substances. The term pollution load comprises excessive organic and inorganic content in the water/effluent. Microbial action on these substances is highly impossible and hence COD method is adopted to estimate "the total pollutant content". It can be described as the amount of O_2 required by $K_2Cr_2O_7$ for the oxidation of organic and oxidizable inorganic matter. In other words it is a chemical oxidation process that brings out the true picture of water quality. The value obtained is approximately 3 times higher than the BOD value. Added, the time period required for COD is only 4 hours. The test is based on the fact that a strong oxidizing agent like potassium dichromate is able to oxidize both oxidizable and inert biological matter. Since the result is the combination of both oxidizable and non-oxidizable organic matter, value of COD is always higher than that of BOD. In case the waste water is rich in organic matter resistant to biodegradation (cellulose, lignin) they remain stable, not interfering with the structure and function of the ecosystem in the receiving water. However the COD value obtained may be high and mislead the analysis to interpret the pollution load of waste water as "too bad". Usage of chemical oxidizing agents for COD determination has been in practice since long time before. Initially potassium permanganate was used and was not up to the mark due to the following reasons.

1. Varied results for different compounds.
2. Variability in degree of oxidation for different concentration of reagents.
3. COD values lesser than BOD values.

Besides ferric sulphate and potassium iodate, potassium dichromate has been found to be most reliable due to its ability to oxidize a wide range of organic substances regardless of their biological assimilability. Added, it is very easy to measure the excess potassium dichromate than any other reagent. To ensure complete oxidation, potassium dichromate should be added with a strongly acidic solution. However, the digestion process enhances the temperature of acidic potassium dichromate to boiling point which ultimately evaporates the volatile materials in the sample. Therefore to prevent this, digestion is done in a reflux condenser. Further, to promote the oxidation of low molecular weight fatty acids, silver ion is added as catalyst. Aromatic hydrocarbons and pyridine are the exceptions which cannot be subjected to oxidation at any cost. For accurate results, potassium dichromate should be dried at 103°C before use.

Since COD results are the direct indication of amount of oxygen consumed, N/8 solution of potassium dichromate is prepared, 8 being the equivalent weight of oxygen. It is reported that N/4 solution of potassium dichromate can be used to estimate large samples. Since the sample contains varied group of organic compounds, the amount of oxidizing agent should be sufficient enough to oxidize them, a step to be taken in the safer side. In case of insufficient amounts, resistible organic compounds may remain unoxidized thus misleading the analyst to add excess amount of oxidizing agent. However at the end of digestion, to evaluate the amount of reduced potassium dichromate, the excess amount should be measured. Ferrous ion prepared from ferrous ammonium sulphate is used as a reducing agent for $K_2Cr_2O_7$. Never expose the reducing agent to air as it would begin to be oxidized by atmospheric oxygen. Titration of ferrous ion with $K_2Cr_2O_7$ is represented as

$$6Fe^{2+} + Cr_2O_7^{2-} + 14H^+ \rightarrow 6Fe_3 + 2Cr^{3+} + 7H_2O$$

Since the additions of reagents and dilution water may bring the organic content from extraneous sources, a blank should be conducted. The end point at which all the oxidized dichromate has been reduced by the titrant ferrous ammonium sulphate (FAS) is indicated by the ferroin indicator. Ferrous 1, 10 phenanthroline sulphate (ferroin) is an excellent indicator to reveal the end point of oxidation–reduction reactions. Colour change is from orange (bluish green–blue) to red and the intermediate bluish shade is due to the presence of Cr^{3+} ions formed by the reduction of dichromate.

3.14.1 Inorganic Interferences

Spent solutions of COD experiment should not be allowed for deliberate discharge as they contain hazardous wastes like H_2SO_4, chromium, silver and mercury. To avoid this, they should be disposed after recovery of silver and mercury. Otherwise reducing the size of the samples will ultimately reduce the amount of reagents used. Using required amount of FAS as titrant will maintain the accuracy of results. Certain inorganic ions like chloride get oxidized and interfere with the COD results. Addition of H_2SO_4 will successfully eliminate this, by combination of mercuric ions with chloride ions to form mercuric chloride. Also the addition of nitrites to nitrates prevents the addition of sulphuric acid to $K_2Cr_2O_7$.

$$6Cl^- + CrO_7^{2-} + 14H^+ \rightarrow 3Cl_2 + 2Cr^{3+} + 7H_2O$$

$$Hg^{2+} + 2Cl^- \rightarrow HgCl_2$$

3.14.2 Applications

1. To evaluate the amount of oxidizable and non-oxidizable wastes in water and waste waters.

2. To evaluate the treatment efficiency of treatment plants.

3. To evaluate the presence of biologically resistant organic substances in water and waste water.

4. To aid in water quality monitoring and to see to it that the wastes discharged to the receiving water are well within the permissible limits in terms of organic content.

3.14.3 Principle

During sample refluxion, potassium dichromate and sulphuric acid bring out the oxidation of organic matter. Excess of $KMnO_4$ is titrated against ferrous ammonium sulphate to determine the organic content of the sample. Addition of silver sulphate promotes the oxidation of complex organic compounds.

$$6Fe^{2+} + Cr_2O_7^- + 14H^+ \rightarrow 6Fe^{3+} + 2Cr^{3+} + 7H_2O$$

Expt. 15   *To determine the COD of the effluent sample*

MATERIALS REQUIRED

Laboratory glassware

COD refluxing unit

It consists of a round-bottomed flask, a reflux condenser and a heating mantle.

REAGENTS

a. Potassium dichromate solution (0.25N)

Dissolve 12.259 g of potassium dichromate powder in 1 l of distilled water.

b. Silver sulphate

Dissolve 1 g of silver sulphate in 100 ml concentrated sulphuric acid.

c. Ferroin Indicator

Dissolve 0.695 g of ferrous sulphate and 1.485 g of 1, 10-phenanthroline in 100 ml of distilled water.

    d.   Ferrous ammonium sulphate solution (0.1N)

Dissolve 39.29 g of ferrous ammonium sulphate in little distilled water. To this add 20 ml of conc. sulphuric acid. Stir well. After cooling make up to 1 l. Prepare fresh before use.

PROCEDURE

1. Take 20 ml of the sample in the round-bottomed flask and of COD reflux unit.

2. For effluents with high COD, reduce the size of sample by diluting it with distilled water.

3. Make sure that the final volume is 20 ml.

4. Similarly for turbid samples, blend the solids using homogenizer and then add the required reagents for digestion.

5. To this, add 10 ml of 0.25N potassium permanganate solution.

6. Shake well and add a pinch of mercuric sulphate powder.

7. Addition of mercuric sulphate should be based on the chloride content of the sample, i.e., it should be added in the ratio of 10:1.

8. See to it that the sample is golden orange in colour.

9. Place the flask in a cool water bath and add 30 ml of sulphuric acid solution.

10. Allow it to stand for 30 minutes.

11. Now digest the content of the flask in a reflux condenser for 2 hours. Following the digestion process, put off the mantle and allow the water to circulate for another 10 minutes.

12. Then detach the unit, dilute it to 150 ml using distilled water. Add 2–3 drops of ferroin indicator and titrate it against 0.1 N ferrous ammonium sulphate solution.

13. Observe for the colour change from bluish green to blue to red.

14. The end point is very sharp and as soon as the solution shows bluish green colour, perform the titration carefully and slowly.

15. Run blank using distilled water.

Procedure for low COD samples

If the COD content is greater, dilution can be made. However it should be remembered that the potassium permanganate solution should also be diluted. This is due to the fact that the quantity of potassium dichromate added to the sample and that remaining after refluxing should be directly proportional, which would otherwise give erroneous results. Always keep the sample away from contamination, and only trained personnel should do the analysis. If the sample ratio of potassium permanganate : sulphuric acid is not maintained at 1:1 it will give a lot of discrepancies in result evaluation. This is because if the ratio is < 1:1, the oxidizing ability of the solution will be absolutely low to degrade the organic content and if it is > 1:1 consumption of $K_2Cr_2O_7$ by blank goes beyond the limit.

CALCULATION

$$\text{COD (mg/l)} = \frac{(B - A) \times N \times 1000 \times 8}{V}$$

where,

B = volume of titrant used against distilled water

A = volume of titrant used against sample

V = volume of the sample

3.15 AMMONIACAL NITROGEN

Major source of ammonia in aquatic bodies is excreta. Since it is highly toxic, it is discharged immediately and in water it forms ammonium hydroxide which further dissociates into ammonium (NH_4^+) and hydroxyl ions (OH^-) ions. Ammonium ions can be as such taken up by plants. Other sources of ammonia are nitrogenous organic matter from sewage and industrial effluents from coke manufacture and petroleum industry.

The four different methods by which ammoniacal nitrogen is analysed are

a. Direct nesslerization Method
b. Direct phenate addition
c. Distillation
d. Volumetric analysis

3.15.1 Direct Nesslerization Method

Nessler's reagent (K_2HgI_4) used in this method is an alkaline solution of potassium mercuric iodide. Addition of this reagent to the pretreated sample favours its combination with the ammonia present to form a yellowish brown colloidal dispersion. The intensity of colour development is directly proportional to the amount of ammonia present.

$$2K_2HgI_4 + NH_3 + 3KOH \longrightarrow \quad + 7KI + 2H_2O$$

$$Hg$$
$$O$$
$$Hg$$
$$NH_2 \text{ Yellow-brown}$$

Colloidal dispersion

The amount of ammonia present in the sample can be estimated either by visual comparison with standard or by photometric methods.

3.15.2 Direct Phenate Addition

This method involves the addition of alkaline phenol solution, hypochloride and manganous salt. Manganese acts as a catalysing agent to speed up the reaction involving phenol, hypochloride and ammonia (to produce indophenol blue colour complex). Based on the concentration of ammonia present in the sample the colour intensity develops. The ammonia in the sample can be determined either by photometry or visual comparison method.

3.15.3 Distillation

Major disadvantage encountered in both the procedure described above is the presence of interfering compounds that tend to approximate the results rather than providing accurate results. To eliminate this, distillation procedure is usually preferred. During distillation, ammonium ion is decomposed to ammonia gas and hydrogen ion

$$NH_4^+ \rightarrow NH_3 + H^+$$

pH higher than 8 will favour this equation but as hydrogen ion starts to accumulate in the residue, the pH of the sample will be dropped thus reversing the reaction automatically. Hence a borate buffer is added to maintain the pH at 9.5. However it should be noted that at high temperature (boiling point) and pH, ammonia in the organic source may expel out and therefore distillation is allowed to run for a limited period of time i.e. till 200 ml of the distillate is collected. The vaporized ammonia is collected in boric acid at the condenser end. Acidic condition transforms the ammonia gas to ammonium ion and prevents its escape. Then the collected distillate is processed for the estimation using Nessler's reagent. While calculating the amount of ammonia, volume of the distillate and the sample size should be taken into consideration.

3.15.4 Principle

At neutral pH, the sample is distilled. The collected distillate when combined with Nessler's reagent produces yellow colour which can be compared with that of standards. Sample with higher ammonia content (75 mg/l) is estimated by distillation with boric acid at the collecting end. The distillate is then titrated against standard sulphuric acid.

Expt. 16   *To determine the ammoniacal nitrogen content of the effluent sample by calorimetric method*

MATERIALS REQUIRED

Laboratory glassware

Spectrophotometer

REAGENTS

a. Phenol nitroprusside solution

    15 g of phenol dissolved in 500 ml distilled water. To this, add 1.5% w/v aqueous solution of sodium nitroprusside. Every time sodium nitroprusside should be prepared fresh.

b. Alkaline hypochlorite solution

    10 g of NaOH dissolved in distilled water. To this add, 2.7 ml of 10% solution of hypochlorite and make up to 500 ml.

c. Standard ammonium chloride solution

    3.82 g of anhydrous ammonium chloride powder dissolved in distilled water. Make up to 1000 ml. From this stock solution (1 g NH_4 N/l) pipette out different concentrations of standard ammonium solutions ranging from 0.1 to 1 mg. Care should be taken to prepare all the solutions in ammonia-free distilled water. Addition of little amount of H_2SO_4 prior to distillation process will expel even traces of ammonia.

PROCEDURE

1. To 20 ml sample, add 2 ml each of phenol nitroprusside solution and alkaline hypochlorite solution.

2. Then make the content of the flask up to 25 ml using ammonia-free distilled water and place the flask in an aluminium box at 25°C to avoid light.

3. After 1 hour, read the absorbance in the spectrophotometer at 635 nm using distilled water as blank.

4. Repeat the same procedure for the different dilutions of standard ammonium chloride solution and measure their respective absorbance.

5. Draw a standard curve using these values and evaluate NH_4N by comparing the absorbance value of the sample with the standard curve.

Expt. 17   *To determine the amount of ammoniacal nitrogen using volumetric method*

MATERIALS REQUIRED

Laboratory glassware

Microkjeldahl distillation assembly

REAGENTS

a. HCl (O.O/N)

   8.34 ml of 12N conc HCl is diluted with 100 ml distilled water. Take 100 ml from this stock solution and make up to 1000 ml (1N) using distilled water (0.1N). Pipette out 100 ml of 0.1N HCl, dilute it with distilled water and make up to 1 l.

b. Boric acid cum indicator solution

   4 g of borax crystals added to 100 ml distilled water. To favour complete dissolution, heat the content. Stir while heating.

PROCEDURE

1. The micro apparatus consists of the following parts
   - Boiling flask
   - Sucking chamber
   - Condenser
   - Sample chamber
   - Conical flask

2. Fill the boiling flask with distilled and water connect the apparatus as described in the schematic diagram.

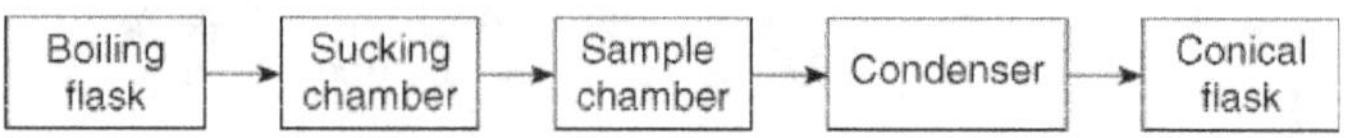

3. Allow the steam to pass to the sample through the sucking chamber.

4. Pour the sample into the sample chamber through the funnel fitted at the top.

5. Keep the conical flask with 5 ml boric acid cum indicator at the lower end of the condenser.

6. The conical flask is positioned in such a way that the end of the condenser is completely immersed in the indicator solution.

7. Now this will vaporize the sample and continue the distillation process and the distillate is collected in the conical flask.

8. About_40 ml of distillate is collected.

9. The presence of ammonia turns the distillate into blue colour.

10. Now titrate the distillate with 0.01N hydrochloric acid. The end point is the colour change from blue to faint pink or brown.

11. Conduct a blank using distilled water.

12. Cool the boiling flask so that the sucking chamber removes off all the wastes which can be withdrawn from the outlet at the bottom.

CALCULATION

Amount of ammoniacal nitrogen in the sample (mg/l)

$$= \frac{(S - B) \times 1000 \times 14}{V}$$

where,

S = Volume of titrant used against sample

B = Volume of titrant used against blank

V = Volume of the sample

3.16 NITRATE NITROGEN

3.16.1 Principle

Four standard methods are available for the estimation of nitrate nitrogen. They are

1. UV spectrophotometry
2. Ion chromatographic method
3. Nitrate electrode method
4. Cadmium reduction method

3.16.2 UV Spectrophotometry

At a wavelength of 220 nm, nitrates absorb ultraviolet radiation and therefore they can be measured in a spectrophotometer that can be operated in a ultraviolet range. The drawback of this method is that certain substances like nitrites, hexavalent chromium and organic compounds absorb ultraviolet radiation at this wavelength and hence the application of this method is restricted to initial screening.

In the spectrophotometric method, the sulphanilamide method and brucine method are the commonly used methods for estimation of nitrate nitrogen. The brucine method utilizes the principle that in acid medium, nitrate ion combines with brucine to form yellow colour, which can be measured spectrophotometrically at 410 nm.

3.16.3 Ion Chromatographic Method

This method can be adopted when the nitrate ion exceeds 0.2 mg/l and the result obtained is accurate and highly reliable. It also favours the simultaneous analysis of other ions in the sample. However the cost and the maintenance of this equipment restricts its commercial application.

3.16.4 Nitrate Electrode Method

Nitrate concentrations up to 1 mg/l can be easily detected by nitrate electrode. It is a liquid membrane electrode and is highly applicable for bio-monitoring of riverine system. Its automated process control favours the rapid estimation even though the initial calibration consumes some time. However chloride and bicarbonate ions interfere with the analysis and it cannot be used to detect low concentrations.

3.16.5 Cadmium Reduction Method

A column of amalgamated cadmium granules is prepared. Prepared sample (filtered sample + ammonium chloride and EDTA reagent) is allowed to pass through the column of cadmium granules. Cadmium reduces the nitrate contents of the sample to nitrite. Then the sample is subjected to diazotization procedure. The major advantage of this method is the detection of nitrate accurately irrespective of the sample size, i.e., it can measure the nitrate concentration as low as 0.01 mg/l and as high as 1 mg/l. Sample size greater than this should be diluted before passing through the column and then appropriately measured since the obtained result evaluates the reduced nitrite and the original nitrite of the sample. For the determination of nitrate, separate analysis of nitrite alone should be done and then subtracted from the nitrate + nitrite results obtained from the cadmium procedure.

3.16.6 Significance

Determination of nitrogen helps to

    i.   perform the disinfection process.

    ii.   check out where the different forms of nitrogen meet the permissible limits.

    iii.   evaluate the efficiency of treatment method.

    iv.   determine the type of treatment method to be adopted.

    v.   estimate the nitrogen content in sludges so as to decide its fertilizer value.

Expt. 18　*To evaluate the nitrate nitrogen of the effluent sample using sulphanilamide method*

MATERIALS REQUIRED

Laboratory glassware

Spectrophotometer

REAGENTS

a.　Phenol disulphonic acid

In a 250-ml conical flask dissolve 25 g of white phenol in 100 ml of concentrated sulphuric acid. Then carefully add 135 ml of concentrated sulphuric acid and stir the contents well. With utmost care, heat it in a water bath for 2 hours. Do not close the flask while heating. Allow it to cool for one hour and store in an amber bottle.

b.　Potassium hydroxide solution (12N)

336.5 g of potassium hydroxide dissolved in 500 ml of distilled water.

c.　Standard nitrite solution

To prepare stock solution dissolve 0.722 g of anhydrous potassium nitrite in distilled water and make up to 1 litre (10 mg NO_3 N/l). From this stock solution, prepare appropriate dilutions with distilled water preferably in the range from 0.1 to 1 mg NO_3N/l.

PROCEDURE

1.　Transfer 25 ml of the sample in to a porcelain basin.

2.　Heat it until it is dried.

3.　Remove it from the hot bath, scrap off the residue and then add 5 ml of distilled water and 1.5 ml phenol disulphonic acid

4.　Stir the flask every time after adding a reagent.

5.　Observe for the formation of yellow colour.

6.　Allow it to stand for 5 minutes.

7.　Then carefully note the absorbance ion in the spectrophotometer at 410 nm. Run a blank using distilled water.

8.　To the standard nitrite solutions, add the reagents as described above and note down the absorbance values.

9.　Draw a standard curve by plotting the concentration of standard nitrate solution against their absorbance values.

10.　Evaluate the total nitrate nitrogen content of the sample by comparing its absorbance value with the standard curve.

Expt. 19　*To evaluate the nitrate nitrogen of effluent sample using brucine method*

MATERIALS REQUIRED

Laboratory glassware

Spectrophotometer

Hot water bath

REAGENTS

a. Brucine Sulphanilic acid solution

   To 50 ml warm distilled water dissolve 1 g of brucine sulphate and 0.1 g of sulphanilic acid. To this add 3 ml of concentrated hydrochloric acid and make up to 100 ml.

b. Sulphuric acid

   Carefully add 500 ml of distilled water and allow it to cool.

c. Sodium chloride solution

   300 g of sodium chloride dissolved in 1 l of distilled water.

d. Standard nitrite solution

   Refer phenol disulphonic acid method.

PROCEDURE

1. Add 2 ml of sodium chloride solution to 10 ml of sample taken in a 25-ml volumetric flask.

2. To these add carefully, 10 ml of concentrated sulphuric acid and 0.5 ml of brucine sulphanilic acid solution.

3. While doing this, place the flask in a cool water bath. Mix thoroughly and then boil it in a hot-water bath for 20 minutes.

4. Allow it to cool and then measure the absorbance (spectrophotometry) using distilled water.

5. To the various concentrations of standard nitrite solutions, add reagents as described above.

6. Draw a standard curve by plotting the concentration of nitrite solutions against absorbance values.

7. Evaluate the nitrite value of the sample by comparing the absorbance value with that of the standard curve.

3.17 TOTAL ORGANIC NITROGEN

Nitrogen is an important constituent of atmosphere and plays a pivotal role in the life processes of all plants and animals. It exists in several oxidation states like $NH_3(-111)$; $N_2(O)$; N_2O (1); $NO(11)$; N_2O_3 (111); NO_2 (IV) and N_2O_5 (V). Out of this, only four forms of N_2 dominate the aquatic systems (NH_3, N_2, N_2O_3 and N_2O_5). The two main sources by which the nitrogen is transferred from the atmosphere to the terrestrial environment are electrical discharge and nitrogen fixation. The principal form of nitrogen in the soil is nitrate and ammonia out of which nitrate is highly preferred by plants for uptake. The next transformation from ammonia/nitrate to protein occurs within the living organisms

$$Producers + NO_3^- \rightarrow amino\ acids \rightarrow proteins$$

$$Producers + NH_3 \rightarrow amino\ acids \rightarrow Proteins$$

Animals totally depend on plants for protein. These proteins are body builders utilized for growth, repair of muscle tissue and at times for energy production. They are released as ammonium carbonate in urine and as organic nitrogen in faeces. After death, the decomposition of animals is carried out by heterotrophs.

$$\text{organic nitrogen} + \text{microbial action} \rightarrow NH_3 + \text{non-digestible detritus/humus}$$

A small portion remains undigested and forms detritus in aquatic system and humus in terrestrial system. The ammonia released is oxidized by autotrophic nitrifying bacteria *Nitrosomonas* to nitrites.

$$2NH_3 + 3O \xrightarrow{\textit{Nitrosomonas}} 2NO_2^- + 2H^+ + 2H_2O$$

Nitrobacter group act on nitrites to form nitrate thus moving the various forms of nitrogen in a cycle

$$2NO_2^- + O_2 \xrightarrow{\textit{Nitrobacter}} 2NO_3^-$$

3.17.1 Denitrification

It is formed in both oxidizing and reducing conditions. Under aerobic conditions nitrates are reduced to nitrites and then to nitrogen, a process termed as denitrification. However this occurs only in the presence of high level of organic matter. Nitrogen beyond the specified level should not be allowed to be discharged in the water as they favour undesirable growth of algae and aquatic plants that disrupt the ecosystem balance. Before 1893, chemists depended on chemical tests to evaluate the sanitary quality in potable as well as polluted waters. Chloride test was one of the chemical tests used by them but unfortunately it provided no information of how recently the pollution has occurred, so, the environmental engineers were able to identify the time of pollution approximately by the form of nitrogen in water. This calculation was totally based on the oxidizing and reducing ability of nitrogen under suitable environmental conditions.

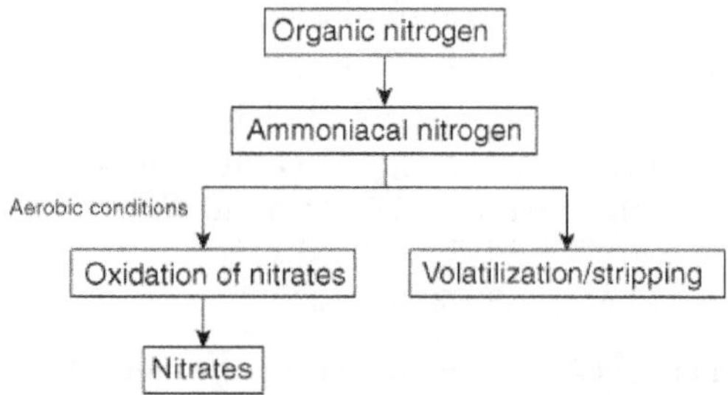

In water, most of the nitrogen will be in the form of organic nitrogen and ammonia. With increasing time they oxidize to ammonia. i.e, in the presence of O_2 and temperature, ammonia is oxidized to nitrites and then to nitrates. Taking these things in to consideration, the time duration of contamination was evaluated.

$$\left.\begin{array}{l}\text{Water with maximum proportion of}\\ \text{organic and ammonia nitrogen}\end{array}\right\} \text{Recently polluted}$$

$$\left.\begin{array}{l}\text{Water with maximum}\\ \text{proportion of Nitrates}\end{array}\right\} \text{Has been polluted long back}$$

Later, the bacteriological test for coliform enumeration was developed, which totally replaced nitrogen analysis in the evaluation of sanitary quality. Since it was found that excess nitrate level leads to methemoglobinemia in infants, deliberate discharge of nitrate into the receiving waters was restricted and maximum permissible limit of nitrate was fixed as 45 mg/l. As nitrogen is an eventual component for living organisms, it is determined in streams and rivers to study their biological productivity. While treating the waste water in biological treatment plants, frequent analysis of different forms of nitrogen, ammonia, nitrate and nitrite is very much essential. This is because, rapid oxidation of organic nitrogen to ammonia depletes the available oxygen. Subsequent conversion of ammonia to nitrate is restricted by two factors a) slow-growing nitrifying bacteria and b) available oxygen. Since the accumulation of ammonia will create anoxic condition, checking the nitrate and ammonia levels at frequent intervals helps to stabilize the waste water treatment process and helps to run it under controlled conditions.

Techniques for the estimation of different forms of nitrogen such as ammonia and nitrate nitrogen emerged out probably by the adverse effects caused by them on the environment. In Kjeldahl method, sulphuric acid is used as an oxidizing agent and the added reagents, mercuric ions and potassium sulphate speed up the oxidation of resistant organic materials. Boiling point of sulphuric acid is 340°C and at this temperature almost all the organic matter is oxidized. To ensure complete oxidation of organic matter, minute changes in the sample during digestion process should be noted. Initially, all the excess water is expelled thus allowing the sulphuric acid to act on the organic matter, when fumes are produced indicating the beginning of digestion. Then the removal of water molecules of organic matter by the acid gives black colour to the mixture followed by the oxidation of carbon. As a result, CO_2 and SO_2 evolve out to show bubbling of sample and finally the completion of digestion is indicated by the clear transparent water-like sample. Even after this, run the digestion for about 20 minutes and collect the distillate in a boric acid containing beaker and then perform the Nessler's method as described above.

3.17.2 Principle

During the distillation of sample, its organic nitrogen is converted to ammonium persulphate in the presence of copper sulphate and ammonium sulphate. Ammonium persulphate, under alkaline conditions, i.e., by the addition of sodium hydroxide liberates ammonia. The distillate is collected with boric acid and then titrated against sulphuric acid.

Expt. 20   *To determine the total organic nitrogen of effluent sample*

MATERIALS REQUIRED

Laboratory glassware

Microkjeldahl distillation assembly

REAGENTS

a.   Digestion mixture

In 200 ml distilled water, add 16.25 g of potassium sulphate. Mix thoroughly and then add 0.4 g of mercuric oxide. To this carefully add 25 ml of concentrated sulphuric acid and make up to 250 ml.

b.   Hypo solution

In 200 ml distilled water dissolve 50 g of sodium hydroxide and 10 g sodium thiosulphate. Make up to 250 ml.

   c.   Boric acid solution

       Add 1 g of boric acid in 100 ml of distilled water.

   d.   Mixed indicator

       Reagent a Dissolve 0.1 g of methyl red in 100 ml ethyl alcohol (95%).

       Reagent b Dissolve 0.5 g bromocresol green in 100 ml ethyl alcohol (95%).

       Mix one part of reagent (a) with 2 parts of reagent (b).

   e.   Hydrochloric acid (0.01N)

       Dilute 8.34 ml of hydrochloric acid in 100 ml from 1N HCl acid solution and make up to 1 l (0.1N). Again take 100 ml from 0.1N HCl acid solution and make up to 1 l (0.001N).

PROCEDURE

1. Transfer 200 ml of sample into a clean dry silica crucible and place it on a hot plate until the contents of the crucible are completely dried.

2. To this, add 4 ml of digestion mixture and 20 ml of distilled water. Mix well and again heat the solution for about 15 minutes.

3. Allow it to cool. Now pour the digested content into the microkjeldahl apparatus.

4. Add 3.5 ml of hypo solution to the sample and conduct the distillation process as described for ammoniacal nitrogen. (The conical flask kept at the end of the condenser consists of 5 ml of boric acid cum indicator solution).

5. Collect the distillate and titrate it against hydrochloric acid. Presence of nitrogen turns the sample to a blue coloured solution.

6. Observe for the colour change from blue to pink colour.

7. Conduct a blank using distilled water.

CALCULATION

Amount of total organic nitrogen in the sample $= \dfrac{(S-B) \times 1000 \times 14}{V}$

where,

    S = Volume of titrant used against sample

    B = Volume of titrant used against blank

    V = Volume of the sample

3.18 DISSOLVED ORGANIC NITROGEN

3.18.1 Procedure

Allow the sample to pass through a (m) millipore filter paper, collect the filtrate and perform the same procedure as that for total organic nitrogen.

3.19 PARTICULATE ORGANIC NITROGEN

Particulate organic nitrogen of the given sample can be evaluated by subtracting the TON from DON.

$$PON\ (mg/l) = TON - DON$$

INORGANIC CONSTITUENTS

3.20 SULPHATE

3.20.1 Principle

Sulphates as such, without any anthropogenic influence are high in arid and semi-arid regions. Apart from this, industrial effluents and sewage when discharged into aquatic system may elevate its sulphate content. Source of sulphate in water is the biological oxidation of sulphur compounds. In excess amount, they cause cathartic effect in living organisms and it is the major anion in water. Gravimetrically it is determined using barium chloride, and formaldehyde is added to the sample to eliminate the interfering ions like SO_3^{2-}. Also, in appreciable amounts they form hard scales in boilers and heat exchangers. In potable waters, the maximum permissible limit is 250 mg/l. Under anoxic conditions, sulphates get reduced to hydrogen sulphide, producing obnoxious odour deterring the aesthetic quality of receiving waters. Nitrate and sulphate act as final electron acceptors and of these two ions sulphate is least preferred. As a result of biochemical oxidation, sulphate ion is reduced to sulphide which combine with the valuable hydrogen ions to form hydrogen sulphide. The existence of different forms of sulphate absolutely depends on pH. At pH >8, sulphate exists as HS^+ and S^{2-} ions while H_2S is present in minimum amount, whereas low pH i.e. <8, induces the formation of unionized shifting the equation to the right and 7 is the optimum pH for the reaction to complete and at this stage H_2S is produced at large level. However it should be noted that this occurs only when sulphate is reduced to yield sulphide ions

$$SO_4^{2+} + \text{organic matter} \xrightarrow{\text{anaerobic action}} S^{2-} + H_2O + CO_2$$

$$S^{2-} + H^+ \rightarrow HS$$

$$HS + H^+ \rightarrow H_2S$$

But the problem is that, the hydrogen sulphide thus produced may escape into the atmosphere which when exceeds beyond 20 ppm causes undesirable effects. Next to carbonic acid, hydrogen sulphide is blamed to causes "crown corrosion" in concrete sewers. This occurs due to the high temperature of domestic waste water associated with anoxic conditions, an excellent environmental condition suitable for the sulphate to get reduced to hydrogen sulphide. This can be avoided by proper ventilation in sewers and by keeping the walls and top of the sewers dry. H_2S-oxidizing bacteria, being common inhabitants of sewers, if provided with optimum conditions (anoxic, moisture and high temperature) will give rise to acid formation and the resultant consequence is "crown corrosion".

$$H_2S + 2O_2 \xrightarrow[\text{Evolution of } H_2S \text{ gas}]{\text{moisture}} H_2SO_4$$

$$SO_4 \xrightarrow{\text{reduction}} H_2S$$

Sulphide minerals in the coal mines and mineral bearing deposits under the action of certain microorganisms are oxidized to sulphate. When this waste water is discharged into the streams or rivers, excess sulphate, high iron content and hydrogen ions that deliberately reduce the pH of the receiving waters will harm the water quality.

Based on the concentration of sulphate in the sample, it can be measured by four different methods.

1. Ion chromatography
2. Gravimetric method
3. Turbidimetric method
4. Automated methyl thymol blue method

3.20.2 Significance

To evaluate the suitability of water for drinking and industrial purposes

i.   to evaluate the condition of water in terms of O_2 i.e. under reduced condition sulphates are converted to H_2S a typical indicator of anaerobic process.

ii. to evaluate the efficiency of anaerobic digestion in waste water treatment plants.

iii. to evaluate the complete dissimilation of organic content in biotreatment plants which can be identified by the release of organic bound sulphur as sulphur ion.

3.20.3 Principle

Addition of barium chloride and hydrochloric acid precipitate sulphate as barium sulphate.

Expt. 21   *To determine the amount of sulphates in the effluent sample*

MATERIALS REQUIRED

Laboratory glassware

REAGENTS

a.   Methyl red indicator

Dissolve 10 g of barium chloride in 100 ml of distilled water. Filter it through Whatman No. 1 filter paper and store for use.

b.   Silver nitrate solution

Dissolve 8.5 g of silver nitrate powder in little nitric acid solution. Make up to 500 ml using distilled water.

c.   Hydrochloric acid (50%)

Add distilled water and conc. hydrochloric acid in the ratio 1:1.

METHOD I (ION CHROMATOGRAPHY)

It is suitable for the sulphate concentration as low as 0.1 mg/L. For further details refer chapter 1

METHOD II (TURBIDIMETRIC METHOD)

Addition of barium chloride to the sample forms poorly insoluble barium sulphate precipitate. The precipitate formation is enhanced by adding acidic buffer solution prepared by the composition of chemicals like magnesium chloride, potassium nitrate, sodium acetate and acetic acid. When the amount of buffer and barium chloride to form the precipitate is standardized, result can be evaluated in a short period of time. To overcome the variables, standard sample of sulphate ion should be used for the samples having similar range of sulphate concentration and then compared to draw a calibration curve.

METHOD III (METHYL THYMOL BLUE)

Acidify the sample using hydrochloric acid and then add barium chloride. Adjust the pH of the sample to 10. Now add methyl thymol blue. It forms a stable complex with barium sulphate whereas the excess barium chloride is in grey colour. The colour formation is measured colorimetrically. Run blank using distilled water.

METHOD IV (GRAVIMETRIC METHOD)

Sulphate concentration up to 10 mg/l can be determined by this method. The method is based on the formation of poorly insoluble barium sulphate. When barium chloride is added in excess to the sample, barium ions of barium chloride combine with sulphate ions to form insoluble barium sulphate. The sample is pretreated with few drops of hydrochloric acid and then boiled. This step is done to remove the barium carbonate precipitate. This is due to the fact that alkaline waters when allowed to reach the boiling point may form $BaCO_3$ precipitate (excess barium chloride gives carbon ions sufficient enough to precipitate sulphate ions). Then the sample is digested to convert the colloidal form of barium sulphate to crystalline form. In the absence of digestion process, the precipitate formed cannot be removed by filtration process. Even after this, filter paper that enables the separation of small crystals of barium sulphate should be used. See to it that all the crystals are being transferred from the sample container to the filter paper. Wash the filter paper many times to remove the excess barium chloride. Sulphate concentration can be evaluated either by combustion or by taking direct weight and subtracting the value from the pre-weighed value of filter paper.

PROCEDURE

1. Transfer 100 ml of sample into a 250-ml volumetric flask.
2. To this, add 2–3 drops of methyl red indicator.
3. Slowly add hydrochloric acid to the flask content and observe for the colour change from red to orange. (This indicates the acidic pH of the solution.)
4. Add still more hydrochloric acid to the flask and boil it.
5. Now add warm barium chloride solution so as to favour the precipitation process.
6. Continue to add it till the precipitation is completed.

7. Again heat it in a hot-water bath for about 2 hours. Do not cool it. Immediately filter it through ash less filter paper.

8. Flush the filter paper with distilled water many times.

9. This should be repeated till the filtrate does not contain traces of chloride.

10. Every time after washing the filter paper, check the presence of chloride in the filtrate by filtrating it against silver nitrate solution. Absence of turbidity indicates nil chloride content and at this stage transfer the filter paper to a silica crucible.

11. Ignite it in a muffle furnace at 80°C for 1 hour. After one hour take the weight of the precipitate.

CALCULATION

$$\text{Sulphate (mg/l)} = \frac{W \times 411.5}{V}$$

where,

W = Weight of the precipitate

V = volume of the sample

3.21 HYDROGEN SULPHIDE

Continuous discharge of organic matter into the water bodies favours the protection of heterotrophic microorganisms. In the presence of oxygen, microorganisms act on H_2S to acquire sulphate and then on sulphide.

1. $H_2S \rightarrow S \rightarrow SO_4$ green bacteria, purple bacteria, colourless bacteria
2. $H_2S \rightarrow SO_4$ aerobic sulphide oxidizers
3. $S \rightarrow SO_4$ Aerobic heterotrophic microorganisms

In due course of time, dissolved oxygen gets depleted and under anoxic conditions, anaerobic microorganisms start to deplete organic matter thereby reducing sulphate into sulphide.

$$SO_4 \xrightarrow[\text{anaerobic condition}]{\textit{Desulfovibrio} \text{ sps.}} H_2S$$

The presence or absence of H_2S acts as an index of pollution status of water and its redox potential.

3.21.1 Principle

The sulphide ions in the sample react with p-amino dimethyl aniline and ferric chloride. The completion of reaction is indicated by the development of blue colour. This method is valid only for determining total sulphides and acid-soluble metallic sulphides.

Expt. 22   *To determine the amount of hydrogen sulphide in the effluent sample*

MATERIALS REQUIRED

Laboratory glassware

REAGENTS

a. Cadmium chloride solution

   Dissolve 2 g of cadmium chloride in 100 ml distilled water.

b. Hydrochloric acid solution (4N)

   Dilute 33.3 ml of hydrochloric acid in 100 ml distilled water.

c. Starch indicator

   Dissolve 1 g of soluble starch in 100 ml distilled water. Heat it under low flame until solution becomes clear. Add a few drops of formaldehyde and store.

d. Sodium thiosulphate solution (0.025N)

   To prepare 0.1N stock solution, dissolve 24.82 g of sodium thiosulphate in 1 l distilled water. Pipette out 250 ml of this stock solution in a volumetric flask and make up to 1 l (0.025N).

PROCEDURE

1. In a glass-stoppered bottle carefully fill the sample up to its rim.
2. Do not entertain air bubbles while filling.
3. Now add 1 ml of cadmium chloride solution to the sample.
4. While doing this, take care to insert the pipette almost to the overflowing of sample.
5. Tightly close the bottle, shake well and let it stand for one full day.
6. Following this, carefully decant the supernatant and to the precipitate formed add little hydrochloric acid.
7. Shake well and titrate the contents against sodium thiosulphate solution using starch indicator. Observe for the change from dark blue to colourless solution.
8. Conduct a distilled water blank in a similar way and determine the H_2S of the sample using the calculation.

CALCULATION

$$H_2S \text{ mg/l} = \frac{(B-A) \times N \times 71 \times 1000}{V}$$

where,

     B = volume of the titrant used against blank

     A = volume of the titrant used against sample

     V = volume of sample (ml)

     N = normality of the titrant

3.22 FLUORIDES

Fluoride is an active component that exists in both simple and complex forms. Only in the year 1930 was it found that when present in excess in water bodies (> 1.0 mg/l), fluoride ions cause mottled enamel commonly called as dental fluorosis. Soon after this investigation, steps were taken to control the fluoride levels in water. In domestic water supplies, commonly used defluoridation methods are

1. sorption and ion methods in which the water is allowed to pass through tricalcium phosphate, bone char, bone meal and activated alumina.
2. lime softening process with the precipitating agent, magnesium hydroxide.
3. coagulation process using alum.

A research study conducted by Dean (1938) revealed that fluorides at optimum level is essential for dental health and the level was found to be 1 mg/l. Deviation from this level creates serious dental problems. For example, if the level drops below 1 mg/l, it leads to dental caries while above 1 mg/l, mottled enamel results.

$$\text{Dental caries} \xleftarrow{<1\,mg/l} \underset{\text{Dental health}}{1\,mg/l} \xrightarrow{>1\,mg/l} \text{mottled enamel}$$

Therefore it becomes necessary to supplement fluorides to the deficient water bodies. The added fluorides act similar to that of natural fluorides and help to control dental caries. Several forms of fluorides added to the water supplies are sodium fluoride, calcium fluoride, hydrogen fluoride, sodium silicofluoride, hydrofluorosilicic acid and ammonium silica fluoride. Before adding fluorides to the water, it is necessary to know the actual content of fluoride in water. In this case continuous monitoring of the water supplies is very essential as the level of fluoride should neither increase nor decrease. Regular sampling of the distribution system should be done.

The main source of fluoride contamination is from the phosphatic fertilizer industry and aluminium processing industry. In the latter, cryolite is used as a solvent which at high temperature releases fluoride gas in the form of smoke. Obviously this smoke escapes and condenses to settle in the nearby environment (soil, vegetation, plant and water.)

In general fluorides are determined by four methods

1. electrode method
2. calorimetric method I using alizarin dye
3. calorimetric method II using SPADNS
4. ion chromatography

Out of these four methods, the calorimetric method I using alizarin dye is the widely accepted one.

3.22.1 Principle

Under suitable conditions, zirconium ion combines with the alizarin dye to form a red coloured product termed as zirconium–alizarin "Lake". Presence of fluorides in the sample favour the bleaching action and the concentration of fluorides determines the intensity of bleaching. Since the bleaching is mediated by the action of ZrF_6^{2-}, concentration of fluoride is inversely proportional to the development of colour intensity.

$$\text{Zr–alizarin Lake} + 6F^- \rightarrow \text{alizarin} + ZrF_6^{2-}$$

Expt. 23   *To determine the concentration of fluorides of the effluent sample*

MATERIALS REQUIRED

Laboratory glassware

Laboratory equipment

    Distillation apparatus

    Calorimeter

REAGENTS

a.   Fluoride stock solution

To a 100-ml volumetric flask, add 0.221 g of anhydrous sodium fluoride. Make up to the mark using distilled water.

b.   Fluoride standard solution

Transfer 100 ml of the stock solution in a 1000-ml volumetric flask using distilled water. Make up to the mark. One ml of this solution contains 0.01 mg fluorine.

c.   Acid zirconium—Alizarin solution

   i.   dissolve 0.7 g of sodium alizarin sulphate in 100 ml distilled water

   ii.   dissolve 0.45 g of zirconyl chloride in 100 ml distilled water.

   iii.   Sulphuric acid solution: carefully add 70 ml of conc. sulphuric acid to 700 ml distilled water. Mix well and cool.

Add solution (ii) to (i), stir well and finally add solution (iii). Make up the mixed solution to a final volume of 1000 ml.

d.   Sodium arsenate solution

To 100 ml distilled water, add 1 g of sodium arsenate, stir well and store it in amber bottle.

SAMPLE PREPARATION

Effluent samples may contain interfering constituents and therefore fluorides have to be separated out. The distillation process serves this purpose.

Distillation apparatus

The apparatus consists of a one-litre round-bottomed flask at one end and a condenser at the other end. A J-shaped glass tube is connected at the neck of the boiling flask. Both the ends of the tube are plugged with rubber stoppers and a thermometer is inserted into the lower end to note down the solution temperature. The upper end of the glass jar and the condenser is attached by a long connecting tube. A 300-ml volumetric flask is placed at the lower end of the condenser.

DISTILLATION

i.   Fill the boiling flask with 400 ml distilled water and 200 ml sulphuric acid.

ii. Mix the solution well. Add few glass beads/boiling chips into the flask.

iii. If the sample contains excess chloride (> 200 mg/l), add silver sulphate crystals to the boiling flask at a rate of 5 mg/l of chloride.

iv. Keep the flask in position and heat it gently.

v.   Slowly increase the temperature until the temperature of the solution reaches 180°C.

vi. See to it that the collected distillate is cold.

vii. Allow the solution of the boiling flask to cool.

viii. Discard the distillate and then add 300 ml of sample to the boiling flask.

ix. Repeat the procedure as described above and collect about 300 ml of the distillate.

PROCEDURE

1. Prepare several dilutions of fluoride standard solution into a series of 50-ml Nessler's tubes with their concentration ranging from 1 ml to 12 ml at 2-ml interval.

2. Shift the pH of each solution to neutral pH (pH 7) and make up the volume of each tube to 50 ml using distilled water.

3. Take 50 ml of distilled water as blank and the prepared sample in separate tubes.

4. Add 1 ml of acid zirconium–alizarin solution to the standards, sample and blank. Stir well and measure their absorbance.

5. Draw a standard graph by plotting their absorbance value against concentrations.

6. From the calibration curve, evaluate the concentration of fluoride in the sample.

Note Residual chlorine interferes with the analysis and can be eliminated using sodium arsenate.

3.23 SULPHIDE

3.23.1 Principle

Sulphides in dissolved forms (H_2S, HS^-, S^{2-}) can be evaluated by this method. P-aminodiethyl aniline and ferric chloride when combined with sulphide give rise to methylene blue coloured product.

3.23.2 Sample Preparation

Exercise great care while collecting the samples. Since oxygen destroys sulphide, sampling should be done in low-oxygen conditions. Addition of 2 ml zinc acetate solution and adjusting the pH to alkaline range (>9) using sodium hydroxide solution. The sample can be preserved for a considerable period of time. However immediate analysis gives accurate results.

Expt. 24 *To determine the concentration of sulphide in the effluent sample*

MATERIALS REQUIRED

Laboratory glassware

REAGENTS

a. Sulphuric acid (1 + 1)

Slowly add 500 ml of sulphuric acid to 500 ml of distilled water. Allow the solution to cool.

b. Amine sulphuric acid stock solution

To 20 ml distilled water, add slowly 50 ml of concentrated sulphuric acid. Mix well and to this solution, add 27 g of p-amino diethyl aniline oxalate. Ensure complete dissolution and make up to 1000 ml using distilled water.

c. Amine sulphuric acid working solution

To 500 ml sulphuric acid solution (1 + 1), add 12.5 ml of amine sulphuric acid stock solution. Mix well.

d. Ferric chloride solution

To 40 ml distilled water, add 100 g of ferric chloride. Stir well.

e. Diammonium hydrogen phosphate solution

Add 100 g of diammonium hydrogen phosphate solution to 200 ml distilled water.

f. Methylene blue solution I

Add 1 g of methylene blue in 1000 ml distilled water.

Standardization of Methylene Blue Solution

- Transfer 100 ml of distilled water into an iodine flask.
- To the flask content, add the following
    i. 10 ml of sodium sulphide
    ii. 5 ml of conc. hydrochloric acid solution
    iii. A known excess of iodine solution
- Mix well and back-titrate the solution with 0.025N sodium thio sulphate solution.
- Use starch as an indicator and the end point is the disappearance of blue colour
- Evaluate the sulphide content in the sample using the following calculation.
- Amount of sulfide present in the sample mg/l = (ml of iodine − ml of thiosulphate solution) × 0.025N × 0.4
- Based on the result, prepare a standard sulphide solution in such a way that one ml of the solution contains 1 mg of sulphide.

- Now transfer 1 ml of this solution to 490 ml of distilled water so that the concentration of sulphide in this solution is 2 mg/l.

- Compare the colour of sulphide solution using 2 drops of methylene blue solution.

- If the colour of the methylene blue solution is intense, perform suitable dilution and if the colour is dull in comparison with the sulfide solution, add methylene blue in drops. Finally see to it that both the methylene blue and sulphide solution match each other.

- In this case one drop of methylene blue solution consists of 1 mg of sulphide.

g. Iodine solution—0.025N

h. Sodium sulphide solution

Transfer 15 g of white coloured sodium sulphide powder into a 1000-ml volumetric flask. Make up to the mark using distilled water. If the powder contains traces of black sulphide, wash it with distilled water and then use.

i. Methylene Blue Solution II

Transfer 10 ml of methylene blue solution I to a separate flask. Make up to 100 ml using distilled water. One ml of this solution contains 0.1 mg of sulphide.

PROCEDURE

1. Take two 10-ml Nessler's tubes. To each tube add 7.5 ml of the sample. Let the tubes be marked as I and II.

2. To tube I, add 0.5 ml of amine sulphuric acid working solution and 3 drops of ferric chloride solution.

3. Similarly add 0.5 ml of sulphuric acid (1 + 1) and 3 drops of ferric chloride to tube II.

4. Hold the tube I with the thumb and invert it once.

5. Appearance of blue colour indicates the presence of sulphide.

6. Let the tubes to stand for three minutes.

7. Now add 1.6 ml of diammonium hydrogen phosphate solution to both tubes. This eliminates the colour of ferric chloride solution.

8. Let the tubes stands for another ten minutes.

9. Add methylene blue solution in drops to tube II until the colour matches with blue I.

10. Select the methylene blue solution (I/II) based on the sulphide content. For sample with sulphide content > 5 mg/l, use methylene blue solution I and for samples with sulphide content <5 mg/l, methylene blue solution II is recommended. At increased levels, dilute the samples accordingly.

CALCULATION

Calculate the sulphide content using the following formula

Amount of sulphide (mg/l) in the sample = $a \times b$

where,

a = no. of drops of methylene blue solution I/II

b = amount of sulphide present in methylene blue solution

(1 for methylene blue I and 0.1 for methylene blue II)

3.24 SULPHITE

Under acidic condition, potassium iodate-iodide solution favours the oxidation of sulphite to sulphate. On addition of starch, the oxidation process is indicated by the appearance of a blue-coloured product.

3.24.1 Sample Collection

Collect the samples with least aeration and temperature > 20°C. Temperature beyond this should be cooled and then used. Conduct the analysis as soon as possible.

Expt. 25   *To determine the amount of sulphite in the effluent sample*

MATERIALS REQUIRED

Laboratory glassware

REAGENTS

a.   Sulphuric acid solution (1+1)

Transfer 100 ml of concentrated sulphuric acid to 200 ml distilled water.

b.   Potassium iodate–iodide solution (0.0125N)

Take 100 ml distilled water in a 1000-ml volumetric flask. To this, add the following.

i.   445.8 mg of potassium iodate

ii.   4.35 mg of potassium iodide

iii.   310 mg of sodium bicarbonate

Dry the potassium iodate powder at 120°C before use. One ml of this solution contains 5 mg sulphite.

c.   Starch indicator

PROCEDURE

1.   Titrate the sample against potassium iodate–iodide solution using starch as an indicator.

2.   End point is the appearance of faint blue colour.

3.   Continue the titration until the colour persists.

4.   Calculate the sulphite level using the formula given below.

CALCULATION

Amount of sulphite present in the sample (mg/l) $= I \times 0.5 \times 1000/S$

where,

I = ml of titrant used

S = volume of sample

3.25 SODIUM AND POTASSIUM

Both the parameters sodium and potassium can be measured using flame photometer.

3.25.1 Principle

Refer chapter 1.

Expt. 26   *To determine the amount of sodium and potassium in effluent sample*

MATERIALS REQUIRED

Laboratory glassware

Laboratory equipments

Hot plate

Flame photometer

REAGENTS

Double-distilled water

Sodium stock solution

PROCEDURE

1. Take approximately 3 g of sodium chloride powder, dry it in hot-air oven at 140°C for about one hour.

2. Then from that amount weigh accurately 2.5422 g and dissolve it in one litre of double-distilled water.

3. One ml of this solution contains 1 mg of sodium.

3.26 CHLORIDES

High concentration of chlorides is the clear-cut index of pollution, the source being the faecal matter containing high quantity of chlorides along with nitrogenous wastes. Direct discharge of untreated industrial wastes may also increase the salt content in receiving waters. Chloride content vary between the samples. Chloride-based health complications have not been reported yet. Beyond 250 mg/l, they give bad taste to water and corrode the metals. Evaluation of chloride content is done by titrating the sample against silver nitrate using potassium chromate indicator. The reaction is as follows.

$$NaCl + AgNO_3 \rightarrow AgCl + NaNO_3$$

$$AgNO_3 + K_2CrO_4 \rightarrow Ag(CrO_4)_2 + 2KNO_3$$

Chloride concentrations in water are directly proportional to mineral content of water. Surface and ground waters have considerable concentration of chlorides than upland and mountain supplies. Human urine is the main source of chlorides and it is estimated that about 6 g of chlorides/person/day is excreted out and therefore sewage contains approximately 15 mg/l of chlorides. It is interesting to note that in areas where water supply is very scarce, chloride level is very high.

3.26.1 Significance

Only chloride level beyond 25 ml is considered to be harmful and it gives a salty taste to water and hence the maximum permissible limit for chlorides is 250 mg/l. This cannot be generalized, in dry areas where water scarcity restricts the water consumption per head, people use water containing 1800–2000 mg/l of chlorides and surprisingly no complication associated with chlorides have been reported. Well before the adoption of bacteriological procedures, chemical test such as chlorides and nitrogen have been is use to detect water contamination. Chloride content evaluation in water used for irrigation gain more importance because salty water after evapo-transpiration (tend to settle in the soil and creates difference in the osmotic pressure of water thus blocking the uptake by the roots. This totally affects the yield of crops and therefore the chloride content is restricted to waters used for irrigating salt-sensitive crops. Therefore estimation of chlorides can be used to

a. evaluate the water quality suitable for drinking.
b. select the waste treatment method that ensures efficient desalting.
c. restrict the use of ground water in the areas nearby oceans.
d. determine the amount of chemicals to be added to eliminate chloride in COD determination.

3.26.2 Procedures for the Estimation of Chlorides

Chlorides can be determined by two types of volumetric methods.

1. Mohr method
2. Mercuric nitrate method

Apart from that it can also be estimated by instrumentation methods given below:

a. Potentiometric method
b. Ion chromatography
c. Calorimetric procedure

3.26.3 Procedure I (Potentiometric Method)

Two electrodes are used in this method a) reference electrode b) silver–silver chloride electrode. Silver nitrate solution is used as a titrant. On titration, chloride forms a complex with silver ions and the complex formation continues as long as the chloride ions are present in the sample. In the absence of free chloride ions, the silver ions tend to increase which in turn is indicated by the sudden rise in the voltage.

3.26.4 Procedure II (Calorimetric procedure)

Here the mercuric thiocyanate is used as a titrant. During the titration process, the mercuric ions combine with the chloride ions in the sample thus liberating the thiocyanate. The thiocyanate complexes with ferric ion to give a deep red coloured solution and the colour intensity is directly proportional to the chloride in the sample. The colour is measured in the colorimeter.

3.26.5 Procedure III (Ion Chromatography)

Refer chapter 1.

3.26.6 Procedure IV (Mercuric Nitrate Method)

It is more advantageous than Mohr's method as it is not affected by interfering compounds. To achieve this, the pH of the sample is shifted to the acidic range i.e. <2.5. The titration process results in the reaction of Hg^{2+} ion with chloride ion to form a soluble $HgCl_2$ complex which shows highly accurate end point.

$$Hg^{2+} + 2Cl^- \rightleftharpoons HgCl_2$$

In the absence of free chloride ions excess Hg^{2+} ions combines with diphenyl carbazone indicator and gives a distinct purple colour. For pH drop, nitric acid should be added and this pH should be maintained in blank standards and sample. The same pH should be maintained throughout the titration and to check it, a xylene cyanol FF pH indicator which is blue green only at a pH of 2.5, is added. As per the standard procedure, about 0.0141 N of mercuric nitrate titrant should be prepared in which 1 ml = 0.5 mg of Cl^-. However to make it more convenient as 1 ml = 1 mg of Cl^-, 0.0282N mercuric nitrate is prepared.

3.26.7 Procedure V (Mohr method)

The titrant used in this method is silver nitrate and it can be standardized using chloride solution. Chloride ion precipitates to form a white coloured silver chloride.

$$Ag^+ + Cl^- \rightleftharpoons AgCl$$

Since the chloride is determined by the amount of excess chloride, potassium chromate is used as indicator. Once all the chloride ions are precipitated, silver ions tend to complex with chromate ion supplied by potassium chromate indicator and the product when it goes beyond the solubility point, forms a reddish brown precipitate.

$$2Ag^+ + CrO_4^{2-} \rightleftharpoons Ag_2CrO_4$$

Since the excess amount of Ag needed to form a precipitate cannot be specified, a blank should be preformed. 100 ml of sample is always preferred. Prior to the titration, adjust the pH to the neutral range (7–8) because both at acidic and alkaline condition the expected silver chromate will not be formed.

$$\text{Acid pH } CrO_4^{2-} \rightarrow Cr_2O_7^{2-}$$

$$\text{Alkaline pH } Ag^+ \rightarrow AgOH$$

Always use 2–3 drops of potassium chromate.

Expt. 27   *To calculate the amount of chloride in the effluent sample*

MATERIALS REQUIRED

Laboratory glassware

Titration assembly

REAGENTS

a.   Silver nitrate (0.02N)

Dissolve 1.7 g of silver nitrate in 500 ml distilled water. Mix well and store it in amber bottles.

b.   Potassium chromate 5%

Dissolve 2.5 g of potassium chromate in 50 ml distilled water.

PROCEDURE

1.   To 50 ml of sample add 2 ml of potassium dichromate solution.

2.   Stir well and titrate it against silver nitrate solution.

3.   End point is the colour change from yellow to red.

CALCULATION

Amount of chlorides present in the sample $= \dfrac{N \times V \times 1000}{S}$

where,

N = Normality of silver nitrate

V = Volume of silver nitrate

S = Volume of sample taken

3.27 RESIDUAL CHLORINE

3.27.1 Principle

Combined and free chlorine residuals in the sample convert the iodide to free iodine. The amount of iodine liberated is indicated by the appearance of blue colour with the addition of starch. Quantity of chlorine residuals in the sample can be evaluated by titrating the blue coloured sample with sodium thiosulphate.

Expt. 28   *To determine the amount of residual chlorine in the effluent sample*

MATERIALS REQUIRED

Laboratory glassware

Titration assembly

REAGENTS

a.   Concentrated acetic acid

b.   Potassium iodide crystals

   c.   Starch indicator

Dissolve 1 g of starch in 100 ml of distilled water. Heat it gently until the solution becomes clear and transparent. While heating, stir it with a glass rod to ensure complete dissolution.

   d.   Sodium thiosulphate solution (0.0025N)

      i.   Sodium thiosulphate solution (0.1N)

Dissolve 2.428 g of $Na_2S_2O_3$. $5H_2O$ in little distilled water and then make up to 100 ml. Use boiled and cooled distilled water.

      ii. $Na_2S_2O_3 \cdot 5H_2O$ (0.0025)

Take 25 ml of reagent (i) and make up to 1 l. One ml of this solution contains 0.08865 mg chlorine.

PROCEDURE

1. Take 100 ml of the sample in the conical flask.

2. By adding suitable acid or alkali, adjust the pH of the sample between 4.5 and 8.0. This will prevent the interference of ions such as ferric iron, nitrite and manganese.

3. Now add 0.1 g of potassium iodide crystals and 1 g of acetic acid to the sample. Stir the flask well.

4. As soon as the solution appears brown in colour, titrate it with 0.0025 N of sodium thiosulphate.

5. When the solution turns colourless, add 2–3 drops of starch indicator.

6. Again titrate it with the sodium thiosulphate and observe for the colour change from blue colour to colourless which is the end point.

7. Note the titrant value and determine the residual chlorine using the calculation given below.

CALCULATION

$$\text{Amount (mg) of chlorine present in the sample} = \frac{N \times 0.0887 \times 1000}{V}$$

where,

N = Normality of the titrant

V = volume of the sample

3.28 NITRITE

Nitrite nitrogen rarely exceeds 1 mg/l and it is less than 0.1 mg/l in surface and ground waters. Therefore the method adopted to measure its content should be highly sensitive.

3.28.1 Principle

Nitrites in the sample react with sulphanilic acid to form a diazo compound which when combined with α–naphthyl amine hydrochloride forms a reddish purple azo dye. For this, calorimetric method is highly preferable. It involves the addition of two reagents

1. Sulphanilamide
2. N-1-naphthyl ethylene diamine dihydrochloride

Nitric ion under acidic condition gets converted to nitrous acid which readily react with amino group of reagent 1 (suphanilamide). The end product is the formation of diazonium salt which when combined with N-1 naphthyl-ethylene diamine dihydrochloride forms a bright coloured pinkish red azo dye. The colour intensity depends on the nitrate concentration in the sample and the amount present can be evaluated by visual comparison with standards. Also the coloured samples can be estimated photometrically.

1. Sulphanilamide + nitrous acid + hydrochloric acid $\rightarrow$ diazonium salt + $2H_2O$
2. Diazonium salt + N-1-naphthyl ethylene diamine dihydrochloride $\xrightarrow{\text{HCl}}$ Red coloured azodye + HCl

---

Expt. 29   *To evaluate the nitrite content of the effluent sample using EDTA method*

MATERIAL REQUIRED

Laboratory glassware

Spectrophotometer

REAGENTS

a. EDTA solution

    500 mg of disodium salt of EDTA dissolved in 100 ml distilled water.

b. Sulphanilic acid solution

    600 mg of sulphanilic acid dissolved in 70 ml distilled water. To ensure complete mixing, slightly heat the distilled water. After cooling add 20 ml conc. HCl and dilute the content to 100 ml with distilled water.

c. α-Naphthalamine hydrochloride solution

    600 mg of α-naphthalamine hydrochloride dissolved in 10 ml of distilled water. Add 1 ml of conc. hydrochloric acid and dilute the content to 100 ml using distilled water.

d. Sodium acetate solution

    27.2 g of sodium acetate dissolved in 100 ml distilled water.

e. Standard nitrite solution

    Prepare 1 litre of sodium nitrite solution by dissolving 1.232 mg of the same in distilled water. Pipette out 4 ml from this solution and make up to 1 litre (1 mg NO_2 N/l). Prepare various concentrations of standard nitrite solution by appropriate dilutions with distilled water preferably from 0.1 to 1 mg NO_2 N/l.

PROCEDURE

1. Add 1 ml each of EDTA solution, sulphanilic acid and α-napthalamine hydrochloride solution one after the other to the various concentrations of nitrite solution.

2. Observe for the absorbance value of these solution on spectrophotometer at 520 nm with distilled waster as blank.

3. Pipette out 50 ml filtered sample in a conical flask and note down its absorbance value after adding all the reagents described above.

4. Draw a standard curve by plotting the concentration of standard nitrite solutions against the absorbance value.

5. Now evaluate the nitrite content of the sample by comparing its absorbance value with the standard curve.

Expt. 30 *To evaluate the nitrite content of the effluent sample using sulphanilamide method*

MATERIALS REQUIRED

Laboratory glassware

Spectrophotometer

REAGENTS

a. Sulphanilamide solution

   1 g of sulphanilamide dissolved in 100 ml of 10% hydrochloric acid.

b. N-1-naphthyl ethylene diamine hydrochloride solution

   0.1 g of N-1-napthyl ethylene diamine hydrochloride solution dissolved in 100 ml distilled water.

c. Standard nitrite solution

   Refer EDTA Method

PROCEDURE

1. Add 1 ml of sulphanilamide solution to 45 ml sample in a 50-ml volumetric flask. Allow it to stand for 5 minutes.

2. Then add 1 ml of N-1 naphthyl ethylene diamine hydrochloride solution. Mix thoroughly and make up to 50 ml.

3. Swirl the flask for 2 minutes and record the absorbance value on spectrophotometer at 543 nm.

4. Run the distilled water as blank.

5. Perform the same procedure to standard nitrite solutions and note down their absorbance value.

6. Draw a standard curve by plotting the concentration of nitrite solution against the absorbance values.

7. Evaluate the nitrite content of the sample by comparing its absorbance value with the standard curve.

3.29 TOTAL PHOSPHORUS

The insoluble nature of rock-bounded phosphorus does not favour its dissolution in water. It occurs in both forms, organic and inorganic, out of which the former constitutes about 85%. The mobility of phosphorus from the aquatic sediments depends on the redox potential value of water i.e. the value indicates oxidizing conditions under which precipitation of phosphorus occurs. But low redox potential value indicates reducing conditions under which it is solubilized to increase the phosphorus content in water. Polyphosphates or inorganic phosphorus are much more important than organically bound phosphorus. Polyphosphates are added to public water supplies to serve as anticorrosive agents and to stabilize calcium carbonate present in hard waters. This increases their level in water deteriorating its quality. The effect of nitrates and phosphates in excessive levels in the aquatic system has been discussed in chapter 2. The maximum permissible limit for phosphorus is 0.005 mg/l. Therefore domestic sewage, agricultural run-off and synthetic detergents are the three major sources of phosphorus in water. Out of these, synthetic detergents contribute more to the inorganic phosphorus content in water. Inorganic phosphorus is released as a result of metabolic breakdown of proteins (in urine) and obviously sewage is rich in phosphorus. However the introduction of synthetic detergents is the major source of phosphorus-rich surface waters. These detergents gained attraction as they did not demand soft water and cleaning efficiency was much more than soap.

They contain phosphates as builders and the composition varies from B – 50%. They contaminate the surface water with inorganic phosphorus three to four times than it was before. Two forms of phosphorus exist in waste water (a) orthophosphates (trisodium phosphate Na_3PO_4; disodium phosphate Na_2HPO_4; monosodium phosphate NaH_2PO_4; diammonium phosphate $(NH_4)_2HPO_4$ b) polyphosphates sodium hexametaphosphate $Na_3(PO_3)_6$; sodium tripolyphosphate NaP_3O_{10}; tetrasodium pyrophosphate $Na_4P_2O_{10}$. To prevent the scale formation in boilers, phosphates are used. At high temperature, complex phosphates are hydrolysed to orthophosphates, even the polyphosphates when entering into the aquatic environment get hydrolysed to orthophosphates. This can be represented as

$$Na_4P_2O_7 + H_2O \rightarrow 2Na_2HPO_4$$

The occurrence of hydrolysis at a faster rate depends on several factors such as high temperature, low pH, enzymes, presence of contaminants in water, and the form in which the phosphorus exists, for example, pyrophosphates hydrolyse better than tripolyphosphates. However this cannot be expected in all cases as it takes days together for the reversion of polyphosphate to orthophosphate. Therefore procedures employed in the determination of total inorganic phosphorus should take into account the polyphosphates. Orthophosphates can be estimated directly as they are stable at a defined pH and temperature. However for the determination of poly and organic phosphorus, their conversion to orthophosphates is essential in domestic and industrial waste waters.

Orthophosphates may be present as phosphoric acid (H_3PO_4); phosphorous acid (H_2PO_4); metaphosphate (HPO_4^{2-}) and phosphate (PO_4^{3-}) fortunately all these forms can be estimated by either of the three methods.

1. Gravimetric method
2. Volumetric method
3. Calorimetric method

3.29.1 Gravimetric Method

This method requires the inorganic phosphorus to be present in large amounts. Stagnant waters continuously receiving fertilizer washout with absolutely minimum or nil chance of dilution may contain phosphorus in excess amount and can be estimated gravimetrically.

3.29.2 Volumetric Method

Waste water with considerable amount of phosphates i.e. not less than 50 mg/l can be determined by this method, and the steps involved in this method are precipitation, filtration, washing up of precipitate and titration. The results acquired are accurate but still the method is not preferred, the disadvantage being its time-consuming procedure.

3.29.3 Calorimetric Acid

This method is commonly in practice in spite of its low accuracy of results. Based on the reagents added, this is again classified into three types However the principle involved is common for all the types.

    a.   Ammonium phosphomolybdate $\rightarrow$ yellow colloidal sol

    b.   Vanadium + ammonium Phosphomolybdate $\rightarrow$ intense yellow colour

    c.   Stannous chloride + ammonium phosphomolybdate $\rightarrow$ heteropoly blue colour

Phosphate complexes with ammonium molybdate to form yellow colloidal sol at intermediate concentrations. At concentration less than 30 mg/l the colour formation is very less and to make it visible, vanadium is added which gives intense yellow colour even at low concentration of phosphate. Also addition of a reducing agent reduces the molybdenum in ammonium phosphomolybdate and produces a blue coloured sol, exactly equal may be stannous chloride or ascorbic acid.

$$(NH_4)_3\,PO_4.12MnO_3 + Sn^{2+} \rightarrow \text{Blue coloured complex} + Sn^{4+}$$

If the waste water consists of interfering compounds, phophomolybdate should be extracted into a benzene isobutanol solution and the extracted sample should be added with the stannous chloride solution.

3.29.4 Polyphosphates

Add sulphuric acid to drop the pH of the sample so as to facilitate the hydrolysis of polyphosphates to orthophosphates. Then boil the sample for about 1.30 hours. Otherwise autoclave the sample at 20 atmospheric pressure. Prior to the addition of ammonium molybdate, the sample should be neutralized to remove the excess acid. Now by following the procedure described above, the hydroysed orthophosphate along with the orthophosphate present in the sample is evaluated and noted as total inorganic phosphate (TIP). Polyphosphate level can be obtained by subtracting orthophosphate from the TIP.

3.29.5 Organic Phosphorus

In waste water analysis, estimation of organic phosphorus is very essential. However this can be achieved by digesting the organic matter so that the phophorus is released as phosphate ion. Wet oxidation method can also be followed by using oxidants like perchloric acid, nitric acid, sulphuric acid and persulphate. Then the sample is estimated for the presence of orthophosphate and the organic phosphate is obtained as follows.

Total phosphorus – inorganic phosphorus = organic phosphorus

3.29.6 Significance

Phosphorus determination is very much essential to

1. checking out the phosphate level in streams.
2. determine the phosphate level to be added in boilers to prevent scale formation and corrosion.
3. assess biological productivity in water.
4. Evaluate the efficiency of waste water treatment methods.

3.29.7 Principle

Phosphates in the sample react with ammonium molybdate to form molybdophosphoric acid. With addition of stannous chloride, molybdophosphoric acid gets reduced to a blue coloured complex.

Expt. 31   *To determine the total phosphorus content of the effluent sample*

MATERIAL REQUIRED

Laboratory glassware

Spectrophotometer

REAGENTS

a.   Perchloric acid (70%)

Dilute 70 ml of perchloric acid with 100 ml of distilled water.

b.   Phenolphthalein indicator

Dissolve 1 g of phenolphthalein in 100 ml of distilled water. Mix well.

c.   Sodium hydroxide solution (M)

Dissolve 4 g of sodium hydroxide powder in 100 ml distilled water.

d.   Ammonium molybdate solution

   i.   Dilute 62 ml of concentrated sulphuric acid in 80 ml distilled water. Allow it to cool.

   ii   Dissolve 5 g of ammonium molybdate in 35 ml of distilled water.

Mix (i) with (ii) and make up to 200 ml.

e.   Stannous chloride solution

Dissolve 0.5 g of stannous chloride in 2 ml of concentrated hydrochloric acid. Make up to 20 ml using distilled water. Do not store the solution. Prepare fresh before use.

f.   Standard phosphate solution

To prepare stock solution, dissolve 4.388 g of dried anhydrous potassium hydrogen phosphate in 1 l distilled water. Then pipette out 10 ml from this solution and make up to 1 l (1 mg/l). By appropriate dilution of stock solution, prepare phosphorus solution preferably from 0.1 to 1 mg/l.

PROCEDURE

1. Transfer 25 ml of the sample in a porcelain basin keeping it in a hot water bath until it is evaporated completely to leave the residue behind.

2. After cooling, scrap off the residue and to this, add 1 ml of perchloric acid. Stir the flask to favour the acid–residue mix.

3. Then heat the content for 5 minutes.

4. When the solution appears colourless, remove it from the flame and let it stand cool.

5. After cooling, add 10 ml of distilled water and 2–3 drops of phenolphthalein indicator.

6. Now using a pipette, add sodium hydroxide in drops to the content of the flask.

7. Continue to add the sodium hydroxide till the solution turns slight pink in colour.

8. Make up to 25 ml using distilled water.

9. To this, add 1 ml of ammonium molybdate solution to get blue colour.

10. Allow it to stand for 5 minutes.

11. Then measure the absorbance value in the spectrophotometer at 690 nm.

12. Conduct a blank using distilled water. Following the same procedure process the standard phosphorus solution.

13. Note down the absorbance value of each concentration.

14. Draw a standard curve by plotting the concentration of phosphorus solution against their absorbance values.

15. Evaluate the total phosphorus content of the sample by comparing its absorbance value with the standard curve.

3.30 DISSOLVED ORGANIC PHOSPHORUS

Allow the sample to pass through a filter. Collect the filtrate and perform the same procedure as described for total phosphorus.

3.31 TOTAL PARTICULATE PHOSPHORUS

Total particulate phosphorus can be evaluated by subtracting the dissolved phosphorus from total phosphorus

$$TPP = TP - TDP$$

where,

TPP = total particulate phosphorus,

TP = total phosphorus and

TDP = total dissolved phosphorus.

Expt. 32   *To evaluate the total inorganic phosphorus content of the effluent sample*

MATERIAL REQUIRED

Laboratory glassware

Spectrophotometer

REAGENTS

Perchloric acid

Phenolphthalein indicator

Sodium hydroxide solution

PROCEDURE

1. To 25 ml of the sample add 1 ml of ammonium molybdate and 3 drops of stannous chloride solution.

2. Observe for the appearance of blue colour.

3. After 10 minutes, to check the persistence of colour produced, measure the absorbance at 690 nm spectrophotometrically.

4. Conduct a blank using distilled water.

5. To the different dilutions of standard phosphate solution (0.1 to 1 mg/l), add 1 ml of ammonium molybdate and 3 drops of stannous chloride solution.

6. Measure the absorbance for each dilution.

7. Draw a standard graph by plotting the concentration of phosphorus solution against the absorbance value.

8. Evaluate the total inorganic phosphorus content of the sample by comparing its absorbance value with the standard curve.

3.32 ORGANIC PHOSPHORUS

Organic phosphorus of the sample can be evaluated by subtracting the inorganic phosphorus from total phosphorus.

$$\text{Organic phosphorus (mg/l)} = TP - IP$$

where,

TP is total phosphorus and

IP is inorganic phosphorus.

3.33 SILICA

3.33.1 Principle

In waters, silica exists as silicate with its concentration ranging from 1–30 mg/l. The aquatic organisms assimilate silicate as they form a structural constituent of many aquatic invertebrates such as diatoms and sponges. It plays an important role in their growth regulation. It returns back these dead organisms to sediments and their solubility depends on several environmental factors such as temperature and pH i.e. higher the temperature and pH, maximum will be the dissolution. It is a common type of impurity in water. When present in small amounts, they can be estimated spectrophotometrically. However when the amount goes beyond 20 mg/l, it is determined gravimetrically.

Expt. 33   *To evaluate the silica content of the effluent sample*

MATERIALS REQUIRED

Laboratory glassware

Spectrophotometer

REAGENTS

a.   Hydrochloric acid (50%)

Dilute 50 ml of conc. hydrochloric acid in 50 ml distilled water.

b.   Ammonium molybdate solution

In 100 ml distilled water, add 10 g of ammonium molybdate. Adjust the pH between 7 and 8 and store it in polyethylene bottles.

c.   Oxalic acid solution

Dissolve 10 g of oxalic acid in 100 ml distilled water.

d.   Standard silica solution

To prepare stock solution with 100 mg silica/l dissolve 971.4 mg of sodium fluorosilicate in 10 ml distilled water. Heat the contents of the flask until it is completely diluted with distilled water; prepare various concentration of standard silica solution preferably from 1 mg to 50 mg SiO_2 at 5-g intervals.

PROCEDURE

1.   Transfer 50 ml of the sample into a 100-ml volumetric flask.

2.   To this add 1 ml of conc. hydrochloric acid and 2 ml of ammonium molybdate solution.

3.   Take care to shake the flask well every time after the reagent is added. Again add slowly 1.5 ml of oxalic acid solution to the flask.

4.   Shake well and measure the absorbance on a spectrophotometer at 410 nm.

5.   Conduct the blank using distilled water.

6. Following the same procedure process the standard silica solution and record the absorbance value for each concentration.

7. Draw a standard curve by plotting the concentration of standard silica solution against their absorbance value.

8. Evaluate the silica content of the sample by comparing its absorbance value with that of the standard curve.

METALLIC CONSTITUENTS

3.34 SAMPLE PREPARATION FOR METAL ANALYSIS

Industrial wastes, in general, are complex with a wide variety of components like organic, inorganic and particulates. In addition to these, majority of effluents possess metals, the most toxic form. They exist in all forms, dissolved or suspended, organic or inorganic. Estimation of metals in the effluent sample cannot be done as the presence of other components especially organic content, absolutely interfere with the results. Therefore, the estimation of metals requires pretreatment of samples and acid digestion methods are commonly practiced. Some of the acid digestion methods are discussed below.

3.34.1 Nitric Acid Digestion

MATERIALS REQUIRED

Laboratory glassware

Hot plate

REAGENTS

Concentrated Nitric acid

PROCEDURE

1. Take 50 ml of acid-preserved sample in a 100-ml volumetric flask.

2. To this, add 5 ml of concentrated nitric acid and a few boiling chips.

3. Place the flask on a hot plate and heat it to the boiling point.

4. Continue to evaporate till the volume of the content is reduced to 10 ml.

5. To avoid drying of the sample during digestion, add conc. HNO_3 in small quantities.

6. The completion of digestion is indicated by the light coloured clear solution.

7. Now allow the solution to pass through Whatman No 1 filter paper.

8. Wash the flask with distilled water twice and filter the same.

9. Collect the filtrate in a 100-ml beaker.

10. Rinse the filter paper with distilled water and collect in the same beaker. Make up to 100 ml, stir well and use it for analysis.

3.34.2 Nitric Acid–Hydrochloric Acid Digestion

MATERIALS REQUIRED

Laboratory glassware

Hot plate

REAGENTS

Concentrated nitric acid

Hydrochloric acid 50%

PROCEDURE

1. Take 50 ml acid-preserved sample in a 100-ml beaker.
2. To this add 3 ml conc. HNO_3 and a few boiling chips.
3. Place the beaker on a hot plate and heat it in such a way that the volume of the content is minimized to 25 ml.
4. Cool the beaker, add 5 ml of conc. HNO_3 .
5. Heat it again in a hot plate, covering the beaker with a watch glass.
6. Raise the temperature of the hot plate and reflux the beaker content.
7. Continue to heat until the digestion is completed, i.e., the beaker content is reduced to 5 ml.
8. Then add 10 ml HCl and 15 ml water per 100 ml of beaker content, heat for another 15 minutes.
9. Filter the content, rinse the beaker and watch glass with distilled water and collect the filtrate in a separate beaker.
10. Centrifuge and leave it to stand overnight.

3.34.3 Nitric Acid—Perchloric Acid Digestion

MATERIALS REQUIRED

Laboratory glassware

Hot plate

REAGENTS

Conc. Nitric Acid

Perchloric acid

Ammonium acetate solution

PROCEDURE

1. Take 50 ml of acid-preserved sample in a beaker.
2. To this add 5 ml of conc. HNO_3 and a few boiling chips.
3. Place it on a hot plate and heat it to the boiling point.
4. Continue to boil until the volume of the beaker is minimized to 15 ml.
5. Now remove the beaker from the hot plate and cool it for 15 minutes.

6. To the cooled solution, add 10 ml each of conc. HNO_3 and $HClO_4$.

7. Again heat the solution gently on a hot plate.

8. As soon as white fumes are formed, add another 10 ml of conc. HNO_3.

9. Prior to the addition of acid, heat it to the boiling point.

10. While heating, cover the beaker with a watch glass.

11. Continue to boil until the solution becomes clear and transparent.

12. Also see to it that the volume is reduced to 5 ml.

13. After the digestion process is over, add 10 ml of 50% HCl and 15 ml of distilled water and heat the solution for another 15 minutes.

14. Ensure that the solution neither precipitates nor forms any residue.

15. Remove the beaker from the hot plate, cool and transfer the solution to a funnel containing Whatman No. 1 filter paper.

16. Rinse the inner walls of the beaker and watch glass with distilled water and pour it into the funnel.

17. Collect the filtrate in a 100-ml volumetric flask, make up to the mark.

18. See to it that the solution is clear.

19. If the solution remains turbid, close the beaker with a watch glass and heat further to get a transparent solution.

20. Now make up to 50 ml using distilled water and allow the solution to pass through Whatman No. 1 filter paper.

21. Rinse the walls of the beaker and the watch glass with distilled water and filter the same. Collect the filtrate in a 100-ml volumetric flask.

22. Make up to the mark. Mix well before use.

23. In the hot medium, $HClO_4$ reacts with organic matter and tends to explode. Therefore, while adding $HClO_4$ ascertain that the content of flask is cooled to room temperature.

3.34.4 Nitric Acid–Perchloric Acid–Hydrofluoric Acid Digestion

MATERIALS REQUIRED

Laboratory glassware

Hot plate

REAGENTS

Nitric acid 50%

Perchloric acid

Hydrofluoric acid 51%

PROCEDURE

1. Take 50 ml of acid-preserved sample in a 100-ml volumetric flask.

2. To this add a few boiling chips and place it on a hot plate and heat gently.

3. Gradually increase the temperature to boiling point.

4. Evaporate till the volume of the content is reduced to 15 ml.

5. To this, add 12 ml of conc. HNO_3 and boil it to evaporate again to near dryness.

6. Remove the flask from the hot plate and allow it to cool.

7. To the cooled solution, add 20 ml of $HClO_4$ and 1 ml of HF and boil until the solution appears clear and transparent.

8. After cooling, add 20 ml of distilled water.

9. Filter it through Whatman No. 1 filter paper.

10. Collect the filtrate in a 100-ml volumetric flask and make up to the mark using distilled water.

11. Mix well before use.

3.34.5 Nitric Acid–Sulphuric Acid Digestion

MATERIALS REQUIRED

Laboratory glassware

Hot plate

REAGENTS

Nitric acid. (concentrated)

Sulphuric acid (concentrated)

PROCEDURE

1. Take 50 ml of acid-preserved sample in a 100-ml beaker.

2. To this, add 5 ml of conc. HNO_3 and a few boiling chips.

3. Place the beaker on a hot plate and heat it gently.

4. Continue to heat until the volume of the beaker is minimized to 15 ml by evaporation.

5. Cool it and then add 15 ml of conc. HNO_3 and 10 ml of conc. H_2SO_4.

6. Heat the contents again.

7. As soon as the solution appears clear and transparent, remove the beaker from the hot plate and cool.

8. If dense white fumes of SO_3 or brown fumes of HNO_3 are developed while heating, extend the boiling time so as to eliminate them from the solution.

9. Dilute the sample to 50 ml using distilled water and filter the same using Whatman No. 1 filter paper.

10. Collect the filtrate in a 100-ml volumetric flask and make up to the mark. Mix well before use.

3.35 ZINC

It is certainly an important trace metal to play a critical role in regulation of tissue growth and repair, digestion and assimilation, eye sight, fertility, cell reproduction and protein synthesis. Also it is said that zinc-supplemented drugs could prevent diarrhoea and pneumonia in infants. Since it is well known for its resistance to moist air, it is used in galvanizing, a process by which iron or steel is coated with zinc and to produce alloys in combination with brass and bronze. Also it is widely used in the preparation of pharmaceuticals and in cosmetic products. Therefore wastes from galvanizing, zinc plating, viscose rayon, dye house, rubber factory, paints, pigments and insecticide manufacturing units contain zinc. Long-term exposures to zinc does not have any deleterious effect but metal fume fever due to inhalation of zinc oxide fume is reported as short-term effect.

3.35.1 Principle

At alkaline pH, zinc combines with dithizone to form a red coloured complex which can be extracted by carbon tetrachloride and measured spectrophotometrically at 535 nm.

Expt. 34   *To determine the amount of zinc in the effluent sample*

MATERIALS REQUIRED

Laboratory glassware

Wash all the glassware with double-distilled water. Rinse them with 50% HNO_3 and sodium citrate solution.

Spectrophotometer

Separatory funnels (150 ml)

REAGENTS

a.   Double-distilled water

b.   Zinc stock solution

Dissolve 100 mg of zinc metal in 50% HCl. Make up to 1000 ml using double-distilled water. 1 ml of this solution contains 100 mg Zn.

c.   Zinc working solution

Carefully siphon 10 ml of zinc stock solution into a 1000-ml volumetric flask. Using double-distilled water make up the solution to the same volume. 1 ml of this solution contains 1 mg Zn.

d.   Potassium cyanide solution

Dissolve 5 g of potassium cyanide (KCN) in 100 ml double-distilled water.

e.   Ammonia solution (concentrated)

f.   Methyl red indicator

Dissolve 100 mg of methyl red powder in a little double-distilled water and then make up to 1000 ml using the same.

g.   Sodium citrate solution

Dissolve 10 g of sodium citrate ($Na_3C_6H_5O_7 \cdot 2H_2O$) in 90 ml double-distilled water. Extract it with 10 ml of dithizone solution. Continue the extraction till the solution remains green. To remove excess dithizone, extract it with carbon tetrachloride.

h.   Acetic acid (concentrated)

i.   Carbon tetrachloride

j.   Bis (2-hydroxy ethyl) dithiocarbamate solution.

   Dissolve 4 g of diethanol amine in 40 ml methyl alcohol. To this add 1 ml of carbon disulphide and mix thoroughly. Prepare freshly before use.

k.   Dithizone solution

   Dissolve 5 mg of dithizone in 100 ml carbon tetrachloride.

l.   Sodium sulphide solution I

   Dissolve 3 g of sodium sulphide in 100 ml double-distilled water.

m.   Sodium sulphide solution II

   Siphon 4 ml of sodium sulphide solution I in 100 ml double-distilled water.

PROCEDURE I

1.   Prepare standard dilutions of zinc working solutions preferably in the range of 1 ml to 10 ml at 1-ml interval.

2.   Make up each dilution to 20 ml using double-distilled water.

3.   Transfer these dilutions into a series of 150-ml separatory funnels.

4.   Simultaneously run a blank (20 ml distilled water) and an aliquot of sample diluted to 20 ml with distilled water (>10 g Zn).

5.   Now add 2 drops of methyl red indicator solution and 2 ml of sodium citrate solution to all the separatory funnels.

6.   Observe for yellow colour formation.

7.   In the absence of colour change, add ammonia solution in drops till the solution appears yellow.

8.   After 5 minutes, add 1 ml of potassium cyanide solution and conc. acetic acid in drops.

9.   Soon after the colour change from yellow to natural peach, add 5 ml of carbon tetrachloride.

10. Wait for the layer separation and discard the carbon tetrachloride layer (Exercise great care while handling CCl_4 as it is highly toxic).

11. To each dilution, add 1 ml of bis 2-hydroxyethyl dithiocarbamate solution.

12. To ensure complete mixing, shake well.

13. Then add 10 ml of dithizone solution. Shake well again.

14. Collect the separated CCl_4 layer in another separatory funnel II.

15. By adding 3–4 drops of dithizone solution to the separatory funnel, carefully extract the aqueous solution and siphon into separatory funnel II. Continue the extraction till the CCl_4 extract remains green.

16. Now add 10 ml of sodium sulphide solution II to the separatory funnel II.

17. Mix well and draw off the aqueous layer.

18. Repeat the process several times until the aqueous layer is separated out and the aqueous solution formation can be identified by distinct pale yellow or colourless layer.

19. Pipette out the remaining CCl_4 layer into a 50-ml volumetric flask.

20. Make up to the same volume using CCl_4.

21. Evaluate the absorbance value for the balance, standard solutions and sample in the spectrophotometer at 535 nm.

Some industrial effluents contain a mixed array of heavy metals such as copper, lead, zinc and cadmium. In such cases, the aforesaid procedure cannot evaluate the exact concentration of zinc present in the effluent, the reason being the interference of other heavy metals. Therefore these heavy metals should be eliminated before proceeding with the experiment.

PROCEDURE II

1. Make up the acid-digested sample to 20 ml using double-distilled water.

2. Adjust the pH of the solution to the acidic range by adding nitric acid (6N) or ammonia solution in drops. (This can be checked out by inserting the pH paper into the solution)

3. Filter this solution in a sintered crucible and wash the filter paper with H_2S solution several times.

4. Finally give a clean thorough wash with hot water and boil the filtrate.

5. Cool it for fifteen minutes.

6. After this, siphon the filtrate into a 150-ml separatory funnel and proceed with the steps described in Procedure I.

3.36 NICKEL

Nickel is a silvery white metal found abundantly in the earth's crust. It is an essential constituent of plant and animal tissues. Good sources of nickel are fruits, vegetables, grains, seafood, etc. Since nickel resists corrosion to a great extent, it is used for a variety of purposes such as finishing coat and undercoat of chromium finish in electroplating industries, ferro-nickel alloys, coins and utensils. Finely powdered nickel acts as a catalyst in the manufacture of solid fats. The effluents of this process, when discharged into the streams, do not cause direct harm as it cannot be absorbed in the gastrointestinal tract. However, blood circulation may be targeted through inhalation or dermal route. Dermatitis is the common phenomena in nickel miners, smelters and refiners. Also it is said that inhalation of nickel dust leads to lung cancer. Out of several nickel compounds, nickel carbonyl ($Ni(CO)_4$) is the most toxic form causing haemorrhagic bronchopneumonia.

3.36.1 Principle

In the alkaline medium, nickel combines with dimethyl glyoxine solution to form a red coloured complex. It is extracted by chloroform and measured spectrophotometrically at 470 nm.

Expt. 35　　*To evaluate the nickel content of the effluent sample*

MATERIALS REQUIRED

Laboratory glassware

Spectrophotometer

REAGENTS

a. Nickel stock solution

Dissolve 447.9 mg of nickel sulphate in distilled water. Make up to 1000 ml. One ml of this solution contains 100 mg Ni.

b. Nickel working solution

Transfer 10 ml of nickel stock solution into a 100-ml volumetric flask. Make up to 100 ml. One ml of this solution contains 10 mg nickel.

c. Hydrochloric acid

Make up 43 ml of conc. hydrochloric acid to 1000 ml using distilled water.

d. Sodium citrate solution

Dissolve 125 g of sodium citrate in 500 ml distilled water.

e. Iodine solution 0.05N

Dissolve 20 g of potassium iodide and 6.4 g iodine in little distilled water and make up to 1000 ml.

f. Dimethyl glyoxine solution 0.5%

Dissolve 1 g of dimethylglyoxine in 100 ml concentrated ammonia solution and dilute it to 200 ml using distilled water. Shake well and if precipitation occurs, filter the solution and use the supernatant.

PROCEDURE I

1. Prepare standard nickel working solution preferably in the range of 0.1 ml to 1 ml (at 0.1-ml interval) in a series of 50-ml volumetric flasks.

2. Take 1 ml of distilled water as blank and suitable aliquot of the acid-digested sample in separate flasks.

3. To all the flasks, add 20 ml of 0.5N HCl.

4. Then slowly add 10 ml of sodium citrate solution.

5. Shake well and add 2 ml of iodine solution and 4 ml of dimethylglyoxine solution.

6. While adding the reagents take care to mix the solution well.

7. Make up the contents of the flask to 50 ml using distilled water. Measure the absorbance value of standard solution, blank and sample.

8. Draw a standard curve by plotting the concentrations of standard solutions against their absorbance values.

9. Evaluate the nickel value of the sample by comparing its absorbance value with that of the standard curve.

Certain effluents may contain heavy metals like copper, iron and manganese other than nickel. Therefore to eliminate them, the steps given in procedure II have to be adopted.

REAGENTS (FOR PROCEDURE II)

In addition to the reagents used for nickel estimation in procedure I, the following reagents are required.

a. Ammonia solution

Make up 10 ml of ammonia to 500 ml using distilled water.

b. Chloroform

PROCEDURE II

1. Transfer the acid-digested sample to a separatory funnel.

2. To the sample, add 10 ml sodium citrate solution and few drops of ammonia. Now add 4 ml of dimethyl glyoxine solution.

3. Mix well and leave it to stand for 3 minutes.

4. During this period, nickel in the sample combines with dimethyl glyoxine to form nickel– dimethyl glyoxine complex.

5. To extract the complex formed, use 30 ml of chloroform.

6. Wait for the layer formation and collect the layered extracts in separatory funnel II.

7. To this add 20 ml diluted ammonia. Mix well and again collect the chloroform layer formed in separatory funnel III.

8. Extract the aqueous layer using 20 ml of 0.5N HCl.

9. Pipette out the aqueous extracts into 50-ml volumetric flask, heat gently and proceed with the steps described in Procedure I.

3.37 MERCURY

3.37.1 Principle

At the acidic pH, mercury ions combine with dithizone to form an orange-yellow coloured complex which is measured spectrophotometrically at 490 nm.

Expt. 36   *To determine the mercury content of the effluent sample*

MATERIALS REQUIRED

Laboratory glassware

Spectrophotometer

REAGENTS

a.　Mercury stock solution

Dissolve 135.4 mg mercuric chloride in 750 ml distilled water. To this add 1.5 ml conc. HNO_3 and make up to 1000 ml. 1 ml of this solution contains 100 mg of Hg.

b.　Mercury working solution

Siphon 10 ml of mercury stock solution into a 1000-ml volumetric flask and make up to the same volume.

c.　Potassium permanganate solution (5%)

Dissolve 5 g of $KMnO_4$ in 100 ml distilled water.

d.　Concentrated sulphuric acid

e.　Potassium persulphate solution 5%

Dissolve 5 g of potassium persulphate in 100 ml distilled water.

f.　Sulphuric acid (0.25N)

Dilute 0.69 ml sulphuric acid in 100 ml distilled water.

g.　Potassium bromide solution (40%)

Dissolve 40 g of potassium bromide solution in 100 ml distilled water.

h.　Hydroxylamine hydrochloride (50%)

Dissolve 50 g of hydroxylamine hydrochloride in 100 ml distilled water.

i.　Dithizone solution

Dissolve 3 mg of dithizone in 500 ml chloroform.

j.　Phosphate carbonate buffer

Dissolve 75 g of disodium hydrogen phosphate and 19 g potassium carbonate in 500 ml distilled water. Purify by dithizone extraction.

k.　Anhydrous sodium sulphat

Except for the mercury stock solutions, all the reagents should be prepared fresh before use.

PROCEDURE

1.　Prepare standard mercury working solutions preferably in the range from 1 ml to 10 ml (at 1 ml interval) in 1000-ml volumetric flasks.

2.　Make up each dilution to 500 ml using distilled water.

3.　Similarly take 500 ml distilled water as blank and suitable aliquot of the acid-digested sample in separate flasks.

4.　To all the flasks add 1 ml of $KMNO_4$ and 10 ml of conc. sulphuric acid.

5.　Shake well and boil the content.

6.　Take care to maintain the permanganate colour in the solution (pink). If the colour disappears, add $KMNO_4$ in drops.

7.　Leave it to stand for 1 hour.

8. Following this, siphon 5 ml of potassium persulphate solution into the flask and give sufficient time for cooling.

9. Now decolourise the solution by adding hydroxylamine hydrochloride solution in drops.

10. Pour the contents of each flask into a series of 1000-ml separatory funnels.

11. Using 25 ml dithizone, extract the funnel contents.

12. Continue to do this, until the colour of the final extract solution reaches the original colour of dithizone (green)

13. To remove the excess dithizone, extract with chloroform. Draw off the separated chloroform layer into a series of separatory funnels II.

14. To this add 50 ml of 0.25 N H_2SO_4. Mix well and collect the extracts in the third series of separatory funnels III.

15. Addition of 50 ml of 0.25N H_2SO_4 and 10 ml potassium bromide solution will separate out the mercury dithizonate complex as aqueous phase, leaving behind the contaminants in organic phase.

16. Remove the lower organic layer and wash the aqueous phase with chloroform several times.

17. Discard the separated chloroform layer and add 20 ml of phosphate carbonate buffer and 10 ml of dithizone solution, mix well and wait for the layer separation. Carefully collect the organic layer and allow to pass through a layer of sodium sulphite in a funnel.

18. Draw a standard curve by plotting the concentration of standard solution against their absorbance values.

19. Evaluate the mercury content in the sample by comparing its absorbance value with the standard curve.

3.38 LEAD

3.38.1 Principle

At alkaline pH, lead combines with dithizone to form lead dithizonate. Since it is soluble in chloroform, it can be drawn off on the chloroform layer leaving behind the lead-free green coloured dithizone.

Expt. 37   *To evaluate the lead content of effluent sample*

MATERIALS REQUIRED

Laboratory glassware

Spectrophotometer

REAGENTS

a. Double-distilled water

Clean all the glassware with double-distilled water and finally rinse them with 50% nitric acid.

b. Ammonia solution

Make up 3.5 ml of ammonia to 100 ml using double-distilled water.

c. Dithizone working solution (0.1%)

Dissolve 100 mg of dithizone in 100 ml of chloroform and store in amber bottles.

d. Dithizone working solution

Pipette out 12 ml of dithizone stock solution into a separatory funnel. To this add 20 ml of 0.5 N ammonia solution, mix well and observe for layer separation. Discard the lower chloroform layer, collect the upper aqueous layer and allow it to pass through a wet filter paper. Store this solution in amber bottles.

e. Sodium hexametaphosphate solution (10%)

Dissolve 10 g of sodium hexametaphosphate solution in 100 ml distilled water. Shift the pH of the solution towards the alkaline side by adding concentrated ammonia . Purify it by extraction with dithizone stock solution. Add 50% HCl to the separatory funnel and proceed with the extraction until the chloroform extract becomes colourless.

f. Hydroxylamine hydrochloride solution ($NH_2OH.HCl$)

Dissolve 1 g of hydroxylamine hydrochloride solution in 100 ml distilled water.

g. Alkaline cyanide solution

Dilute 340 ml of ammonia in 680 ml of water. Add 3 g of sodium sulphate and 30 ml of 1% potassium cyanide solution to this solution. Shake well and store it. Exercise great care while handling.

h. Lead stock solution

Dissolve 1.599 g of lead nitrate in 50 ml double-distilled water. To this add 10 ml concentrated nitric acid and make up to 1 l. One ml of this solution contains 1 mg Pb.

i. Lead intermediate solution

Transfer 10 ml of lead stock solution in a 1000-ml volumetric flask and make up to the volume of the same. One ml of this solution contains 10 mg Pb.

j. Lead working solution

Pipette out 10 ml of lead intermediate solution into a 100-ml volumetric flask and make up to the mark. One ml of the solution contains 1.0 mg Pb.

PROCEDURE I

1. Transfer different dilutions of lead working solution (from 1 ml to 10 ml) into a series of separatory funnels.

2. Make up each dilution to 50 ml.

3. Similarly take 50 ml double-distilled water as blank and suitable aliquot of digested sample in separatory funnels (if the effluent is free of organic matter skip the digestion procedure).

4. Always maintain the pH of the sample in neutral range.

5. To all the flasks, add 1 ml each of sodium hexametaphosphate solution and hydroxylamine hydrochloride solution.

6. After thorough mixing, add 30 ml of alkaline cyanide solution and 0.5 ml dithizone working solution and again mix the content well.

7. Finally pour 10 ml of chloroform and shake the funnel vigorously. Observe for layer formation.

8. Carefully collect the chloroform layer and measure the absorbance value in the spectrophotometer at 510 nm.

9. Prepare a standard curve by plotting the concentration of standard solutions against their absorbance value.

10. Evaluate the lead content in the sample by comparing its absorbance against the standard curve.

PROCEDURE II

1. Presence of tin and bismuth in the effluent interface with lead extraction. Therefore to eliminate tin and bismuth, the following steps have to be adopted.

2. To the sample add 50% HCl and adjust the pH below 3. Now pour the sample content into the separatory funnel.

3. Add 20 ml of 0.1% dithizone solution, mix well and draw off the chloroform layer.

4. Continue the extraction till the dithizone retains green colour. Collect the aqueous layer, adjust its pH to neutral range and proceed with the steps described in Procedure I.

3.39 COPPER

In nature, copper exists as oxides and sulphides. It is essential for all living organisms to mobilize iron and to form haemoglobin. Therefore intake of copper in the range of 1–100 mg does not pose any danger. However in excessive amounts it causes vomiting, diarrhoea, stomach cramps and nausea. Continuous accumulation of copper in the body leads to a disease known as Wilson's disease. In fish it cause epizootic ulcerative syndrome. It is extensively used in making utensils and household items. Also it is found in copper pickling liquors and ammonium rayon wastes. Since certain bacteria, fungi and algae are sensitive to copper, it is sprayed in water and used as a fungicide.

3.39.1 Principle

Addition of hydroxylamine hydrochloride solution reduces copper present in the sample. In the acidic medium, reduced copper combines with neocuproin to form a yellow coloured complex. The colour obtained is extracted by chloroform which can be determined spectrometrically at 457 nm. This method is valid up to 0.8 mg copper/100 ml.

Expt. 38 *To determine the copper content of effluent sample*

MATERIALS REQUIRED

Laboratory glassware

Calorimeter

Copper-free distilled water

REAGENTS

a. Conc. hydrochloric acid

b. Hydroxylamine hydrochloride solution

    Dissolve 50 g of hydroxylamine hydrochloride solution in 250 ml double-distilled water.

c. Neocuproine solution

    Dissolve 0.2 g of neocuproin in 200 ml methyl alcohol. Store it in a cool place.

d. Sodium citrate solution 25%

    Dissolve 125 g of sodium citrate dihydrate in 500 ml double-distilled water. To eliminate the impurities present in it, pour the contents into a 1000-ml separatory funnel. To this add 10 ml each of hydroxylamine and neocuproin solution and mix well. Again add 50 ml of chloroform and draw off the lower layer of chloroform extracts. Stores the aqueous layer in tightly stoppered bottle.

e. Copper stock solution

    Dissolve 0.1 g of copper in 10 ml double-distilled water. Add 3 ml of conc. nitric acid in drops, shake well and let it to stand for 5 minutes. Heat the solution in a hot plate. Since complete dissolution occurs only at boiling temperature, allow it to boil but take care not to let the solution spatter. After cooling, make up to 1000 ml. One ml of this solution contains 100 mg Cu.

f. Copper working solution

    Siphon out 50 ml of the stock solution into a 1000-ml volumetric flask and make up to the marked volume. One ml of this solution contains 1 mg Cu.

PROCEDURE

1. Prepare various dilutions of copper working solution (1–10 ml) into a series of separatory funnels.

2. Make up to 50 ml using double-distilled water.

3. Add 2–3 drops of conc. HCl and 5 ml of hydroxylamine hydrochloride to all the separatory funnels.

4. Shake vigorously and again add 10 ml each of sodium citrate solution and neocuproin solution.

5. Mix well and pour 20 ml of chloroform . Leave it for 10 minutes.

6. Draw off the chloroform extract in a 100-ml beaker.

7. Continue the extraction process using 20 ml chloroform and collect the chloroform layer in the same beaker.

8. Dilute the content to 50 ml using isopropyl alcohol. Measure the absorbance value at 457 nm in the spectrophotometer.

9. Run the blank and the acid-digested sample as described above.

10. Draw a standard graph by plotting the concentration of standard of solution against their absorbance value.

11. Determine the copper content in the sample by comparing its absorbance value with that of standard curve.

3.40 HEXAVALENT CHROMIUM

Chromium is a brightly coloured lustrous metal. In greek, "chrome" means colour and hence the name. Being an essential trace element, it is involved in sugar metabolism and obviously insufficient chromium content leads to loss of weight and rise in blood sugar. The main source of chromium is whole grains, dried beans and peanuts. However when their level in the body exceeds the needed limit, more deleterious will be the effect. Also it should be noted that the valency of the metal radical determines its toxicity. For instance, trivalent chromium is not harmful and is considered as an essential food nutrient, but, hexavalent chromium compounds, chromates and dichromates are highly toxic causing gastrointestinal ulcers and kidney dysfunction. Their movement into the environment has serious effects on aquatic and terrestrial life. Hexavalent chromium compounds are used in industrial processes like metal picking, electroplating, aluminium anodizing and leather tanning and as anticorrodents in water-cooling systems. Also they are used in the manufacture of paints, dyes, paper explosives and ceramics. They are released in bulk form as effluents and evaluation of their quantity and form of existence (hexavalent/trivalent) will aid in selecting the apt treatment technology. Two different methods for each form of metal are given below.

3.40.1 Principle

In acid medium, hexavalent chromium combines with S-diphenyl carbazide to form red-violet coloured complex, which is measured spectrophotometrically at 540 nm.

In acidic medium, chromate gets converted to dichromate which when titrated against sodium thiosulphate/ferrous ammonium sulphate reveals the chromium content in the sample.

Expt. 39 *To determine the chromium content of effluent sample by calorimetric method*

MATERIALS REQUIRED

Laboratory glassware

Calorimeter

Double-distilled water

REAGENTS

a. Chromium stock solution

    Dissolve 283 mg dried $K_2Cr_2O_7$ is 100 ml distilled water. Make up to 1000 ml. One ml of this solution contains 100 mg chromium.

b. Chromium working solution

    Transfer 10 ml chromium stock solution into a 1000-ml volumetric flask. Make up to the same volume. 1 ml of this solution contains 1 mg of Cr.

c. Diphenyl carbazide reagents

    i. Sulphuric acid: Dilute one part of sulphuric acid with 9 parts of distilled water.

    ii. Dissolve 0.2 g of diphenyl carbazide in 100 ml of 95% ethyl alcohol. Mix (i) with (ii) and store it in a cool place.

PROCEDURE

1. Prepare various standards of chromium working solution (preferably in the range of 1 ml to 20 ml at 2-ml interval) in a series of 50-ml volumetric flasks. Make up to 50 ml.

2. To the flasks, add 2.5 ml of diphenyl carbazide reagent and shake well.

3. After ten minutes measure the absorbance at 540 nm in a spectrophotometer. Run a blank using distilled water.

4. For determination of chromium, use clear samples.

5. If the sample is turbid, centrifuge it and take the supernatant for analysis.

6. Adjust the pH to around 7 and then follow the procedure as described above.

7. Draw a standard graph by plotting the absorbance values of standard solutions against their concentrations.

8. Evaluate the chromium content in the sample.

Expt. 40   *To determine available chromium content of effluent sample titrimetrically*

MATERIALS REQUIRED

Laboratory glassware

REAGENTS

10N H_2SO_4

Ammonium bifluoride

Sulphuric acid

Potassium iodide

Sodium thiosulphate (0.1N)

0.1N FAS

Ferroin indicator

PROCEDURE

1. To 100 ml of the sample add 2 ml of 10N H_2SO_4 and mix well.

2. Add 1 g each of ammonium bifluoride and sulphuric acid.

3. Ensure complete mixing of the flask content by swirling the flask every time after reagent addition.

4. Finally add 2 g of potassium iodide and shake well.

5. Keep it away from light for 30 minutes. Following this, add 1–2 drops of starch indicator.

6. Titrate it against 0.1N sodium thiosulphate solution.

7. The end point is the disappearance of blue colour. Determine the available chromium by using the following calculations.

Titration can also be performed by using ferrous ammonium sulphate using ferroin indicator. The end point is the colour change from blue or bluish green to red.

CALCULATION

$$\mathrm{Cr} = \frac{V \times N \times 17.332 \times 1000}{S}$$

$$\mathrm{CrO}_4^{2-} = \frac{V \times N \times 38.66 \times 1000}{S}$$

where,

V = Volume of the titrant

N = Normality of the titrant

S = Volume of the sample

3.41 TOTAL CHROMIUM

3.41.1 Principle

Potassium permanganate solution oxidizes the chromium to chromate and this when allowed to react with S-diphenyl carbazide gives a violet coloured complex. Such colour extract can be measured photometrically at 540 nm.

Alternatively boiling the sample with persulphate and silver nitrate catalyst oxidizes the trivalent chromium to dichromate. Furthermore, heating the sample at boiling point decomposes the excess which when titrated against ferrous ammonium sulphate reveals the dichromate content in the sample.

Expt. 41 *To determine the total chromium content of effluent sample photometrically*

MATERIALS REQUIRED

Laboratory glassware

REAGENTS

    a.   Reagents used for the estimation of total chromium.

    b.   Potassium permanganate solution 0.1N

        Dissolve 316 mg of $K_2Cr_2O_7$ in 100 ml distilled water.

    c.   Sodium azide solution

        Dissolve 0.5 g of sodium azide in 100 ml distilled water.

PROCEDURE

1. Transfer 10 ml of the sample into a 100-ml flask.

2. To this add 10 ml of 10% sulphuric acid.

3. Slowly pipette out $KMnO_4$ into the flask till the solution appears pink.

4. To the coloured solution, add 30 ml of distilled water and heat it in hot water bath for half an hour.

5. While heating add sodium azide till the solution become colourless.

6. Then pour the content into a beaker and add 2.5 ml diphenyl carbazide solution.

7. Mix well and make up to 50 ml and measure the absorbance at 510 nm.

8. For samples enriched with organic matter perform the digestion process with nitric acid and sulphuric acid.

9. Adjust the pH of the distilled sample to neutral range by addition of ammonia.

Expt. 42   ***To determine the total chromium content of the effluent sample titrimetrically***

MATERIALS REQUIRED

Laboratory glassware

Titration assembly

REAGENTS

    a.   Silver nitrate solution

        Dissolve 1.7 g of silver nitrate powder in 100 ml distilled water.

    b.   Ammonium persulphate.

    c.   10N sulphuric acid

    d.   Ferrous ammonium sulphate

    e.   Ferroin indicator

PROCEDURE

1. To 100 ml of sample, add 4 ml of silver nitrate and 2 g of ammonium persulphate. Mix well and heat the flask content to boiling point for 30 minutes.

2. Remove the flask from the flame and let it stand for another 30 minutes.

3. Now add 20 ml of 10N sulphuric acid and a few drops of ferroin indicator.

4. Titrate the content against 0.1N ferrous ammonium sulphate and observe for the colour change from bluish green to red.

5. Determine the total chromium by using the calculation.

CALCULATION

Cr (mg/l) = $V \times N \times 17.332 \times 1000$/ml of sample

where,

V = Volume of titrant

N = Normality of titrant

3.42 CADMIUM

3.42.1 Principle

Cadmium, when combined with dithizone, forms a pink to red coloured complex. The colour is extracted using chloroform and measured spectrometrically at 518 nm.

Expt. 43   *To determine the amount of cadmium in the effluent sample*

MATERIALS REQUIRED

Laboratory glassware

Spectrophotometer

Separatory funnel

REAGENTS

a. Water

Use double-distilled water for reagent and solution preparation.

b. Stock cadmium solution

Dissolve 100 mg Cd metal in 20 ml double-distilled water. To this add 5 ml of conc. hydrochloric acid and heat the content to favour thorough mixing. Make up to 1000 ml. This stock solution can be used to prepare standard cadmium solution (1 ml = 100 mg Cd).

c. Standard cadmium solution

Transfer 1 ml of cadmium stock solution into a 100-ml conical flask. Add 1 ml conc. hydrochloric acid and make up to 100 ml (1 ml=1 mg Cd). Prepare fresh.

d. Sodium potassium tartrate solution

Dissolve 250 mg of sodium potassium tartrate in distilled water. Make up to 1 l.

e. Sodium hydroxide–potassium cyanide solution.

Solution (a) Dissolve 400 g NaOH and 10 g KCN in double-distilled water. Make up to 1 l.

Solution (b) Dissolve 400 g NaOH and 0.5 g KCN in double-distilled water. Make up to 1 l.

f. Hydroxylamine hydrochloride solution

Dissolve 20 ml of hydroxylamine hydrochloride in 100 ml distilled water.

g. Dithizone solution

h. Chloroform

To test its suitability, add a drop of dithizone to the chloroform taken in a test tube. The positive result is the appearance of faint green colour.

i. Tartaric acid solution

Dissolve 20 g of tartaric acid in double-distilled water. Make up to 1 l. Keep this solution in refrigerator.

j. Conc. HCl

k. Thymol blue indicator

Dissolve 0.4 g thymo sulphnaphthalein in 100 ml double-distilled water.

l. Sodium hydroxide 6N

PROCEDURE

1. Prepare various standards of cadmium working solution (preferably in the range of 2 μml to 20 μml at 2-μml interval) in a series of 50-ml of volumetric flasks.

2. Conduct a blank using distilled water.

3. Pipette the sample into a separating funnel. The concentration of the sample should not exceed 20 μg Cd. If the concentration is >20 μg Cd, dilute it.

4. Maintain acidic pH (2.8) in all the separatory funnels using thymol blue indicator and make up the volume of each separator tube to 25 ml.

5. Then add 1 ml each of sodium potassium tartrate solution and hydroxylamine hydrochloride solution; 5 ml of sodium hydroxide–potassium cyanide solution I and 15 ml dithizone stock solution.

6. Shake well after the addition of each reagent. During this time, let the vapour pressure to let out through the stopper.

7. Take a second set of separatory funnels. For convenience sake, let us assume them as F(11). Siphon into each funnel 25 ml of tartaric acid solution.

8. Collect the extracted chloroform layer from separatory funnels I into separatory funnels II.

9. Repeat the extraction using 10 ml chloroform and then collect the same into separatory funnels II.

10. Shake well and allow for layer formation. After 5 minutes, discard the lower chloroform layer.

11. Using 5 ml chloroform, wash the layers once again.

12. Also, remove floating drops of chloroform by blowing air.

13. Now, to the Cd in the tartaric acid content, add 0.25 ml of hydroxylamine hydrochloride solution, 15 ml of dithizone working solution and 5 ml of sodium hydroxide–potassium cyanide solution II.

14. Mix well and wait for the layers to separate.

15. Allow the chloroform layer to pass through a plug of cotton wool.

16. Measure the OD value of the solution spectrophotometrically at 518 nm.

17. Process the blank using distilled water.

3.43 CALCIUM

3.43.1 Principle

Adjusting the sample pH to alkaline condition allows the magnesium to precipitate as hydroxide and then the calcium is evaluated by EDTA method using murexide indicator.

Expt. 44   *To determine the calcium hardness in the effluent sample*

MATERIALS REQUIRED

Laboratory glassware

REAGENTS

a. NaOH solution (8%)

8 g of NaOH dissolved in 100 ml of distilled water.

b. Murexide indicator

Add 0.2 g of ammonium purpurate to 100 g of sodium chloride. Grind it well in a mortar and pestle.

c. EDTA solution (0.01M)

3.723 g of disodium salt of EDTA dissolved in distilled water. Make up to 1000 ml.

PROCEDURE

1. In 50 ml sample, add 1 ml of sodium hydroxide solution and a pinch of murexide indicator.

2. Swirl the flask to ensure complete mixing.

3. The colour of the solution turns pink.

4. Now titrate it against EDTA solution till the colour changes from pink to purple.

5. Note down the end point. Repeat the procedure till you attain the constant weight.

6. Determine the calcium using the following formula.

7. Express calcium hardness as $CaCO_3$ mg/l and Ca mg/l.

CALCULATION

$$\text{Calcium hardness (Ca mg/l)} = \frac{E \times 400.5 \times 1.05}{V}$$

$$\text{Calcium hardness (CaCO}_3 \text{ mg/l)} = \frac{E \times 1000 \times 1.05}{V}$$

where,

E = volume of EDTA titrant (E) in ml

V = volume of sample

3.44 MAGNESIUM

3.44.1 Principle

Like calcium, rock is the source of magnesium. However due to its low solubility, it is not so abundant as calcium. Its presence in water gains more significance because it is an important constituent of the photosynthetic pigment, chlorophyll. It is usually determined by titration with EDTA with Erichrome black 'T' indicator.

Expt. 45   ***To determine the magnesium hardness in effluent sample***

The total hardness and calcium hardness are determined.

CALCULATION

Magnesium (mg/l) = (Total hardness − Calcium hardness) × 0.243

3.45 TOTAL IRON

Iron helps in carrying oxygen from the lungs to blood cells. Its deficiency leads to anaemic condition, suppression in growth and development in infants, lethargy, defective immune function and loss of memory. Meat, fish, poultry products, legumes, broccoli, green leafy vegetables and nuts contain more iron. Although iron is inevitable for body activities, its intake in excess causes deleterious effects such as siderosis. Effluents from iron picking and mines consist of more iron in the form of iron hydroxide.

3.45.1 Principle

Iron in its oxidized state (ferri) is very essential for several enzymatic and redox processes in aquatic organisms whereas certain microorganisms like *Creaothrix* spp. and *Leptothrix* spp. utilize reduced form of iron as an energy source. Therefore in aquatic systems it is in both forms and its state in water is determined by the availability of CO_2. Under anoxic condition, it occurs as ferrous bicarbonate whereas in O_2-rich environment, ferrous bicarbonate is oxidized to brown coloured insoluble ferric hydroxide evolving CO_2. Water flowing through the basaltic region has high level of iron in potable waters. Its prescribed limit is 1 mg/l and when it exceeds this limit, it causes staining of laundry and ceramics. Growth of iron-specific bacteria raises the level of iron in water. Phosphate or polyphosphate are the interfering compounds.

Expt. 46   *To determine the iron content of the effluent sample*

MATERIALS REQUIRED

Laboratory glassware

Spectrophotometer

REAGENTS

a.   Concentrated hydrochloric acid (12N)

b. $KMnO_4$ (0.lN)

3.16 g of $KMnO_4$ dissolved in distilled water and make up to 1 l.

c.   Hydroxylamine hydrochloride solution

10 g of hydroxylamine hydrochloride dissolved in distilled water. Make up to 100 ml.

d.   Ammonium acetate buffer solution

100 g of ammonium acetate dissolved in 60 ml distilled water. To this, add 280 ml of glacial acetic acid.

e.   Phenanthroline solution

50 mg of 1,10-phenanthroline monohydrate dissolved in 50 ml distilled water. Ensure complete mixing and heat up to 80°C in a water bath.

f.   Standard iron solution

Dissolve 1.404 g of FAS in 20 ml of sulphuric acid diluted with 50 ml of distilled water. Add $KMnO_4$ solution in drops till faint pink colour appears (See to it that the colour retains). Make this solution to 1 l. Keep this stock solution (200 mg Fe/l) to prepare standard iron solution from 1 to 5 mg.

PROCEDURE

1.   In 50 ml of sample, add 2 ml of HCl and 1 ml of hydroxylamine hydrochloride solution.

2.   Mix well and heat the contents till boiling point.

3.   Put off the flame when the content is half of its original volume.

4.   After cooling, add 2 ml each of ammonium acetate buffer solution and phenanthroline solution.

5.   Dilute the content with distilled water and make up to 100 ml.

6.   Keep this flask for ten minutes and measure the absorbance on spectrophotometer (510 nm) using distilled water as blank.

7.   Repeat the same procedure for standard iron solution series and note down their respective absorbance at 510 nm.

8.   Using these values draw a standard curve.

9.   Evaluate the total iron content by comparing sample value with the different dilutions of known standard solution.

3.46 MANGANESE

3.46.1 Principle

Persulphate in acid solution oxidizes the soluble manganese compounds to permanganate. The resultant colour of the product can be measured spectrophotometrically at 545 nm.

$$2MnSO_4 + 5(NH_4)_2S_2O_8 + 8H_2O + 10HNO_3 \rightarrow 2HMnO_4 + 12H_2SO_4 + 10NH_4NO_3$$

Expt. 47   *To determine the manganese content of the effluent sample*

MATERIALS REQUIRED

Laboratory glassware

Laboratory equipment

Hot plate

Spectrophotometer

REAGENTS

a.   Manganese stock solution

Prepare 1% sulphuric acid solution. To this, add 1 g of pure manganese metal. Stir well and ensure complete dissolution. Make up this solution to one litre. One ml of this solution contains 1 mg manganese.

b.   Manganese working solution

Transfer 5 ml of manganese stock solution to a 500-ml volumetric flask. Make up to the mark using distilled water. One ml of this solution contains 0.01 mg of manganese.

c.   Special reagent

Mix 200 ml of concentrated nitric acid to 100 ml of distilled water. To this, add 37.5 g of mercuric sulphate, 100 ml of 85% phosphoric acid and 17.5 mg of silver nitrate. Stir well and make up to 500 ml using distilled water.

d.   Ammonium persulphate crystals

PROCEDURE I

1.   Prepare several dilutions of manganese working solution in a series of beakers with their concentration ranging from 5 ml to 50 ml at 5-ml interval. Make up the content of each beaker to 100 ml using distilled water.

2.   Take 100 ml of distilled water as blank.

3.   Similarly pipette out 100 ml sample in a beaker.

4.   To all the beakers, add 5 ml of special reagent.

5. Continue to heat until the content of each beaker is reduced to 40 ml.

6. Now add 1 g of ammonium persulphate to each beaker and boil for another minute.

7. Carefully remove the beakers from the hot plate and cool them under running tap water.

8. Measure the absorbance value of all the solutions, sample, standards and blank in spectrophotometer at 545 nm.

9. Draw a standard curve by plotting the absorbent value of standards against their concentrations.

10. From the standard curve, evaluate the concentration of manganese in the sample and express the results as mg/l.

PROCEDURE II (for turbid and organic-rich samples)

1. Transfer 50 ml of the sample in a silica crucible, add 5 ml of concentrated hydrochloric acid and continue to boil until the size of the sample is reduced to half of its original volume.

2. See to it that the suspended particulates are completely dissolved and once confirmed, continue to boil till all the sample evaporates to dryness.

3. By following the digestion procedure (nitric acid + sulphuric acid), digest the sample.

4. Process the digested sample as described in procedure I.

ACCESSORY CONSTITUENTS

3.47 CYANIDE

3.47.1 Principle

During distillation, sodium hydroxide absorbs the cyanide in the sample which when treated with bromine water and pyridine pyrazolone reagent converts the cyanide to cyanogen bromide and then to a blue dye.

$$\text{Simple cyanides} \xrightarrow[\text{distillation}]{\text{acid}} \text{HCN}$$

Alkali ferro- and ferricyanide + $CuCl_2/MgCl_2$ → simple cyanides

Expt. 48   *To evaluate the cyanide content of the effluent sample*

MATERIALS REQUIRED

Laboratory glassware

Titration assembly

REAGENTS

a. Acid cuprous chloride solution

    i. Hydrochloride acid (5 N)

       Dissolve 43 ml of conc. HCl in 100 ml of distilled water.

    ii. Cuprous chloride powder 2 g

       Dissolve (ii) in 100 ml of (i).

b. Magnesium chloride solution

Dissolve 5 g of magnesium chloride in 100 ml of distilled water.

c. Sodium hydroxide solution 1N

Dissolve 4 g of sodium hydroxide in 100 ml of distilled water.

d. Sodium hydroxide solution 0.5N

Dilute 1N of sodium hydroxide solution with the same volume of distilled water.

e. Sodium hydroxide solution 0.1N

Take 10 ml of 1N sodium hydroxide solution and dilute to 100 ml using distilled water.

f. Cyanide stock solution

Dissolve 1.25 g KCN in little distilled water. To this add 100 ml of 1N NaOH and make up to 500 ml using distilled water. Do not store the solution for a very long time. Standardize it against silver nitrate solution (0.1N) to get a sharp colour change from canary yellow to salmon pink. 1 ml of this solution contains 1 mg CN.

g. Cyanide intermediate solution

Dilute 10 ml of cyanide stock solution to 100 ml using distilled water. To this add 100 ml of 1N sodium hydroxide solution. Mix well. Always prepare fresh. One ml of this solution contains 10 mg CN.

h. Cyanide working solution

To 100 ml of cyanide intermediate solution, add 1 ml of 1N sodium hydroxide solution and then make up to 100 ml. Prepare fresh. One ml of this solution contains 1 mg CN.

i. Saturated bromine water

j. Arsenous acid solution

Dissolve 2 g of arsenous oxide in 100 ml boiled distilled water

k. 1-phenyl 3-methyl 5-pyrazolone solution

To 100 ml boiled distilled water, add 1 g pyrazolone and stir well. Prepare fresh.

l. Mixed Pyridine–Pyrazolone Reagent

Dissolve 25 ml bispyrazolone in 25 ml pyridine. To this, add 125 ml of phenyl 3-methyl 5-pyrazolone solution and mix well. Prepare fresh.

PROCEDURE

Pretreatment

Based on the effluent composition, they should be pretreated and the steps involved in pretreatment vary between the chemicals. For example the presence of sulphide at alkaline pH, oxidizing agents, fatty acids, and aldehydes in the sample interfere with cyanide analysis.

1. Take 50 ml of the sample in a distillation flask.

2. Make up to 200 ml and to this add 10 ml acid cuprous chloride solution.

3. If the solution contains more than 0.5 mg/l thiocyanates, replace cuprous chloride solution with 20 ml magnesium chloride.

4. To the flask attach the reflux condenser and at its receiving end, place 0.5N sodium hydroxide solution.

5. Take care to immerse the tip of the condenser in NaOH solution.

6. Continue to run the distillation till 100 ml of distillate is collected.

7. Following distillation process, pour the distillate into a conical flask and make up to 100 ml.

8. From this distillate, siphon 20 ml into a 50-ml volumetric flask.

9. To this, add 0.5 ml each of conc. HCl and saturated bromine water.

10. Shake the flask well and stopper it.

11. Leave it to stand for 10 minutes.

12. Then add 0.5 ml arsenous acid solution and shake well. This will remove excess bromine.

13. Finally to the content of the flask, add 5 ml of pyridine–pyrazolone reagent.

14. Observe for colour development.

15. In the absence of colour formation, leave the flask undisturbed for about 30 minutes and then make up to 50 ml.

16. Run blank using 20 ml 0.1 N NaOH.

17. Similarly prepare several dilutions of standard cyanide working solution (1–10 ml) and make up each dilution to 20 ml using distilled water.

18. Add all the reagents to the standard solutions as in the blank described above.

19. Measure the absorbance of blank, standards and sample at 620 nm in spectrophotometer.

20. Draw a standard graph by plotting the concentration of standard solution against their absorbance value.

21. Evaluate the cyanide content in the sample by comparing the standard graph with its absorbance value. Express the result as cyanide mgl^{-1}.

3.48 OIL AND GREASE

Oil and grease removal is very important because of two factors poor— solubility in water and tendency to separate itself from the aqueous phase. Interestingly, it is these characteristics that favour their separation when the waste water is subjected to treatment. Unless properly removed at the preliminary treatment stage itself they complicate the transportation process of waste, collapse the biological treatment and finally interfere with the nutrient cycling and destroy aquatic life in receiving waters.

The major sources of oil and grease are slaughterhouses, sewage and food processing industries. In the preliminary treatment methods, they are separated as scum and then transferred with the settled solids; the process is termed as skimming. High atmospheric pressure is applied at the bottom of the tank. Poor solubility and low specific gravity make the oil and grease to float on the surface as dense scum layers. However, these methods do not remove the entire oil and grease content as considerable amounts remain in the waste water by attaching themselves to the micro particles of emulsifying agents. When the waste water is subjected to secondary treatment, i.e., biological treatment, microorganisms act on the emulsifying agent thus leaving free the oil and grease content. Again they adhere to large-sized particles of the waste water and therefore settle at the bottom. This can be confirmed by the presence of grease balls in the final settling tanks.

The above-said process occurs only when the waste water entering the biological treatment has 5–10% of oil and grease. If there are excess amounts, they interfere with the microbial process and collapse the treatment tank. For instance, it has been observed that they exhibit a smothering effect. That is they form a slime layer on the microbial biomass preventing the exchange of O_2 between the medium and the microbes. However the problem has been overcome in advanced high-rated reactors. Short retention time favours the microbial biomass to act on emulsifying agents but does not permit them to adhere to the released grease or to oxidize the freed particles of oil and grease. This can be done by immediately drawing the oil and grease out.

Oil and grease involves hydrocarbons, esters, oil, fats, waxes and high molecular weight fatty acids. Standard procedures to estimate the oil and grease content involve solvent extraction as a major step. Low molecular weight hydrocarbons do not undergo efficient partition into the solvent, and therefore will not be included in the result. Also, a few methods suggest going in for sample drying at 103°C, a preliminary step before extraction. The problem is that substances with low boiling point and vapour pressure at 103°C will escape. Also in the presence of drying oils in waste water, these procedures favour the oxidation of the unsaturated linkages and make them absolutely insoluble. In spite of all these difficulties, the methods employed for the determination of oil and grease can totally be relied upon as most of the materials that come under the term oil and grease can be measured. In the case of petroleum industrial wastes, the drying procedure should not be adopted as most of its constituents have low boiling points and the analytical chemist should have a thorough knowledge about the chemical composition of the effluent to be treated in terms of its volatile and nonvolatile materials. Determination of oil comes under routine analysis. It is performed only when the effluent has an oil source.

Waste water	Contaminants
Domestic wastes	oil, fats, wax, fatty acid
Industrial wastes	low molecular weight hydrocabons, heavy fuel, lubricating oils, glycerides of vegetable and animal origin, calcium and magnesium soaps

Oil and grease do not dissolve in solvent and for the determination, the samples should be pretreated with HCl. At a low pH (1.0), the free fatty acids are released.

$$Ca(C_{17}H_{35})_2 + 2H^+ \rightarrow 2C_{17}H_{35}COOH + Ca^{2+}$$

In general, four methods are available to measure the oil and grease content

1. Partition gravimetric method
2. Partition infrared method
3. Partition gravimetric method with pretreatment
4. Hydrocarbon analysis

3.48.1 Partition Gravimetric Method

The samples to be determined are extracted by a suitable solvent, CFC-113. They are separated and used to measure oil and grease.

3.48.2 Partition Infrared Method

The samples extracted by the solvent CFC-113 are scanned under infrared light to measure oil and grease. Oil and grease standards are prepared according to the composition of oil and grease content in the sample used for calibration and naturally the reliability of the method depends on the standards used.

3.48.3 Partition Gravimetric Method with Pretreatment

The pretreatment of the sample involves acidification with hydrochloric acid and then filtration. The main objective of the method is to include the volatile hydrocarbons in the determination of oil and grease. However, it is a tedious and time-consuming procedure. After acidification, it is filtered and dried. Filtration retains the insoluble, harmful, high-molecular-weight oily and greasy substances. Dried samples are extracted by CFC-113 as considerable amount of water has been removed by drying and solvent extraction is made easier.

3.48.4 Hydrocarbon Analysis

This method is highly preferred for industrial wastes containing hydrocarbons. Therefore to exclude fatty acids and other fatty materials, silica gel is added with CFC-113 while extracting the sample. This favours the removal of fatty materials and the extracted sample is analysed by any of the three procedures described above. In the case of oil and grease determination in sludges, drying should be done as a preliminary step; however it takes a long period of time. In laboratories, the problem is tackled in a different way i.e., by application of chemicals, free water molecules are converted to a chemically bounded form, which makes its separation easier. The technique is termed as chemical dehydration technique where $MgSO_4 \cdot H_2O$ is added to the sample after acidification. This combines with the free water molecules of the sample to form $MgSO_4 \cdot H_2O$, and then the sample is crushed and then selected for extraction by CFC-113.

3.48.5 Significance

i.   Determination of oil and grease in domestic and industrial wastes helps to choose the suitable treatment method for its removal.

ii. Determination of oil and grease in treated effluents helps to evaluate the efficiency of the treatment employed.

iii. Determination of oil and grease in raw sludge helps to evaluate the efficiency of anaerobic digestion. Also, the level estimated in treated sludge is very essential to determine its commercial value as a fertilizer.

3.48.6 Principle

Since oil and grease are soluble in petroleum ether, during extraction, it is separated as ether layer. Evaporation of ether layer leaves behind the oil and grease as residue which can be weighed.

Expt. 49   *To determine the amount of oil and grease in the effluent sample*

METHOD I

MATERIALS REQUIRED

Laboratory glassware

REAGENTS

a.   Flocculating agents

    i.   Ferric chloride (1%)

       Dissolve 1 g of ferric chloride in 100 ml distilled water.

    ii. Ammonium hydroxide (1%)

       Dissolve 1 g of ammonium hydroxide in 100 ml distilled water. Other than this any flocculating agent given in the table below can be used.

Flocculating agents	Concentration
Magnesium sulphate	1%
Lime	2%
Aluminium sulphate	1%
Zinc acetate	10%
Sodium carbonate	5%

b.   Hydrochloric acid

    Dilute 50 ml of hydrochloric acid with 150 ml of distilled water.

c.   Petroleum ether

d.   Sodium sulphate

PROCEDURE

1. To determine oil and grease, always collect samples in separate bottles.
2. Prior to analysis, shake the bottle well.
3. Transfer 500 ml of the sample into a beaker.
4. Add 5 ml of 1% ferric chloride solution and stir well.
5. Immediately add ammonium hydroxide in drops till the suspended particulates agglomerate to form floc drops.
6. Stir vigorously while adding and leave the solution to stand for 10 minutes (following this, if the effluent does not show a distinct bottom layer of precipitate, repeat the procedure with a fresh sample).
7. After 10 minutes, see to it that the effluent shows a precipitate layer at the supernatant and decant it.
8. To prevent sucking the precipitate along with the clear layer do not pipette out the supernatant till the end.
9. Pour 5 ml of dilute hydrochloric acid into the beaker and dissolve the precipitate. Transfer the whole content into a separatory funnel.
10. Shake well and extract the aqueous layer with petroleum ether.
11. Collect the extracts in a beaker.
12. To this, add 2 g of sodium sulphate and stir it with a glass rod. Keep it undisturbed for 30 minutes.
13. Now allow the contents to pass through a Whatman filter paper (No 42) filled with sodium sulphate and collect the filtrate in a pre-weighed evaporating dish.
14. Before use, wipe the cone of the funnel with cotton immersed in petroleum ether.
15. Finally wash the filter paper with 20 ml of petroleum ether and collect it in the water bath. Allow the solvent to evaporate.
16. In the presence of water droplets after evaporation, add few ml of acetone and evaporate the residue once again.
17. Repeat this procedure until the residue is clear of water droplets.
18. Remove the evaporating dish, wipe its outside and weigh.

CALCULATION

Amount of oil and grease in the sample (mg/l) $= I - F$

where,

$I = $ initial weight of the dish

$F = $ final weight of the dish

METHOD II

MATERIALS REQUIRED

Laboratory glassware
Hot-water bath

REAGENTS

a. Sulphuric acid

Dilute 50 ml of concentrated sulphuric acid in 100 ml of distilled water.

b. Petroleum ether

c. Absolute ethyl alcohol

PROCEDURE

1. Take 250 ml of sample in a 500-ml volumetric flask.
2. To this, add 10 ml of concentrated sulphuric acid and 50 ml of petroleum ether.
3. Mix well and transfer the whole content to a separatory funnel. Allow it to stand for 30 minutes.
4. Do not disturb the funnel till two distinct layers, upper petroleum ether layer and lower water layer are formed. Carefully drain out the lower water layer.
5. Allow the petroleum layer to pass through a filter paper before use.
6. Immerse the filter paper in petroleum ether for 30 minutes and collect the filtrate in a pre-weighed glass beaker (W_1).
7. After the filtration is completed, flush the filter paper with petroleum ether 2 or 3 times.
8. Make sure that the filter paper is free of oil and grease.
9. Now heat the contents in a hot-water bath until all the petroleum ether has evaporated.
10. Allow it to cool and note down its weight (W_2).

CALCULATION

Amount of oil and grease in the sample $(\text{mgl}^{-1}) = W_1 - W_2$

where,

W_1 = Initial weight of the beaker

W_2 = Final weight of the beaker

3.49 PHENOLS

3.49.1 Principle

In the alkaline medium, phenol combines with 4-amino antipyrine and potassium ferricyanide to form a red antipyrine dye, which can be measured spectrophotometrically at 460 nm.

Expt. 50　　*To evaluate the amount of phenol present in the effluent sample*

MATERIALS REQUIRED

Laboratory glassware

Titration assembly

REAGENTS

a. Phenol stock solution

Dissolve 1 g of phenol in distilled water in a 1000-ml volumetric flask and make up to the volume of the same.

Standardization

- Dilute 25 ml of phenol stock solution to 100 ml and transfer it into an iodine flask.

- Siphon 20 ml of bromate–bromide solution into the flask.

- Seal the lid with 10 ml of potassium iodide (10%), stopper it and keep the flask away from light for 1 hour.

- After this, remove the lid and wash the sealed potassium iodide into the flask by adding another 10 ml of the same.

- Mix the content well and titrate it against sodium thiosulphate (0.25N) using starch as an indicator.

- Run blank using distilled water.

Calculation

$$\text{Amount of phenol present (mg/l)} = \frac{(V_1 - V_2) \times 3.921 \times 1000}{S}$$

where,

V_1 = ml of titrant used for blank

V_2 = ml of the titrant used for sample

S = volume of the sample

b. Phenol intermediate solution

Pipette out 10 ml of phenol stock solution and dilute to 100 ml using distilled water (1 ml contains 10 mg phenol)

c. Phenol working solution

Pipette out 100 ml of phenol intermediate solution and make up to 1000 ml using distilled water.

d. Ammonia–ammonium chloride buffer solution

Dissolve 16.9 g ammonium chloride in 143 ml concentrated ammonia solution and make up to 250 ml using distilled water.

e. 4-Amino antipyrine solution

Dissolve 2 g of 4-amino antipyrine in a little distilled water and make up to 100 ml.

f. Potassium ferricyanide solution

Dissolve 8 g of potassium ferricyanide in a little distilled water and make up to 100 ml.

g. Chloroform

h. Sodium sulphate (granular)

i.   Sodium arsenite

j.   Phosphoric acid $(1+9)$

Dilute 10 ml of H_3PO_4 with 90 ml of distilled water.

PROCEDURE

For phenol determination, collect the sample in separate bottles. Do not store the sample beyond 24 hours. For better results, perform the analysis as early as possible to avoid the interference of organic substances such as oxidizing agents and sulphur compounds in the sample and carry out the following steps.

Step 1

Eliminate the oxidizing agents by adding little amount of sodium arsenite and mix well.

Step 2

i.   To eliminate the sulphur compounds adjust the pH of the sample towards acidic range (>5) and stir vigorously.

ii. To this add copper sulphate solution so as to precipitate the sulphide as cupric sulphide (0.1 g/100 ml of sample).

iii. Now transfer 500 ml of sample in a round-bottomed flask and run the distillation process.

iv. Collect the distillate in a beaker placed at the receiving end of reflux end.

v.   Continue to run the distillation till the distillate collected is equal to the volume of the sample.

vi. If the distillate collected is insufficient, add distilled water to the flask.

vii. Run the distillate till the expected volume is reached. If the collected distillate continues to remain turbid, the sample should be subjected to pretreatment as described below.

Pretreatment of the Sample

- Adjust the pH of the sample (7.5) to the acidic range using 1N H_2SO_4.

- Add a few drops of methyl orange indicator.

- Add NaCl to the sample (30 g/100 ml sample)

- Transfer the whole content into a separatory funnel (I) and purify by extraction with 40 ml chloroform.

- Continue the extraction using another 25 ml of chloroform.

- Collect the extracts in the separatory funnel (II) and again purify by extraction with 2.5N NaOH (10 g/100 ml).

- Repeat the extraction three times with minimum quantity of NaOH (75 ml).

- Collect the extract in an evaporating dish and place it in a water bath until all the chloroform is removed.

- Cool the residue and make up to 500 ml using distilled water and proceed with the distillation process.

1. Prepare several dilutions of phenol working solution preferably in the range of 1 ml to 50 ml at 5-ml intervals.

2. Transfer each dilution into a series of separatory funnels.

3. Shift the pH of the dilution to the alkaline side (<9) by adding 10 ml of ammonia–ammonium chloride buffer solution.

4. To this add 3 ml each of 4-amino antipyrine solution and potassium ferricyanide.

5. Remember to shake the funnel well after each addition.

6. Leave it to stand for 5 minutes.

7. Extract the content with 25 ml of chloroform.

8. Allow the collected extract to pass through a filter paper containing sodium sulphate (5 g dissolved in a little distilled water) and the sample following the same procedure as described above.

9. Draw a standard curve by plotting the concentration of standard solutions against their absorbance value.

10. Evaluate the amount of phenol content in the sample by comparing its absorbance with that of the standard graph.

3.50 SURFACTANTS

3.50.1 Principle

Anionic surfactants combine with methylene blue to form a blue coloured salt, which can be measured spectrophotometrically at 650 nm.

Expt. 51   *To evaluate the surfactant content of the effluent sample*

MATERIALS REQUIRED

Laboratory glassware

Separating funnel

REAGENTS

a. Aniline phosphate solution (1%)

Dissolve 10 g of anhydrous disodium hydrogen phosphate in 1000 ml distilled water. The solutions should have alkaline pH (10), for which appropriate amount of sodium hydroxide is added.

b. Neutral methylene blue solution

Dissolve 150 mg of methylene blue in 500 ml distilled water.

c. Chloroform

    d.   Acid methylene blue solution

Dissolve 35 mg methylene blue in little distilled water. To this add 6.5 ml of conc. H_2SO_4 and make up to 1000 ml.

    e.   Linear alkyl sulphonate (LAS) stock solution

Dissolve 0.1 g of Manoxol OT in a little distilled water and make up to 1 litre. One ml of this solution contains 100 mg manoxol OT.

    f.   Linear alkyl sulphonate working solution

Siphon 10 ml from LAS stock solution and make up to 100 ml using distilled water. One ml of this solution contains 10 mg manoxol OT.

    g.   Hydrogen peroxide

PROCEDURE

1. Prepare several dilutions of phenol working solution preferably in the range of 2 ml to 20 ml at 2-ml intervals.

2. Make up each dilution to 100 ml using distilled water.

3. Keep a blank (100 ml distilled water) and sample (20 ml of sample made up to 100 ml using distilled water) in separate beakers.

4. Transfer the standard solutions, blank and sample into a series of separating funnels.

5. To the contents of the funnel add 10 ml of alkaline phosphate solution, 5 ml of natural methylene blue solution and 15 ml of chloroform and mix well.

6. Do not shake vigorously as rapid mixing leads to the formation of emulsion.

7. Leave it undisturbed for 30 seconds and wait for layer separation.

8. Collect the chloroform layer in a separating funnel II already filled with 110 ml of distilled water and 5 ml of acid methylene blue solution.

9. Repeat the extraction using chloroform for 3–4 times and collect the extracts in the same separating funnel (II).

10. After distinct layers are formed, draw off the chloroform layer and allow it to pass through a small funnel.

11. Prior to use, plug the funnel cone with chloroform-saturated cotton.

12. Collect the extract in a beaker, wash the funnel with 20 ml of chloroform and collect it in the same breaker.

13. Make up to 50 ml and measure the absorbance value at 650 nm in a spectrophotometer.

14. Draw a standard graph by plotting the concentration of the standard solutions against their absorbance values.

15. Evaluate the surfactant content in the sample by comparing its absorbance with the standard graph.

3.51 TANNIN AND LIGNIN

3.51.1 Principle

Tannin and lignin reduce tungstophosphoric acid and molybdophosphoric acid to form a blue coloured complex.

Expt. 52   *To evaluate the tannin and lignin content of the effluent sample*

MATERIALS REQUIRED

Laboratory glassware

Separating funnel

REAGENTS

a. Tannin lignin reagent

i.   Sodium tungstate   25 g

ii. Phosphomolybdic acid   5 g

iii. Phosphoric acid (85%)   12.5 ml

Dissolve (i), (ii) and (iii) in 200 ml distilled water. Reflux the mixture for 2 hours. After cooling make up to 250 ml using distilled water.

b. Sodium carbonate solution

Dissolve 100 g of sodium carbonate in 500 ml distilled water. Warm the distilled water prior to use.

c. Tannin stock solution

Dissolve 1 g of tannic acid or lignin compound in 1000 ml distilled water. 1 ml of the solution contains 1 mg tannin/lignin compound.

d. Tannin working solution

Siphon 10 ml from the tannin stock solution and make up to one litre. One ml of this solution contains 0.01 mg tannin or lignin compound.

PROCEDURE

1. Prepare several dilutions of tannin or lignin (1–10 ml) working solution.

2. Make up each dilution to 50 ml using distilled water.

3. Take 50 ml distilled water as blank and filtered sample in separate beakers.

4. To the sample, blank and standard dilutions, add the following reagents, one after the other. Take care to mix the content of the beaker after each addition.

    • Sodium carbonate solution   10 ml

    • Tannin lignin reagent   2 ml

5. Wait for the maximum colour development and measure the absorbance value in spectrophotometer at 700 nm.

6.   Draw a standard graph by plotting the concentration of standard solution against their absorbance values.

7.   Evaluate the tannin lignin content of the sample by comparing its absorbance value with that of the standard graph.

3.52 BORON

3.52.1 Principle

Acidification and evaporation of boron with the added curcumin reagent gives out a red coloured, alcohol-soluble rosocyanin. The product obtained can be measured calorimetrically at 540 nm.

Expt. 53   *To determine the content of boron in the effluent sample*

MATERIALS REQUIRED

Laboratory glassware

Laboratory equipment

Water bath

Spectrophotometer

REAGENTS

a.   Boron stock solution

To the 100-ml volumetric flask, add 57.16 mg anhydrous boric acid and make up to the volume using distilled water. One ml of this solution contains 0.1 mg of boron.

b.   Boron working solution

Transfer 10 ml of boron stock solution to the volumetric flask containing 100 ml distilled water. Mix well. One ml of this solution contains 0.01 mg of boron.

c.   Ethyl alcohol 95%

d.   Curcumin reagent

To 80 ml of 95% ethyl alcohol, add 40 mg of homogenized curcumin and 5.0 mg of oxalic acid. Mix well. Add 42 ml of hydrochloric acid and make up the flask content to 100 ml using ethyl alcohol.

e.   Sodium hydroxide 1N

Add 4 g of sodium hydroxide in a beaker containing 100 ml of distilled water. Stir well.

a.   Hydrochloric acid

Transfer 10 ml of concentrated hydrochloric acid into a beaker containing 110 ml distilled water.

SAMPLE PREPARATION

1.   Take one ml of sample in an evaporating dish, add 2 ml of 1N sodium hydroxide, mix well and evaporate in a water bath (55°C) to dryness.

2. Following this, ignite the content of the dish at 550°C. Allow the dish to cool.

3. Then add 2.5 ml of hydrochloric acid (1 + 11) to the dish, mix well.

4. Centrifuge the mixed content and pipette out the supernatant into 25 ml of volumetric flask

5. Make up to the mark using distilled water.

6. From the standard curve, evaluate the concentration of boron in the sample and express the results as mg/l.

PROCEDURE I

1. Prepare several dilutions of boron standard solutions into a series of evaporating dishes with their concentration ranging from 0.1 to 1 ml at 0.2-ml interval.

2. Take 1 ml of distilled water as blank in a separate dish.

3. To each dish add 4 ml of curcumin reagent.

4. Gently rotate the dishes so that the reagent is mixed well with the solution in the dish.

5. Evaporate the dish contents by placing them on a water bath maintained at 55°C.

6. Do not remove the dishes from the water bath until the odour of the acid is eliminated completely.

7. Following the evaporation, cool the dishes and slowly add 10 ml of ethyl alcohol to each dish.

8. See to it that the acid dissolves the residue of the dish.

9. After the dissolution is confirmed, transfer the contents of the dishes in to a series of 25-ml volumetric flasks.

10. Make sure that the content of the dish is transferred completely and make up each flask to 25 ml using alcohol.

11. Now measure the absorbance value of standards and blank in a spectrophotometer at 540 nm.

12. Draw a standard curve by plotting the absorbance value of standards against their concentrations.

PROCEDURE II (For samples with hardness >100 mg/l)

1. Immerse moderate amount of ion exchange resin in a H5 hydrochloric acid solution.

2. After the slurry is formed, carefully pack it in a glass column. (50 cm length; 1.2 cm breadth).

3. Ensure even packing and do not entertain air bubbles while packing.

4. Now allow the 1 + 5 HCl solution to pass through the column.

5. Then repeat the procedure using distilled water.

6. After the column is ready, transfer 25 ml of prepared sample into the column.

7. Adjust the bottom stopper in such a way that the sample comes out of the column at a rate of two drops per second.

8. Draw off the eluted sample in a conical flask.

9. Wash the column with distilled water several times until the eluted sample is about 50 ml..

10. Take one ml of the sample and process it as described in procedure I.

3.53 MICROBIOLOGICAL CHARACTERISTICS OF SEWAGE

Sewage composition varies with the source which subsequently reflects its microbial composition. However presence of more than 70% of organic matter encourages the growth of several groups of microorganisms like algae, fungi, protozoa, bacteria and viruses. Therefore, not surprisingly, a drop of sewage contains bacteria in lakhs. Among the different types, soil microorganisms predominate and they include *B. subtilis, B. megaterium, B. mycoides, P. fluorescens, Achromobacter* spp. and *Micrococcus* spp. Also, coliforms like *E. coli, Proteus* spp. and *Serratia* spp. are present in large numbers. Presence of pathogens like *Vibrio cholerae, Salmonella typhi, Salmonella paratyphi* and *Shigella dysenteriae* in sewage makes its treatment complicated. Occasionally viruses causing poliomyelitis, hepatitis are said to be present in the sewage, the source being the faeces of the infected host.

Sewerage refers to the conveyance system that brings sewage from various places either to the treatment plant or to the place of disposal. Based on the type of sewage it carries, sewerage is classified into three types/divisions.

1. Sanitary sewers which convey domestic/industrial waste water.
2. Storm sewers which convey surface run-off and storm water.
3. Combined sewers which convey all types of waste water.

3.54 SEWAGE TREATMENT

In the sewage treatment, purification is done is steps. Since several treatment methods (both aerobic and anaerobic) by which sewage can be treated have been discussed in detail in Chapter 2, treatment methods specific to sewage are dealt with in this chapter.

Treatment can also be done on a local scale (i.e.) from individual houses and small firms like hotels, hospitals, shopping complexes and theatres. Sewage from these areas can be collected and treated by two basic anaerobic methods:

1. Septic Tank
2. Imhoff Tank

3.54.1 Septic Tank

These are small tanks built at ground level in the backyard of individual houses. They are kept closed except for a small vent from which the fermented gases are allowed to escape. The basic mechanism behind these tanks are sedimentation and biological degradation of organic solids. Soon after the entrance of sewage into the tank, solids are retained, thus giving sufficient time for biological action. Lack of oxygen favours the proliferation of anaerobic microorganisms which leads to the production of organic

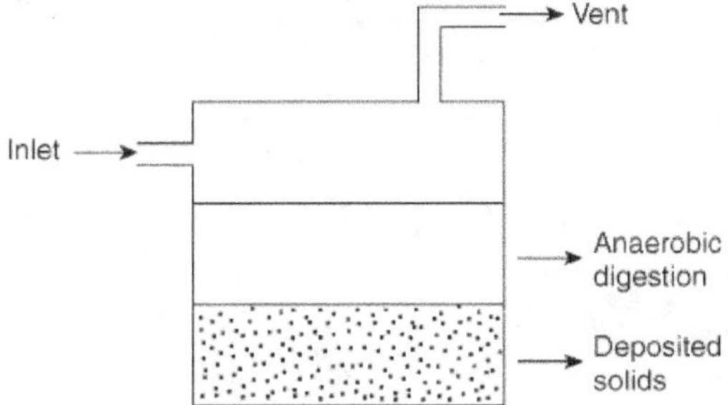

Figure 3.1   Diagrammatic representation of septic tank

acids and hydrogen sulphide. Overflowing effluent is drained out for further aerobic treatment. After a considerable period of time, sedimented sludge is removed and dried, and can be used as a fertilizer. Satisfactory BOD and solids reduction are guaranteed in this method. It is a cheap and best method at local level. However, this technology does not eliminate the pathogenic microorganisms from the sewage and therefore the effluent requires further treatment.

3.54.2 Imhoff Tank

The principle behind the imhoff tank is similar to that of septic tank except for a slight modification in its physical design. The body of the tank is divided into two divisions – upper and lower section. Sedimentation occurs in the upper section, whereas anaerobic decomposition takes place in the lower section. After settling, the solids move into the lower division where the condition is absolutely anoxic. Backward movement of solids from this layer is prevented. Large size of the tank favours complete digestion in a short period of time (2-4 hours).

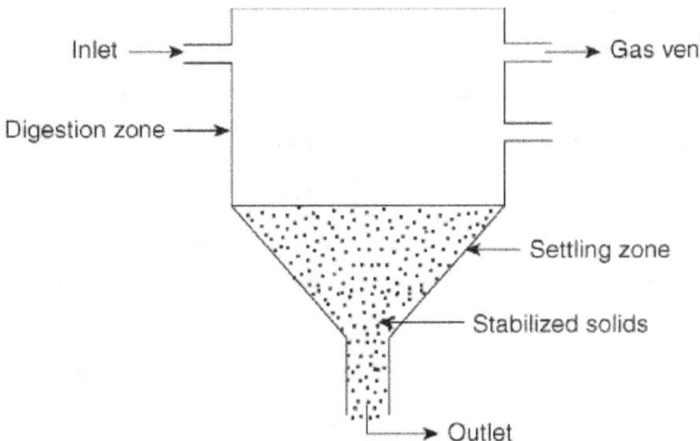

Figure 3.1   Diagrammatic representation of Imhoff tank

Microorganisms utilized in the imhoff tank are as follows:

Bacteria	Fungi	Protozoa
Pseudomonas, Nitrosomonas	*Zoophagus*	Phytoflagellates
Zooglea, Sphaerotilus	*Arthrobotrys, Geotrichum*	Zooflagellates Amoebae
Beggiatoa, Achromobacter	*Pullalaria*	Ciliates
Flavobacterium, Escherichia	*Alternaria*	
Nocardia, Bacillus, Leptothrix	*Penicillium Cephalosporium Candida*	

3.54.3 Biotransformation in Sewage Treatment

Aerobic treatment methods   Nitrification and denitrification are the two predominant processes in aerobic treatment.

Anaerobic treatment methods   These methods include hydrolysis, fermentation, acetogenesis and methanogenesis.

3.55 DISINFECTION

Sewage disinfection is a process by which the pathogenic microorganisms are completely eliminated. The main aim of this process is to prevent the spread of diseases and is adopted to purify public water supplies, swimming pools, recreation areas and shell fish growing areas. Several factors that influence the disinfection process are

1. Composition and temperature of the sewage to be analysed
2. pH of the sewage
3. Type of microbial species to be destroyed
4. Number of microorganisms in the sewage
5. Selection of disinfection
6. Concentration of disinfection
7. Contact period between disinfection and waste water

Disinfection can be done by both physical and chemical means.

Physical means   Application of gamma radiation, ultraviolet radiation and heat.

Chemical means   Use of chemical agents like chlorine, chlorine dioxide, ozone, combination of these agents.

Out of these two methods, chemical means are found to be cheap and effective. Among several chemical agents, chlorine application serves best to decontaminate waste water. Advantages of chlorine application are

1. Cheap and best method to kill pathogens.
2. In moderate levels it does not impart unpalatable taste or colour to water.
3. It is readily soluble and beyond saturation level, it leaves residue.

4. Even at high concentration it does not render harm to environment and organisms.

5. The technology is easy to practice and any error can be identified easily.

6. It exists in all three forms, solid, liquid and gas.

7. It is potentially toxic to a wide range of microorganisms.

8. It is highly efficient in the treatment of cyanide.

3.55.1 Chlorination

Application of chlorine to the sewage system is called as chlorination. The primary aim of chlorination is to control odour formation, hydrogen sulphide accumulation and septicity. The mechanism behind the chlorination-induced death of pathogenic microorganisms is the disruption of metabolic activities. Blocking the vital enzymes of microorganisms leads to their death.

The immediate reaction of chloride with water is the formation of hypochlorous acid $(HOCl)$. Since hypochlorous acid is categorized under weak acids, under alkaline conditions it gets dissociated to protons (H^+) and chlorite ions (OCl^-). This condition is not favourable for disinfection because $HOCl$ is 100 times more toxic than chloride ions. This essentially stresses the importance of pH during disinfection process.

On the other hand, when they stand free (chlorine, hypochlorous acid and chlorite ion), they are more effective and termed as free chlorine residuals or free available chlorine residuals. However this is not the case under normal conditions. This is because, in free form, they are unstable and therefore combine with ammonia to mono-, di- and trichloramines. The problem lies in the fact that these combined chlorine residuals are less effective.

3.55.2 Different Methods of Chlorine Application

Based on the type of waste water, several methods are adopted for chlorine application and the three important processes are discussed below.

1. Break-point chlorination
2. Superchlorination
3. Chloramination

Break-point chlorination   This method is adopted for waste waters with significant amount of ammonia. Limited amount of chlorine added will combine with ammonia to form monochloramines and continuous addition of chlorine will lead to the formation of di- and trichloramines. Hence formation of di- and trichloramines can be prevented by keeping the chlorine–nitrogen ratio as low as possible. Already formed mono- and dichloramines in mixtures fail to remain stable. Similarly trichloramines and monochloramines combine together to dissociate into nitrogen and hydrochloric acid. If this process continues, at one point the ammonia, monochloramine and free available chlorine are completely eliminated. When chlorine is added in excess at this stage, free available chlorine will detoxify the pollutants. This process of application is called break-point chlorination.

Superchlorination   If the sewage is suspected to contain resistant microorganisms, excess of chlorine is added. In addition, the contact period is prolonged and following the contact period, excess chlorine can be removed by using sulphur dioxide.

$$Cl_2 + SO_2 + 2H_2O \rightarrow 2H_2SO_4 + 2HCl$$

Chloramination   Chlorine when applied to organic-rich waste waters, leads to undesirable odour. To overcome this problem, chlorine in combination with ammonia is added. Chloramines thus formed are reactive to organic matter and both the disinfection process and odour control can be achieved by this method.

3.56 SEWAGE DISPOSAL

Disposal of sewage can be classified into two major divisions, namely sludge disposal and effluent disposal.

3.56.1 Sludge Disposal

In a sewage treatment plant, sludge arises from a variety of sources—sludge from primary treatment (sedimentation), sludge from secondary treatment (digested form from aerobic and anaerobic sewage treatment plants) and sludge from tertiary treatment (sludge from coagulation, and sedimentation process of softening plants). These sludges based on their quality can be subjected to several methods and a few methods are discussed below.

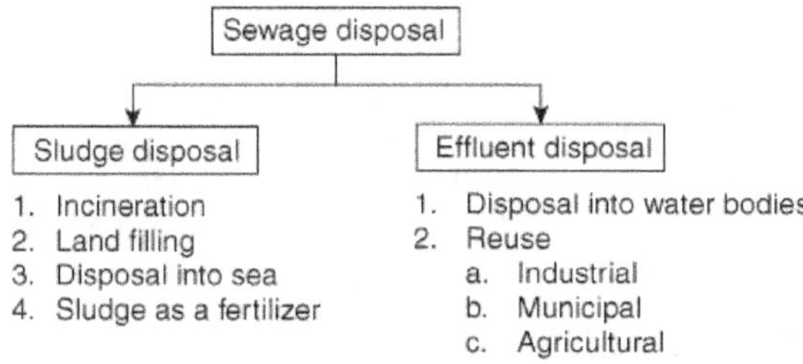

3.56.2 Processing of Sludge

Sludge disposal requires the elimination of bound water and strong agglomeration of its solid particles. To favour this, several methods can be adopted. They include, thickening, chemical conditioning, elutriation and heat treatment. In the thickening process, the sludge tank is stirred by a low-speed clarifier. This increases the thickening of solids and decantation of supernatant can be done after thickening. Addition of chemical agents like aluminium sulphate, iron salts and aluminium carbonate serves best to process the sludge. Elutriation is a process applied to sludge with alkaline pH. In this process, excess alkalinity is lowered by dilution and then allowing the sludge to settle. Dewatering of sludge is done by four methods. They are drying beds, pressure filtration, vacuum filtration and centrifugation. In drying beds, sludge is spread on rectangular beds with porous bottom made up of ash. Water gradually moves to the porous bottom where it is removed. Bed height of 125–250 mm is sufficient enough to give way to sun's rays and wind throughout its layer. Partially dried sludge cracks which further allow the penetration of the sun's rays and wind. Completely dried sludge is removed either manually or mechanically. Pressure filtration is a method in which the sludge is dewatered by mechanical action. A metal chamber with its inner layer lined by filter cloth is used for this process. The metal back of the chamber has grooves. During the process, the sludge is pressurized at 600–850 kPa and then pumped into the metal chamber. High pressure throws the sludge on to the surface of the filter cloth while the water gets drained through the grooves. Vacuum filtration process involves a rotating drum that is kept submerged in the sludge tank. A vacuum of about 90 kPa is applied to the submerged segments of the rotating drum. Sucking of sludge on the filter cloth occurs naturally which subsequently is thrown out when the drum rotates.

Of the several disposal methods, incineration is quite expensive and cannot be adopted in all cases. However sludge with toxic chemicals cannot be disposed by other methods and incineration is the only option. Similarly land fill method serves best to dispose waste with low organic content. In the case of sludge with high organic content, it can be used as a fertilizer.

EXERCISES

1. ______________ are the sole contributors of methyl orange acidity?

2. pH value of a solution depends on its ______________.

3. Under natural conditions, sulphate content is rich in ______________ and ______________ regions.

4. Sulphate in excess causes ______________ effect on living organisms.

5. Under anoxic conditions iron is present as ______________ and under oxic condition is present as ______________.

6. In the presence of O_2 and temperature, ammonia is oxidized to ______________ and then ______________.

7. Mobility of phosphorus from aquatic sediments depends on the ______________ of water.

8. Silica forms ______________ of ______________ and ______________.

9. One molecule of sodium thiosulphate is equal to one atom of ______________.

10. Conductivity is influenced by ______________.

11. Chromium is involved in ______________.

12. Structural composition of manganese resembles ______________.

13. Manganese poisoning leads to ______________ disease.

14. In nature, copper exists as ______________ and ______________.

15. Continuous accumulation of copper in organisms leads to ______________.

16. Beyond ______________ mg/l, chlorides give bad taste to water.

17. Natural colour-producing substances exist as ______________.

18. Addition of boron with curcumin reagent gives rise to an end product called ______________.

19. Fluorides below 1 mg/l of water leads to ______________.

20. A drop of sewage contains bacteria in ______________.

21. ______________ and ______________ are the two processes involved in septic tank.

22. Anaerobic microbial processes lead to the production of ______________ and ______________ in septic tanks.

23. Chlorine combines with water to form ______________.

24. Presence of complicated substances in sewage is attributed to ______________.

25. What is the principle behind COD estimation?

26. Give reasons for the inefficiency of potassium permanganate to act as an oxidizing agent in COD estimation.

27. In COD estimation, why should the waste water be digested?

28. What is the role of silver ion in COD estimation?

29. Ferrous ammonium sulphate titrant used in COD estimation should not be exposed to air. Why?

30. What is the necessity of blank in COD estimation?

31. In the oxidation–reduction reaction of COD estimation, what is the reason for the appearance of intermediate bluish shade?

32. What should be the acid and sample ratio in COD estimation?

33. Apart from atmospheric diffusion, what is the source of CO_2 in aquatic environment?

34. What is phenolphthalein acidity?

35. How will you take samples for CO_2 estimation?

36. At which concentration does calcium produces scales on conduit tubes and pipes?

37. Why should the samples be titrated twice for CO_2 estimation?

38. Combustion products of fossil fuel power plants leads to mineral acidity. How?

39. What is the ultimate effect of acidic effluent on aquatic organisms?

40. How is CO_2 concentration in aquatic system controlled by the sufficient amount of calcium and magnesium?

41. Discuss the chemical reactions by which iron pyrites and sulphur acidify running waters.

42. Define pH of the medium.

43. List out the influencing factors that induce pH change in a medium.

44. pH of the medium is the best environmental indicator—justify.

45. How will you measure hydrogen ion activity?

46. Addition of any acid to water will increase its hydrogen ion concentration. How?

47. List out the differences between hydrogen and glass electrode.

48. How will you select electrodes to determine the pH range of a solution?

49. What is the use of Imhoff cone in total solids estimation?

50. How will you estimate solids in sludges?

51. What is the fate of sulphate under anoxic condition?

52. Discuss the pH-dependent oxidation reduction reaction of sulphate.

53. Explain crown corrosion.

54. Which is the major source of magnesium in waters?

55. Addition of O_2 to the ground water rich in iron and manganese elevates its sulphate content. Why?

56. Which form of iron is essential for enzymatic redox processes in aquatic organisms?

57. What is the maximum acceptable concentration of iron and manganese in aquatic bodies?

58. In the presence of phosphate and heavy metal content, how will you estimate iron and manganese?

59. In manganese estimation, how will you eliminate interfering agents?
60. Name two constituents that impart temporary hardness to water.
61. List out the hardness-causing cations and anions.
62. Explain the source of carbonates and bicarbonates in water.
63. Which constituent renders permanent hardness to water.
64. Major source of ammonia in aquatic bodies is by excretion. Discuss.
65. Give an account of nitrification process in aquatic bodies.
66. What happens when nitrogen is present in excess in aquatic bodies?
67. Which form is more toxic, ionised or unionised ammonia?
68. In highly polluted waters which form of organic nitrogen predominates?
69. List out the different forms of organic nitrogen present in water.
70. Nitrate electrode method is not suitable for samples with low nitrate concentrations. Why?
71. What are the main sources of phosphorus in water?
72. Inorganic phosphorus addition to water is more than that of organically bound phosphorus. Why?
73. What are the suitable environmental conditions for silica dissolution in water.
74. What are the two methods by which silica can be determined?
75. Name the two forms of phosphorus in water.
76. Explain with reaction, conversion of polyphosphates to orthophosphates in aquatic environment.
77. How will you estimate organic phosphorous in water?
78. What is skimming?
79. What is the average DO concentration in fresh water?
80. Why is dichromate standard less preferred in DO estimation?
81. How will you preserve sodium thiosulphate titrant?
82. What is the role of manganous sulphate in DO determination?
83. How will you estimate DO under field conditions?
84. Name a few substances that impart turbidity to water.
85. How was turbidity measured in olden days?
86. In Jackson candle nephelometric method, which substance was kept as an arbitrary standard?
87. Which property of chromium determines its toxicity?
88. Which form of chromium is used for a wide range of industrial purposes?
89. What is the role of magnesium in plants?
90. What is the functional role of iron in organisms?
91. Give an account of the effect of iron deficiency in living organisms.
92. What are the sources of nickel?
93. Nickel in polluted waters when consumed cannot harm organisms. Why?
94. What is the role of zinc in man?
95. Define total alkalinity.

96. What are the different forms of alkalinity in water.

97. How will you estimate carbonate alkalinity?

98. Name two volumetric methods by which chlorides can be determined.

99. How will you estimate phenolphthalein alkalinity?

100. In which form can coagulants precipitate the colloidal and suspended particulates in waste water?

101. What is the source of colour in water?

102. Mention a few toxic properties of colouring compounds in water.

103. Write down the principle of manganese estimation in effluent.

104. What is the source of fluorides in water?

105. Write down the principle of sulphide estimation in effluent.

106. Write down the effluent sources of heavy metals given below.

| Chromium | Copper | Nickel |
| Iron | Manganese | Zinc |

107. List out the interfering compounds in boron estimation.

108. Discuss the significance of COD estimation in organic-rich waste waters.

109. Name the standard colour in colour measurement?

110. In chloride estimation, which method is more preferable?

111. What is the importance of turbidity measurement in waste water analysis?

112. In BOD estimation, growth of nitrifiers should be suppressed. Why?

113. How will you eliminate the interfering compounds in BOD estimation?

114. In BOD estimation of effluent samples, seeding of microorganism is necessary. Justify.

115. What is the source of microbial seed in BOD estimation?

116. Why should a blank be maintained for BOD estimation?

117. How will you determine dilutions for high strength waste waters.

118. Sewage composition determines its microbial composition. Justify.

119. List out predominant microbial groups in sewage.

120. What aspect makes the sewage treatment complicated?

121. What is a sewage?

122. What are the three types of sewers?

123. What is the major drawback associated with septic and imhoff tank?

124. Differentiate between septic and imhoff tank.

125. How will you disinfect sewage waters?

126. Explain the chemical means of disinfection in sewage.

127. What is chlorination?

128. When does HOCl get transformed to chlorite ions?

129. Explain free chlorine residuals.

130. What is superchlorination?

131. In waste waters with more ammonia, chlorine is added in excess. Why?

132. Explain the condition under which chloramines are more suitable than chlorine.
133. What are the sources of sludge?
134. What are the chemical agents added to condition sludge?
135. Write down the process by which sludge can be dewatered.
136. Incineration-based sludge disposal method cannot be adopted in all cases. Why?
137. Describe the composition of total solids in sewage.
138. What is the significance of sewage treatment?
139. What happens when ferric sulphate and lime are added to sewage?
140. What is the necessity for tertiary treatment in sewage?
141. Write down a few tertiary treatment methods that can be employed for desalination of sewage.
142. Ocean dumping of sewage is discouraged. Justify.
143. How will you collect sewage samples?
144. Give an account of microbes involved in composting.
145. Which is better, composting or sanitary land filling?
146. pH of the sewage is more important for any kind of treatment. Discuss.
147. Sewage composition is always unstable. Why?
148. How do season and weather influence sewage composition?
149. At what level does sewage interfere with the self-purifying ability of aquatic bodies?

（**4**）

SOIL ANALYSIS

Soil is the basic foothold of terrestrial ecosystems that supports plant growth and serves as a habitat for a wide range of flora and fauna. The nutrient cycling process will not be completed without soil decomposers. Precisely, soil is the land surface of the earth.

4.1 SOIL STRUCTURE

It refers to the typical arrangement of soil particles. In general, at least a part of the soil is aggregated, and the stability of the soil structure depends on its aggregation. Non-aggregated soil can either be consolidated or massive. Well-developed soil structure constitutes aggregates of 1–5 mm in size and the interconnecting voids do not stagnate water. Aggregates are of different types.

1. *Fragments* Peds broken up naturally
2. *Clods* Soil lumps formed by tillage
3. *Nodules* Soil aggregates formed by insoluble cementing
4. *Pans* Soil particles aggregated to form horizontal regions
5. *Pores* voids of cylindrical shape

4.2 SOIL PROFILE

A vertical section of the soil can be divided into several layers called horizons. They occur as a mix and it is very difficult to separate them. However the composition of these components differs between soil groups. The source of inorganic materials in the soil is the regolith, surface layer of rocks. This layer when subjected to physical and chemical weathering, action of wind/water is broken down into pieces to form soil particles. The different horizons of soil include:

1. Upper layer of organic matter (L horizon)
2. A layer that is prone to washing (A horizon)
3. A layer that receives the washed materials (B horizon)
4. A layer of weathered parent material (C horizon)
5. A layer of unweathered bed rock (R horizon)

Figure 4.1 Soil profile

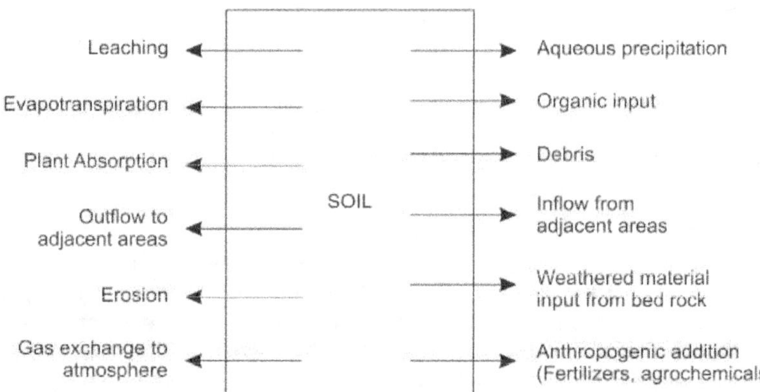

Figure 4.2 Changes in the soil ecosystem

4.3 PHYSICAL CHARACTERISTICS OF SOIL

4.3.1 Soil Texture

The relative proportion of clay, silt, gravel and sand determine the texture of soil. Accordingly they are divided into three types. They are

1. Coarse-textured
2. Medium-textured
3. Fine-textured

It is this particle size that determines the sustenance of plant life. In the coarse-textured soil, the particle size is considerably big to allow the water to drain. Also it fails to hold the nutrients. Fine-textured particles, in contrast, hold high moisture and nutrients. Similarly four different sizes are recognized—clay (<0.002 mm), silt (0.002–0.05 mm), sand (0.05–2 mm), gravel (>2 mm) and accordingly they are allocated to the appropriate textural class (Table 5.1). Medium-textured soil with 20% clay, 40% sand and 40% silt is considered to be suitable for plant growth. Substantially large-sized particles are categorized as gravel. If the proportion of all the particles are found to be more or less equal, it is called loamy soil. The soil texture determination therefore requires a thorough knowledge about the particle size and distribution. Moreover, by mere experience, a rough estimation can be done, i.e., by rubbing the moist soil between two fingers, the soil texture can be assessed.

Sand particles	Feel by fingers
Coarse	Gritty
Fine sand	Silky
Silt	Smooth and non-sticky

Table 4.1   Size of particles of different soil types (VSDA system of classification)

Soil Particles	Size in mm
Gravel	>2.0
Very coarse sand	1.0–2.0
Coarse sand	0.5–1.0
Medium sand	0.1–0.5
Fine sand	0.05–0.1
Silt	0.002–0.05
Clay	< 0.002

Assessment of soil texture gains importance as the proportion of clay in the soil and its mineral composition determine its ability to adsorb cations from the soil solution and the amount and nature of organic content. Also the water-holding capacity of the soil depends on the pore space, which in turn depends on the structural organization of organic and inorganic matter. Soils with unattached particles are said to be structureless or single-grained and their maximum proportion would be coarse particles. If the size of the soil is massive, it aggregates to form a characteristic size and shape. However the aggregation differs between size, stability and internal porosity. In general, clay particles tend to flocculate except for alkaline soil where sodium ions do not favour its aggregation. The clay particles are arranged in stacks known as domains. Based on the addition of sand or silt particles the organic molecules interlink these domains to form micro- or macro-aggregates. It is reported that the neutral and calcareous soils are said to have stable aggregates. Binding agents such as humus, iron and aluminum oxides are more effective and their size ranges from 20–250 µm diameter.

4.3.2 Bulk Density and Pore Space

Soil when dried has a bulk density (P_b) exactly half that of its original value. This is because water in its pore space is replaced by air that it can be defined as the ratio of the mass of dried soil to its volume. Soil porosity, fractional pore space (e) can be evaluated from bulk density (P_b) and particle density P_s.

$$\text{Soil porosity } (e) = 1 - \frac{\text{Bulk density } (P_b)}{\text{Particle density } (P_s)}$$

$$\text{Percent pore space} = 100 \left[1 - \frac{\text{Bulk density } (P_b)}{\text{Particle density } (P_s)} \right]$$

4.4 COMPONENTS OF SOIL

Soil is an open ecosystem consisting of several components like soil water (20–30%), soil CO_2 (20–30%), weathered rock, minerals (45%) and organic matter (5%).

4.4.1 Soil Water

Water in soil determines the soil's energy to do work. It is expressed as per unit quantity of water. Water that escapes when the soil is kept at 105°C for 24 hours is termed as soil water. Only the water that is not bound within the structure of soil minerals (water of crystallization) will be vaporized. Soil water accompanied by dissolved salts is called soil solution. Both the soil water and water of crystallization is not available to plants. In the absence of available water, the plant wilts and the soil at this stage is called *wilting point*. *Field capacity* of the soil is the soil water minus the water lost by drainage and the difference between the field capacity (FC) and wilting point (WP) is called *available water capacity* (AWC). Three different forces facilitate the movement of water in the soil and they are capillary force, gravitational force and osmotic force. The sum total of all these forces is the *water potential*. Water potential is the energy difference between available water and soil water. Soil solution is not pure and it consists of dissolved gases and solutes. Solute concentration varies between soil groups. For example in saline soils, it ranges from 50 to100 mol m^{-3} whereas in non-saline soils, the level is not too high (1–20 mol per cubic metre of the soil solution). The solute concentration also depends on factors such as parent rock composition; environmental conditions that favour the rock formation; addition of fertilizers, manure and lime; quality of water used for irrigation; and rate of deposition of salt and gases from the atmosphere. The general composition of solute in soil with pH >5 is as follows: Ca^{2+}, Mg^{2+}, K^+, Na^+, NO_3^-, Cl^-, SO_4^{2-} and HCO_3^-. In acid soils Al^{3+}, $AlOH^{2+}$ and Mn^{2+} predominate.

4.4.2 Soil Air

The air present in soil is called soil air and the composition of air in soil is different from that of atmosphere because of two reasons.

1.   Slow exchange of gases between atmosphere and soil.
2.   Respiratory activity of soil organisms.

This difference becomes severe when the organic content of the soil is enhanced. Decomposition of organic matter elevates the CO_2 content and decreases the O_2 content. Similarly, waterlogged conditions bring down the O_2 level even to zero. Under this condition, CO_2 content may go up to 20. Immediate change in the soil organisms is the death of sensitive organisms and domination of anaerobes. This results in the production of reduced form of nitrogen, nitrous oxide, iron, manganese and sulphur. Accumulation of volatile organic species such as methane, ethane and fatty acids takes place.

The proportion of air in the soil depends on the total fractional pore space and the water content of the soil respectively. In general, the CO_2 content is more in soil and O_2 content is low. This can be attributed to several reasons: a) reduced diffusion of atmospheric air, b) root penetration, c) respiratory activity of soil organisms and d) decomposition of organic matter.

$$C(H_2O)_n + O_2 \rightarrow CO_2 + H_2O$$

The respiratory quotient (RQ) defined as the ratio of moles of CO_2 produced to moles of O_2 consumed is 1 in high aerated soils and the atmospheric exchange of gases depends on the differences in gas concentration between soil and atmosphere. However for the exchange to occur, conditions should be favourable in soil. Among several factors, water is the prime factor. If the water is more, gas will be less and this depletes the available O_2 in due course of time.

4.4.3 Mineral Components

Factors that determine the mineral components of the sand and silt fractions of soil are the mineral composition of the parent material and the degree of weathering. Out of several minerals, quartz (SiO_2) predominates. In addition strongly weathered soils have minerals such as ilmenite, zircon and haematite. The proportion of sand and silt determines the physical properties of soil. For example, components of soil with maximum silt content have minimum pore space. Also, to some extent they influence the chemical property of the soil. This can be understood by the fact that the presence of mica and feldspar in the soil restrict the easy release of calcium, magnesium and potassium during weathering. Therefore the minerals of the sand and silt fractions are the direct outcome of weathering and hence termed as primary minerals. On the other hand, the minerals of the clay fraction are the resultant products of primary minerals and hence termed as secondary minerals/clay minerals. Their main composition is aluminosilicates and hydrated oxides. These minerals by their physical and chemical properties influence the nature of soil.

4.4.4 Chemical Elements in Soil

The earth's crust constitutes a wide variety of elements (O, Si, Al, Fe, C, Ca, K, Na, Mg, Ti, N and S). Except for soil potassium, all elements irrespective of their essentiality to plant growth, are present in concentrations less than 0.1% whereas the soil potassium is more than 1.5%. Of several elements, aluminosilicates form the major proportion. The composition of soil is different from that of the parent rock as it consists of carbon, nitrogen and sulphate in excess amounts, the reason for this being the presence of organic matter in the soil and their decomposition. The concentration of each soil component influences its property. For example let us take the pore space of the soil. The minimum the pore space, more will be the water holding capacity, which ultimately reduces the O_2 content in soil. Moreover waterlogged condition brings about changes in the decomposition rate. On the other hand, in the absence of organic input, the decomposition rate brings down the organic content of the soil to zero. Therefore organic content of the soil depends on the decomposition rate and the quality of organic carbon that is added newly to the soil ecosystem. In polluted soils, inorganic content is more. The possible sources are radioactive wastes from a nuclear explosion; hospital waste; sewage and intensive agricultural practices.

4.4.5 Organic Matter

Materials of biological origin (partially decayed plant and animal remains and products synthesized by microorganisms) constitute soil organic matter. It supplies sulphur and phosphorus to soil organisms. Also it gives the nitrogen when needed. The decomposition process of organic matter is not complete as a minimum portion of the organic matter is resistant to decay. This portion is termed as humus. In general, humus constitutes to only 5% of the total soil. However it is responsible for the moisture-holding capacity of medium-textured soils. This is due to the fact that it adheres to the inorganic particles of the soil to form pellets called "Robusts". This results in the formation of voids thus giving way to outlet of water and inlet of air. Humus occupies the upper layer, i.e., up to 10 cm from the top. Organic compounds in this layer constitute about 1–3% carbon. However, this level gears up to 30% in peat soil. Sources of carbon compounds in the soil are numerous (roots, stone leaves, woody material, animal residues, faeces, sewage, sludge). The decomposition rate depends on the available form of substrate. For instance, leaves and stems decompose at a faster rate than wood. The end product of decomposition is carbon dioxide and humus. Under X-ray diffraction, it is a dark-coloured material with collapsed structure. It is made up of loosely connected high molecular weight substances.

Humus formed after decomposition cannot be expected to be of similar structure and composition. Based on the substrate humified, it shows different properties. Very few types are soluble in acids and the majority are soluble in alkaline solutions. On acidification of the extract at low pH (<2), it precipitates to form humic acid. Yet, a small portion remains in the solution and it is termed as fulvic acid. Analysis of humus compounds show that it consists of a variety of chemical compounds. It has been found that polysaccharides of humus consist of 10–15% of humus carbon. The sugar units in this portion are known as uronic acids and it helps to give the inorganic mineral constituents in soil. When the humus is subjected to several chemical processes like partial oxidation, reduction and acid hydrolysis, it gives aromatic and aliphatic compounds. It also has amino acids, sulphates, inositol phosphates, phospholipids and nucleic acids. The acid fraction of humus (humic, fulvic) has carboxylic and phenol groups in considerable amounts. At suitable pH, they dissociate to acquire negative charge, i.e., at alkaline pH, phenolic groups are got rid of and between pH 4.5 and 7, carboxylic groups are removed.

4.4.6 Nutrient Content of Humus

The nutrient content of the humus is considerably more (C/N ratio is 10–14; C/P ratio is 100; C/S ratio is 80–100) and therefore when mineralized by the microorganisms, it releases the nutrients which can then be absorbed by plants. It can complex with a variety of metal ions such as copper, lead, iron, nickel, manganese, cobalt, calcium, zinc and magnesium. Above all, humus is very much essential for plant growth as it is responsible for the formation of stable aggregates in soil.

4.5 PROCESSING OF SOIL

Soil formation is a process which takes long periods of time and occurs in steps. The weathering process is initiated as soon as small plants start to develop on a mineral substrate. The mineral substrate need not be necessarily formed from the weathering of rock surface. It may be heap of debris formed by any motive force like wind, ice or river or it may be a partly weathered rock. Continuous growth of plant on the mineral substrate contributes to soil organic matter. Naturally by the process of decomposition, organic matter is transformed into humus, which forms an excellent binding agent between clay aggregates. Soil organisms like earthworms facilitate the dispersion of both organic and inorganic content, which ultimately change its structural organization. This is why the weathered natural debris exhibit themselves in different arrangement. Apart from soil fauna, several factors bring about the transformation of minerals from one layer to the other. Changes in the soil ecosystem depend on the input and output of materials. Physical, chemical and biological processes in the soil also translocate the material from one layer to the other. These include

- Biological activity
- Weathering
- Secondary mineral formation
- Leaching
- Gleying
- Acidification
- Lessivage
- Podzolisation
- Reduction
- Alkalination and salinization
- Desalinization
- Erosion and deposition
- Parent material
- Upward translocation
- Physical mixing
- Surface relief
- Climate
- Clay eluviation
- Soil microorganisms

4.5.1 Biological Activity

Decomposition of plant and animal materials results in the production of water, carbon dioxide and a residue called humus. The decomposition process is induced by several factors and they are

a. Mechanical breakdown
b. Chemical reactions
c. Soil fauna activity
d. Action of decomposers

Mechanical breakdown may be caused by wind and water and the chemical reactions include processes like hydrolysis, oxidation and hydration. Soil fauna like wood lice, earthworms and eelworms act on the organic matter to break them into pieces. Decomposers such as bacteria and fungi according to their need, break down the organic substrates and utilize them. The residue obtained by the decomposition process is an insoluble, condensed heterogeneous polymeric material. Amino acids accompanied with the oxidized aromatic compounds form the humus material. Oxidation of aromatic compounds is microbe-based and the humus with its high density phenolic, carbonyl and carboxylic functional groups acts as an excellent binding agent. It interlinks metal ions by chelation and pH-dependent cation exchange capacity. The humus particles have a size of <2 μcm and are found to bind with inorganic compounds of the clay component of the soil. The attractive force behind the clay–humus complex are the hydrogen bonds, van der Waals forces and complexations with cations bound to the mineral portion. Well-dispersed humified organic matter in a mineral soil is called as Mull humus. It is the characteristic of moist, oxidized and mildly acidic soils. i.e. the soil provides favourable condition for the humification process which does not allow the accumulation of organic matter on the surface. To say clearly, these soils are rich in microorganisms and detritivores necessary for decomposition. If the growth and survival of these organisms are limited, surface accumulation of the organic matter may occur. For the aforesaid process to occur, the soil should possess the following characteristics: high acidity, low temperature and waterlogged conditions. This leads to the formation of peat, the organic layer separated out in the top layer as Mor humus.

4.5.2 Weathering

Rock formation demands specific environment conditions like high temperature and pressure, absence of air, water and biological activity. Since these conditions do not combat with surface soil conditions, rocks are said to be thermodynamically unstable and hence subjected to change and the process is called as weathering. Basically three types of weathering are recognized.

1. Physical weathering
2. Chemical weathering
3. Biological weathering

Physical weathering   It brings about a change physically i.e. the rocks are broken down merely into fragments. This is mediated by several processes. They are a) soil erosion, b) heating and c) trapping of water inside the rock. Soil erosion washes out materials from the rock, which cause the rock to expand. During daytime, heating up of rock may expand them whereas at night, sudden cooling of the environment makes them to contract which induces them to weather. Similarly water trapped inside the rock due to extreme cold conditions get converted to crystals which create pressure on the rock and make them to weather. Above all, strong winds or running water may break the rock into fragments.

Chemical weathering   Here the chemical composition of the rock changes, which in turn results in weathering. Some of the chemical reactions involved in chemical weathering are oxidation, dissolution and hydrolysis.

i.   *Oxidation*   Ingress of air into the cracks and crevices of rock oxidize some of its reduced species such as $Fe^{(11)}$ or $S^{(11)}$.

ii.   *Dissolution*   Water dissolves the ionic bonds that link the minerals. This is a common phenomenon for univalent ions, e.g. weathering of sodium chloride.

iii.   *Hydrolysis*   The end products of this process are a weak acid and a strong base, which makes the solution alkaline. A typical example is the hydrolysis of calcium carbonate to calcium hydroxide and carbonic acid. When the hydrolysis is favoured by the acidified species, the process is said to be acid hydrolysis. Acid species carbonic acid, is formed by the combination of carbon dioxide and water which on reaction with calcium carbonate give rise to calcium ions and bicarbonate ions. This is a neutralization reaction that occurs with the change in pH.

Biological weathering   Root penetration plays a dominant role in weathering of rocks. The action of soil microorganisms may change the chemical composition of rocks. For example, CO_2 released by their respiratory activity induce the carbonic-acid-mediated acid hydrolysis. Similarly, secretion of root exudates such as carboxyl acids and phenols induce acid hydrolysis. Also, microorganisms-based oxidation–reduction reactions change the chemical composition of rocks.

4.5.3 Secondary Mineral Formation

Minerals formed from the weathering process are called secondary minerals. They are more stable than primary minerals and therefore form the components for horizon development. These minerals are generally termed as clays. Most of them are hydrated iron and aluminum oxides or silicate minerals known as clay minerals. Due to the chemical or biological weathering, primary minerals within the parent rock tend to dissolve. Even in the incomplete dissolution state, they exist in a different chemical form. If they are dissolved completely they recrystallize to acquire a new form. Both forms of primary minerals are now termed as secondary minerals. Clay minerals in general, have silica tetrahedral layer (phyllosilicates) and octahedral layers of aluminum or magnesium. Both the layers are held together by shared oxygen atoms. Clay minerals exchange ions or atoms from their lattice with the ions or atoms of several elements. This property is called as isomorphous replacement.

$$\text{Tetrahedral layer } Si^{4+} \rightleftharpoons Al_3^{+}$$

$$\text{Octahedral layer } Al_3^{+} \rightleftharpoons M^{2+} \text{ ions}$$

This change occurs during the mineral formation and hence they acquire a negative charge which is permanent. Under these conditions, they attract cations which can be exchanged with other cations in the solution. This is called cation exchange capacity i.e. it is because of this property that the divalent cations such as K^+, Ca^{2+}, Mg^{2+}, NH_4^+ and Na^+ are retained in the soil which would otherwise get leached from the soil.

4.5.4 Leaching

Leaching process involves two stages. Percolating water is the motive force for material movement. In the first stage, movement is upward and in the second stage it removes the matter entirely from the soil. The process depends on the soluble property of the material and the rainfall. Maximum their solubility, more are they subjected to leaching. This can be easily understood from the fact that sodium chloride due to its extreme solubility gets leached away easily when compared to calcium carbonate that is tightly held in soil even in areas of heavy rainfall. Apart from these two factors, the form in which the particular compound exists decides its leaching i.e. calcium hydrogen carbonate is more sensitive to leaching than calcium carbonate.

4.5.5 Acidification

Water percolation through the soil profile, shifts the soluble substances between layers. If the water happens to be acidic it turns the soil acidic too. Alteration in soil pH changes the soil fauna also. Only resistant soil species survive. Soil fungi predominate under this condition. Earthworms are highly sensitive and therefore get reduced in number. This reflects in the decomposition process too, which will be very slow. As a result they accumulate in the top layer to form Mor humus.

4.5.6 Gleying

Change in chemical speciation may induce solubility. This is a characteristic property of waterlogged mineral soils. Naturally, the O_2 content drops to create anoxic condition and the anaerobes predominate. Iron (III), which is highly insoluble may be reduced to soluble iron (II) in the absence of O_2. Iron (II) exists as blue-green-grey coloured layer, a mixture of iron (II) and (III) hydroxides. This layer is apparent and is called as gleyed horizon. Due to leaching they may be transferred to aerobic areas i.e. near root regions where oxygen supply is sufficient. Again the oxidation process occurs to transform iron (II) to Iron (III). This results in the formation of orange brown coloured insoluble complex.

4.5.7 Lessivage

"A" horizon is a stable layer with soil aggregates cemented together by iron (III) oxide. Anoxic conditions induce the conversion of iron (III) oxide to iron (II) oxide. This will ultimately loosen the layer "A" thus washing out the clay and iron oxides and the process is called as lessivage. It is a purely mechanical process and in due course of time, phenolic leachates complex with iron (II) oxide.

4.5.8 Podzolisation

It occurs in areas with poor water-holding capacity. Coniferous forests are the characteristic example for acid soil. Low molecular weight organic acid, fulvic acid and polyphenols of plant leaves form soluble complex with aluminum and clay. The movement of these complexes from 'A' horizon to 'B' horizon results in the formation of bleached ash grey colour in the 'A' horizon. In the "A" horizon, aluminum cations and iron (II) oxide are formed. These compounds complex with polycarboxylic acid, humic compounds and polyphenolic leaf leachates and get leached to 'B' horizon. Complete removal of coloured iron organic matter gives a peculiar pale colour to the 'A' horizon and hence it is termed as 'E' horizon or Elution horizon whereas the 'B' horizon with the accumulation of iron (II) and organic matter is thickened. Processes occurring in this horizon are:

a.   Breaking up of iron complex which again converts the iron (II) to iron (III).

b.   Precipitation of aluminium

c.   Microbes-based oxidation process that oxidizes organic chelate agents.

d.   Polyvalent cations (Ca^{2+}, Mg^{2+}, Al^{3+}) in excess concentration form flocs.

The resultant product of all these processes is podzol with a characteristic pale 'E' horizon and darkened 'B' horizon.

4.5.9 Reduction

Soil with minimum pore space holds large amount of water than air. This limits the supply of O_2 and enhances the level of CO_2. As a result, organic matter builds up and forms peat. On the other hand, microorganisms reduce Fe^{3+} to Fe^{2+} to give a characteristic grey or bluish grey colour to the soil. If the existing condition is temporary, ingress of O_2 oxidizes the Fe^{2+} to Fe^{3+} once again to show a mottled appearance i.e. brown colour in oxidized and grey colour in reduced areas.

4.5.10 Salinization and Alkalination

Salinization refers to the increased sulphate and chloride content of soil and alkalination refers to the increased sodium content on the cation exchange sites. Weathering processes, inward movement of salt from oceans, and lands irrigated with ground water may be the possible sources. When salinity is caused by sodium ions, calcium and magnesium ions precipitate as carbonates leaving sodium to adsorb on cation exchange sites. This suddenly raises the soil pH to 8 and above, and collapses the entire soil structure.

4.5.11 Desalination

Under tropical conditions, silica is leached out resulting in highly porous soil rich in iron oxide. These soils are generally termed as oxisols.

4.5.12 Erosion and Deposition

Wind and water are the two motive forces for soil erosion. When transferred to a new place they are influenced by those environmental conditions and there they undergo a variety of chemical processes and change their structure and character accordingly.

4.5.13 Parent Materials

Influence of parent material is more in young soil. After the soil processes begin to occur, the effect of parent material on the soil property is less pronounced. However, the nature of the parent material determines the chemical process. For example, granite rocks cannot be expected to be sandy in the final form. Their chemical process is clear-cut. Since they have low pH, they exhibit podzolisation. Similarly basic igneous rocks with ferromagnesium minerals give rise to a soil of clayey or loamy textural class.

4.5.14 Upward Translocation

In the arid regions, the precipitation process does not combat with the evapo-transpiration process i.e. the latter exceeds the former. As a result, water tends to move upward due to capillary forces. This brings both the soluble and sparingly soluble materials to the surface. When the water gets evaporated, the dissolved salts are left behind making the soil saline. Out of several salts, sodium compounds are highly soluble and therefore accumulate in high concentration on the surface thus making it saline.

4.5.15 Physical Mixing

Mixing of organic and inorganic components in the soil may be induced by several factors. a) biological, b) chemical, c) physical, d) human activity. Under the influence of these factors, the soil components disperse themselves to bring out a characteristic soil structure.

4.5.16 Surface Relief

Movement of soil materials downhill allows the shallow soil to accumulate on the upper slope. Moisture content of the soil in valleys and low levels are more than those on slopes. The reason behind is the nearness of water table in valleys and depressions. Moreover here, the water flow is restricted and the water stagnates irrespective of the soil's water holding capacity. Therefore, apart from climate and parent material, altitude also determines, the soil pattern.

4.5.17 Climate

Soils that are influenced by climate are called zonal soils and the characteristics of intrazonal soils depends on the nature of the parent material. Soil formation is restricted to azonal soils.

4.5.18 Clay Eluviations

Water movement in pores facilitates the movement of clay from the 'A' horizon to the 'B' horizon. In the 'B' horizon they get deposited as clayskins on pore walls. Reduction in the water flow and changes in the chemical conditions that do not favour the clay suspension are the two factors that cause clay deposition.

In general microorganisms are classified into five groups—bacteria, virus, fungi actinomycetes, protozoa. The characteristic features of these microbial groups are as follows.

4.5.19 Soil Microorganisms

Soil microorganisms play a predominant role in the decomposition process. The major soil microorganisms include the following:

Bacteria   They are either spherical (cocci) or rod shaped (bacilli). Some cocci are subjected to change when they shift their habitats. Under unfavorable conditions, they form spores and protect themselves from drying. Their population in the soil depends on the availability of organic matter. That is why their population near roots is 100 times greater than the rest of soil. Since the soil habitat favours the growth of a variety of microorganisms, decomposition process is easier.

Actinomycetes   The size of actinomycetes is equal to that of bacteria. They have a filamentous structure like fungi; most of their cellular characteristics resemble bacteria. They are heterotrophic in nature and utilize the nutrient source from plant residues in soil and composts. They are widely distributed and found to be dominant in warmer temperature.

Fungi   Cellular filaments of fungi are termed as hyphae and their size range is 5–20 μm diameter and a few cm to several metres in length. In the presence of abundant nutrient (leaf litter and compost), their hyphae structure aggregate together to form rhizomorphs. Based on their nutritional requirements, they may be classified into several groups such as heterotrophs, saprophytes, plant pathogens and plant root invaders. They have well developed enzyme systems to decompose all components of plant material including the most resistible tannin, lignin and cellulose compounds. Since they adapt themselves to acid pH, they are the dominant decomposers of acid soils. Some specific groups of fungi exhibit a mutual relationship with plant roots to form a structural association called mycorrhiza.

Algae   They are photoautotrophs whose size ranges within 10–40 mm diameter. They have coloured pigments and therefore exist in different colours, blue-green, brown, red and green. They are the initial colonizers of barren land supplying organic materials for the subsequent invaders. Polysaccharides secreted by them act as binding agents and thus they agglomerate the soil particles and stabilize them. In association with fungus (lichen), they act as pollution indicators in places where the atmosphere is contaminated with sulphur dioxide. Cyanobacteria is the typical example of algae with a characteristic ability to fix nitrogen and photosynthesize. This makes them as an ideal candidate to maintain the fertility of the soil ecosystem.

Protozoa   They are either photoautotrophic or saprophytic. They exist in single cells (50 μm). They move between soil pores in search of moisture. Under unfavourable conditions i.e. dry condition, they form eggs and feed on bacteria and fungi and have a check on their population. It is reported that while feeding on bacteria and fungi they liberate the nutrients and therefore help in the nutrient cycling in the soil environment.

The type of microbial community present in soil depends on the nature of soil. This is because, microorganisms utilize a wide variety of substrates for energy source and based on the energy source they depend on, they are classified into two types, autotrophs and heterotrophs. Again based on the need of O_2, they are classified into three types (a) aerobes, (b) anaerobes and (c) facultative anaerobes. These organisms secrete enzymes which brings about decomposition and chemical transformation processes in soil. Therefore measurement of enzyme activity gives an indication of the extent of specific biochemical processes in soil which in turn gives a real picture about the soil fertility.

4.6 ANALYSIS OF SOIL

In general, soil is analysed to evaluate its status in terms of

- Fertility
- Pollution load
- Ability to support antibiotic-producing microorganisms

For the above objectives, the physicochemical and biological parameters of the soil are analysed.

4.6.1 Sample Collection

Samples should be collected in polyethylene bags and then transferred to the laboratory. Soon after the collection, the soil is labeled as given below.

Name of the sampling site	: _________________________
Method of sampling	: _________________________
Parameters to be analysed	: _________________________
Time and date of collection	: _________________________

Certain parameters like pH, redox potential, organic nitrogen and phosphorous are to be evaluated immediately and for the remaining parameters the soil should be allowed to air-dry. Drying temperature varies with the parameter analysed.

4.6.2 Sample Preparation

Collected soil should be separated from unwanted substances such as stones, roots, twigs, etc. Then it should be homogenized to get a powdered form. If needed it can be sieved (0.5–2 mm) and then used.

PHYSICAL PARAMETERS

Physical parameters of the soil include bulk density, specific gravity, moisture content, water holding capacity, cation exchange capacity and loss of ignition.

4.7 BULK DENSITY

4.7.1 Principle

It is the mass per unit volume of the soil, which varies with different soil types. For example coarse-textured soil shows higher values of bulk density whereas fine-textured soil shows low values. Determination of bulk density is considered to be important because it restricts infiltration and makes soil impermeable. Also it prevents root penetration into deeper regions. The specific gravity of the soil is directly related to its bulk density. It is determined by drying the soil volume to a constant weight at 105°C.

Expt. 54　　*To determine the bulk density of soil sample*

MATERIALS REQUIRED

Laboratory glassware

　Measuring cylinder

　Silica crucible

Spatula

　Laboratory equipment

　　Hot-air oven

　　Desiccators

　　Balance

PROCEDURE

1. Homogenize the soil sample before use.
2. Then dry it in a hot-air oven at 105°C and record its dry weight.
3. Repeat this procedure several times till a constant weight is achieved.
4. Transfer the dried soil sample to a measuring cylinder.
5. For convenience sake, keep the volume as 100 cm in the measuring cylinder.
6. Now weigh the soil sample at this volume and note down the value.

CALCULATION

$$\text{Bulk density } (g/cm^3) = \frac{\text{Weight of the soil in grams}}{\text{Volume of the soil in } cm^3}$$

4.8 SPECIFIC GRAVITY

The relative weight of the given volume of soil compared with an equal volume of distilled water at a given temperature is called specific gravity. The average specific gravity of soil is about 2.65 mg/m^3.

Expt. 55　　*To determine the specific gravity of soil sample*

MATERIALS REQUIRED

Laboratory glassware

　Sample bottles

　Measuring cylinder

Spatula

Laboratory equipment

　Balance

PROCEDURE

1. Before estimation, homogenize the soil sample and dry it in a hot-air oven at 105°C.

2. Repeat this until a constant weight is achieved.

3. Take two wide-mouthed glass bottles and record their initial weight.

4. Now transfer the dried soil sample to one of the bottles up to a fixed volume.

5. Fill the second bottle with distilled water to the same volume.

6. Take the weight of both the bottle with soil and distilled water.

7. Using the formula given below, determine the specific gravity

CALCULATION

$$\text{Specific gravity of soil} = \frac{y - y_1}{z - z_1}$$

where,

y = final weight of bottle with soil

y_1 = Initial weight of bottle used for soil

z = Final weight of bottle with distilled water

z_1 = Initial weight of bottle used for water

4.9 MOISTURE CONTENT

4.9.1 Principle

Moisture content of the soil is its water content. Presence of water in soil determines its texture and compactness which ultimately reflects its suitability to support life. Source of moisture in the soil is infiltration and irrigation. However, persistence of moisture in soil depends on several factors such as water holding capacity, evaporation and plant uptake. Moisture content of the soil is determined by subtracting the final weight of the dried soil sample from the initial weight of the soil sample as such. Since drying up the soil at 105°C removes its moisture, this method is followed.

Expt. 56   *To determine the moisture content of soil sample*

MATERIALS **R**EQUIRED

Laboratory glassware

Sample bottle

Silica crucible

Spatula

Laboratory equipment

Hot-air oven

Desiccators

PROCEDURE

1. Dry the homogenized soil sample in a hot-air oven at 105°C till a constant weight is achieved.

2. Let that be the initial weight of the sample (I).

3. Cool it in a desiccator and record its final weight (F).

4. Now determine the moisture content of the sample using the formula given below.

CALCULATION

$$\text{Moisture content (\%) of the soil sample} = \frac{(I - F) \times 100}{I}$$

where,

I = Initial weight of the sample (in g)

F = Final weight of the sample (in g)

4.10 WATER HOLDING CAPACITY (WHC)

4.10.1 Principle

Physical and chemical characteristics of soil determine its ability to retain water. When the soil is flooded with water, considerable amount gets infiltered and fills all the pore space leaving no space for the ingress of atmospheric air. This is the stage at which the soil is said to be at its maximum water holding capacity.

Expt. 57 *To determine the water holding capacity of soil sample*

MATERIALS REQUIRED

Laboratory glassware

Petri dish

Filter paper (Whatman No.1)

Spatula

Laboratory equipment

Hot-air oven

Perforated soil boxes

Balance

PROCEDURE

1. For WHC determination, bottom-perforated rounded soil boxes of about 5.6 cm and 1.6 cm diameter are used.

2. The diameter of the bottom holes should not be greater than 0.75 mm.

3. Record the initial weight of the empty box.

4. Homogenize the soil sample to be analysed before drying it at 105°C.

5.   Keep a filter paper (preferably Whatman No.1) above the perforated bottom of the soil box.

6.   Now fill the box with dried soil and record its final weight (F_1).

7.   Place the soil box in a petri dish filled with water and leave the whole set-up undisturbed for about 12 hours.

8.   This allows the water in the petri dish to enter into the soil box and ultimately saturates it.

9.   After 12 hours, record the weight of the box (F_2).

10. Take care to wipe the box dry on the outside before weighing.

CALCULATION

$$\text{WHC } (\%) = \frac{(F_2 - F_1) - (F_1 - I) \times 100}{(F_1 - I)}$$

where,

I = Initial weight of soil box (in g)

F_1 = Final weight of soil box with soil (in g)

F_2 = Final weight of the soil box with water-saturated soil (in g)

4.11 DETERMINATION OF CATION EXCHANGE CAPACITY

4.11.1 Principle

Refer flame photometer in chapter 1.

Expt. 58   *To determine the cation exchange capacity of soil sample*

MATERIALS REQUIRED

Laboratory glassware

Laboratory equipment

Flame photometer

Centrifuge

Shaker

REAGENTS

a.   Ethanol 95%

b.   Sodium acetate (IN)

Weigh 136 g of sodium acetate trihydrate and transfer into a 1000-ml volumetric flask containing little distilled water. Make up to the mark using distilled water. Adjust the pH of the solution to 8.2.

c.   Ammonium acetate (IN)

To a 2-litre volumetric flask, add 154 g of ammonium acetate and dilute to 1.8 litre. Adjust the pH of the solution to 7.0 by adding diluted ammonium hydroxide or acetic acid. Finally make up to the volume using distilled water.

PROCEDURE

1. Take 5 g of soil in a beaker. To this add 25 ml of sodium acetate solution (IN) and stir well.

2. Carefully transfer the beaker content into a centrifuge tube. See to it that the beaker content is completely transferred into the centrifuge tube.

3. Shake the tube for about 5 minutes and then centrifuge it at 2000 rpm for about 10 minutes.

4. Collect the supernatant in a 1-litre volumetric flask.

5. Repeat the procedure four times. Every time use 25 ml of IN sodium acetate and collect the supernatant.

6. Now add 25 ml of 95% ethanol to the tube, shake well, centrifuge and collect the supernatant.

7. Repeat the procedure three times.

8. Again add 25 ml of IN ammonium acetate, shake well and centrifuge at 2000 rpm to collect the clear liquid.

9. Repeat the procedure another two times.

10. Dilute the supernatant to the marked volume using distilled water. This liquid consists of sodium ions replaced from the soil.

11. Determine the CEC capacity of the soil by using flame photometer.

12. Go through the operation manual of flame photometer. After setting the sodium filter, initiate the compressor and light the burner of the instrument. Set the air pressure at 5 lbs and gas feeder to produce blue sharp flame cones.

13. Using highest sodium standard solution, adjust the instrument to show full reading. Similarly using extract solution, set the zero reading.

14. Begin from the lowest to the highest standard and record their emission value.

15. Draw a standard curve by plotting the reading of standard against their concentration.

16. From the calibration curve, evaluate the concentration of sodium in the solution.

17. Calculate the CEC of the soil solution using the following formula.

CALCULATION

Cation exchange capacity in 100 g of soil (ml) $= \dfrac{E \times 100}{W} \times \dfrac{V}{1000}$

where,

E = Sodium concentration in the extract

V = Volume of extract (100 ml)

W = Weight of soil (g)

4.12 ESTIMATION OF LOSS OF IGNITION

4.12.1 Principle

At 600°C, organic matter, calcium carbonate, magnesium carbonate and moisture content of the soil are completely lost. Therefore, the final weight of the soil shows a considerable reduction from its initial weight.

Expt. 59   _To estimate the loss of ignition of soil sample_

MATERIALS REQUIRED

Laboratory glassware

    Petri dish

    Silica crucible

Spatula

Laboratory equipment

    Desiccator

    Bunsen burner

    Balance

    Homogenizer

PROCEDURE

1. Take a clean silica crucible with a lid.
2. Heat the crucible in a red-hot flame for about 30 minutes.
3. Cool it in a desiccator and estimate its initial weight (I_1).
4. Transfer the pre-weighed soil sample (20 g) in the crucible.
5. Replace the lid and again weigh it (I_2).
6. Homogenize the soil sample prior to use.
7. Now ignite the silica crucible in a red-hot flame for 8 hours thus giving sufficient time for the organic content and moisture content to be expelled.
8. After this, cool the crucible in the desiccator and take its final weight (F)

CALCULATION

$$\text{Loss of ignition (moisture free)} = \frac{I_2 - F}{I_2 - I_1} \times 100 \times \frac{100}{100 - M}$$

where,

$$I_1 = \text{Initial weight of the silica crucible}$$

$$I_2 = \text{Initial weight of silica crucible } (I_1) + \text{ soil}$$

$$F = \text{Final weight of silica crucible } + \text{ soil}$$

$$M = \text{Moisture content of soil}$$

CHEMICAL PARAMETERS

Soil is a complex mixture of organic and inorganic constituents. Major inorganic constituents of the soil include Al, Si, Ca, Mg, Fe, K and Na and humus contributes to the organic content of soil. The composition of these constituents depends on several factors like the nature of parent matter, climatic conditions, biological and human activity. In general for soil analysis, pH, EC, alkalinity, chlorides, sulphate, nitrogen, phosphorus and organic carbon are taken into consideration and analytical procedures for these parameters are given below.

For the analysis of the chemical parameters namely pH, EC, alkalinity, chlorides and sulphates, a soil suspension is to be prepared. For the preparation of soil suspension (1:10 w/v), add 10 g of soil in 100 ml of distilled water and stir well.

4.13 pH

4.13.1 Principle

It is an important criterion for the evaluation of soil quality measurement as the microorganisms and higher plants respond markedly to the chemical environment. In general soil pH indicates the acidity, alkalinity and neutrality of soil solution. Alkaline soils have higher concentration of Na^+ ions on the cation exchange site whereas acidic soils result from the formation of organic and inorganic acids. During the process of organic matter decomposition, leaching also contributes significantly to acidity. Based on the concentration of H^+ ions in the solution, glass electrodes of the pH meter acquire electrode potential which can then be measured potentiometrically against a calomel electrode. The potential difference between these two electrodes gives the pH value of the given solution.

4.13.2 Calomel Electrode

It consists of two layers a) inner layer which is a mixture of mercuric chloride and mercury paste and b) outer layer which is a saturated solution of potassium chloride. The inner tube is connected to the outer tube through a small opening. It gives the standard voltage with a resistance of 2000–3000 ohms.

4.13.3 Glass Electrode

The inside of the glass electrode tube consists of a wax-insulated lead wire made up of silver coated with silver chloride. The bottom of the tube has a glass bulb that acquires a potential on contact with H^+ ions. On operation, potassium chloride escapes out through the opening and favours the current passage between the electrodes by means of an ionic bridge. This allows the electric potential to develop on both the electrodes. Since the absorption of H^+ ions vary the potential of glass electrodes, the difference is measured as pH of the solution.

Expt. 60   *To evaluate the pH of the soil sample*

MATERIALS REQUIRED

Laboratory glassware

Laboratory equipment

pH meter

Blotting paper

Distilled water

Soil sample

REAGENTS

Buffer solution (4.0, 9.2)

PROCEDURE

1. Warm up the instrument for 15 minutes before use.
2. Adjust the pH knob and zero adjustment knob to zero position and set the temperature control knob to the solution temperature.
3. Immerse the electrodes in distilled water and see to it that the pH knob and zero adjustment knob are pointing towards zero.
4. Now wipe the electrodes with a blotting paper, immerse the electrode into the buffer solution of pH 4.0.
5. Keep the functional switch to the pH range of the buffer solution and adjust the pH knob to the exact pH of the buffer solution.
6. Similarly, calibrate the pH for buffer solution 9.2.
7. Now take 10 g of soil in a 100 ml beaker.
8. Stir well and allow the soil suspension to settle.
9. Immerse the electrodes carefully into the soil suspension.
10. Take care to stir well the solution before taking the reading.
11. Adjust the functional switch to the particular pH range and record the pH.

4.14 ELECTRICAL CONDUCTIVITY

4.14.1 Principle

Electrical conductivity of soil represents its total soluble salts. It is an important physicochemical characteristic of soil indicating the nutrient availability of soil for plant growth. This clearly shows that the plants can absorb ions only when they are present in free form. According to several researchers, effluent-flooded soil consists of more soluble salts than the control soil. Electrical conductivity is directly proportional to the amount of salts in the solution. It is the measurement of resistance offered to the current flow and the resistance increases with the decreased solution concentration. Therefore the resistance is recorded by immersing two parallel electrodes in the soil solution. The method is based on the formula

$$R = r \times L/A$$

where,

R = Electrical resistivity

L = Distance between the electrodes (1 cm)

A = Area of the conductor (1 cm^2).

r = Proportionality constant called specific conductance.

Since conductivity is the reciprocal of resistance it is expressed as mhos/cm, where mho is reciprocal of ohm.

Expt. 61 *To measure the electrical conductivity of soil sample*

MATERIALS REQUIRED

Laboratory glassware

Laboratory equipment

Conductivity meter

Soil sample

REAGENTS

Calcium sulphate solution

0.1N Potassium chloride solution

PROCEDURE

1. Switch on the instrument and leave it undisturbed for about 15 minutes.
2. Check accuracy of the instrument by using a solution of known EC.
3. Commonly, saturated calcium sulphate solution and 0.1N potassium chloride solution are used. Their EC values should be 2.2 and 1.41 dsm^{-1} respectively.
4. Dissolve 20 g of soil in a 50-ml beaker containing distilled water.
5. Stir well and then allow the soil to settle.
6. Wash the electrodes with double-distilled water, blot them dry and carefully immerse them into the soil solution.
7. Adjust the temperature knob and measure the EC of the soil solution.
8. To acquire the specific conductivity of the soil solutions multiply this by the cell constant.

4.15 ALKALINITY

4.15.1 Principle

Refer alkalinity in chapter 3.

Expt. 62 *To determine the alkalinity of soil sample*

MATERIALS REQUIRED

Refer alkalinity in chapter 3

PROCEDURE

1. In 50 ml of soil suspension, add 2–3 drops of phenolphthalein indicator.

2. If the solution shows pink colour, titrate it against sulphuric acid.

3. Appearance of slight pink colour indicates the presence of hydroxides or carbonates whereas a colourless sample confirms the presence of free CO_2.

4. Note down the colourless end point as "p".

5. Now add 2–3 drops of methyl orange indicator to the same flask and proceed with the titration till the solution turns from yellow to orange.

6. Record the value as "t" and calculate the alkalinity using the formula.

CALCULATION

Amount of alkalinity in the sample (mg/g) $= \dfrac{F}{10} \times \dfrac{V}{W} \times \dfrac{1}{(100 - M)}$

where,

F = Alkalinity of filtrate (mg/l)

V = Total volume of suspension (ml)

W = Weight of soil used in suspension (g)

M = Moisture content of soil (in %)

4.16 SULPHATES

4.16.1 Principle

Refer sulphates in chapter 3.

Expt. 63   *To estimate the amount of sulphate in soil sample*

MATERIALS REQUIRED

Refer sulphate in chapter 3.

PROCEDURE

1. Transfer 100 ml of sample in 250 ml volumetric flask.

2. To this, add 2–3 drops of methyl red indicator.

3. Slowly add hydrochloric acid to the content of the flask and observe for the colour change from red to orange (This indicates the acidic pH of the solution).

4. Add still more hydrochloric acid to the flask and boil it.

5. Now add warm barium chloride solution so as to favour the precipitation process.

6. Continue to add it till the precipitation is completed.

7. Again heat it in a hot water bath for about 2 hours. Do not cool it. Immediately filter it through ashless filter paper.

8. Flush the filter paper with distilled water many times.

9. This should be repeated till the filtrate does not contain even traces of chloride.

10. Every time after washing the filter paper, check the presence of chloride in the filtrate by titrating it against silver nitrate solution. Absence of turbidity indicates nil chloride content and at this stage transfer the filter paper to a silica crucible.

11. Ignite it in muffle furnace at 800°C for 1 hour. After one hour, weigh the precipitate.

CALCULATION

$$\text{Amount of sulphates in the sample (mg/g)} = \frac{F}{10} \times \frac{V}{W} \times \frac{1}{(100 - M)}$$

where,

F = Sulphate content of filtrate (mg/l)

V = Total volume of suspension (ml)

W = Weight of soil used in suspension (g)

M = Moisture content of soil (in %)

4.17 AVAILABLE SULPHUR

Inorganic sulphate is the available form of sulphur for crop growth. In reduced forms (elemental sulphur and sulphide) they become relatively insoluble and soluble sulphate can be determined by barium sulphate precipitation method.

4.17.1 Principle

Heat treatment hydrolyse the inorganic and organically held sulphate sulphur which can be easily extracted by sodium chloride. Addition of barium chloride crystals to the extracted sample favour the conversion of sulphate sulphur into barium sulphate. This results in the turbidity formation and therefore can be determined in a spectrophotometer.

Expt. 64   *To estimate the amount of available sulphate in soil sample*

MATERIALS REQUIRED

Laboratory glassware

Laboratory equipment

    Water bath

    Hot-air oven

    Spectrophotometer

REAGENTS

a. Sodium chloride solution (1%)

   Dissolve 10 g of sodium chloride in 1000 ml distilled water. Mix well.

b. Barium chloride dihydrate crystals

c. Stabilizing solution

   To 250 ml distilled water, add 75 g sodium chloride, stir well and then add 30 ml conc. hydrochloric acid, 100 ml absolute ethanol and 50 ml glycerol. Ensure complete dissolution of sodium chloride. Make up the volume to 500 ml using distilled water.

d. Standard sulphate solution

   Add 1.080 g potassium sulphate in 100 ml distilled water. Stir well. One ml of this solution contains 2 mg of sulphate.

PROCEDURE

1. Transfer 5 g of soil into a silica basin containing 20 ml distilled water.

2. Heat it in boiling water bath till all its water content evaporates.

3. Following this, heat it in a hot-air over at 102°C for 1 hour. After one hour incubation period, allow it to cool.

4. Now transfer the crucible content into a centrifuge tube containing 33 ml of 1% sodium chloride.

5. Rotate it at 2000 rpm for about 5 minutes and collect the clear liquid in a silica crucible.

6. Repeat the procedure till 25 ml of the extract V_2 is collected.

7. Now add 2 ml of 3% H_2O_2 to the silica crucible and heat it in a boiling water bath to eliminate the organic matter.

8. Again to remove the excess water, place it in a hot-air oven at 102°C for 1 hour.

9. Transfer the residue to a centrifuge tube containing 25 ml water, rotate it at 2000 rpm and collect the supernatant.

10. Pour 10 ml of this extract into a conical flask and make up the volume to 30 ml using distilled water.

11. Similarly prepare several dilutions of standard sulphate solution into a series of beakers with their concentrations ranging from 5 ml to 50 ml at 5-ml interval. Make up the volume of each beaker to 30 ml using distilled water.

12. Add 2.5 ml of stabilizing solution and 0.3 g of barium crystals to the standards and soil extract solution.

13. Mix the contents of each flask well and observe for the turbid formation.

14. Measure the absorbance of both the standards and extract solution in a spectrophotometer at 340 nm.

15. Prepare a standard curve by plotting the concentration of standards against their absorbance values.

16. From the calibration curve evaluate the sulphate content in the extract solution. Let this be *a*.

CALCULATION

$$\text{Amount of sulphate mg/kg} = \frac{a \times V_1}{V_2 \times W}$$

where,

a = amount of sulphate (mg/l) in the soil extract.

V_1 = Volume of extract (10 ml)

V_2 = Volume of aliquot (25 ml)

W = Weight of soil

4.18 CHLORIDES

4.18.1 Principle

Refer chlorides in chapter 3.

Expt. 65   *To estimate the amount of chlorides in soil sample*

MATERIALS REQUIRED

Refer chlorides in chapter 3.

PROCEDURE

1. To 50 ml of soil suspension add 2 ml of potassium dichromate solution.

2. Stir well and titrate it against silver nitrate solution.

3. End point is the colour change from yellow to red.

CALCULATION

$$\text{Amount of chlorides in the sample (mg/g)} = \frac{F}{10} \times \frac{V}{W} \times \frac{1}{(100 - M)}$$

where,

F = Chloride content of filtrate (mg/l)

V = Total volume of suspension (ml)

W = Weight of soil used in suspension (g)

M = Moisture content of soil (in %)

4.19 HYDROCHLORIC ACID EXTRACT

4.19.1 Principle

Mineral constituents of soil can be evaluated by subjecting the sample to hydrochloric acid digestion which releases the minerals that have been locked up within the soil particles.

Expt. 66   *To prepare hydrochloric acid extract and to estimate the amount of sesquioxide in soil sample*

MATERIALS REQUIRED

Laboratory glassware

Laboratory equipment

Sand bath

REAGENTS

HCl (1:1)

PROCEDURE

1. Take a known weight of the ignited soil in a 250-ml conical flask.
2. Add a few drops of distilled water, 60 ml of 1:1 hydrochloric acid and digest the contents in a sand bath for about 8 hours. During digestion, place a funnel over the conical flask.
3. After digestion, dilute the digested content with 200 ml distilled water [keep it undisturbed for about 10 hours].
4. Filter the content through a No. 3 filter paper in a 500-ml volumetric flask.
5. Apply a thin layer of vaseline over the mouth of the conical flask before proceeding with the filtration so that the digested content will be filtered to the conical flask safely.
6. Also transfer the soil to the filter paper.
7. Wash it several times with hot water until the soil content is free of chlorides and collect the filtrate.
8. Now remove the flask, make up to 500 ml with cold distilled water.

CALCULATION

Amount of sesquioxides on moisture free basis

$$= (b-a) \times \frac{500}{50} \times \frac{100}{W} \times \frac{100}{(100-M)}$$

where,

M = Moisture content of soil

50 = Volume of HCl extract used (ml)

500 = Total volume of HCl extract (ml)

W = Weight of soil

b = Weight of sesquioxide in the silica crucible

a = Weight of silica crucible

4.20 IRON AND ALUMINIUM OXIDES

4.20.1 Principle

For the estimation of iron and aluminium oxides, ammonia is added as a precipitating agent. Precipitation followed by ignition, converts aluminum and iron hydroxides to oxides which is then estimated.

Expt. 67　　*To estimate the amount of iron and aluminium oxides in the soil sample*

MATERIAL REQUIRED

    Laboratory glassware

    Laboratory equipment

      Water bath

      Oven

REAGENTS

    Solid ammonium chloride

    Ammonium hydroxide

    Silver nitrate

    Litmus paper

PROCEDURE

1. To 50 ml HCl extract, add 2 g of solid ammonium chloride in a 250-ml beaker.
2. Boil the solution, take care to cover the beaker. Before boiling put off the flame.
3. With great care add diluted ammonium hydroxide to the hot solution.
4. While doing this, stir the content to ensure complete mixing.
5. The purpose is to turn the solution alkaline and this can be confirmed by placing the litmus paper in the solution.
6. Heat this again to the boiling point for 5 minutes and cool it for few minutes. Care should be taken to transfer the solution along with the precipitate.
7. Wash with hot water and then filter through No. 3 filter paper in a 500-ml beaker.
8. After the filtration is over, remove the precipitate, add some distilled water and boil again.
9. Then follow the same procedure as described above.
10. Repeat this several times until the filtrate runs free of chloride.
11. Evaporate the collected filtrate over a water bath until the filtrate is reduced to 250 ml which can be used to estimate Ca, Mg and K.
12. Heat and dry the precipitate in a hot-air oven for 15 min and place it in a pre-weighed silica crucible. Ignite the filter paper and record its weight.

4.21 CALCIUM

4.21.1 Principle

Calcium gets precipitated to calcium oxalate on adding ammonium oxalate in acid medium. This when titrated against standard potassium permanganate quantify the calcium.

$$CaCl_2 + (NH_4)_2C_2O_4 \rightarrow CaC_2O_4 + 2NH_4Cl$$

Calcium chloride — Ammonium oxalate — Calcium oxalate — Ammonium chloride

$$2KMnO_4 + 8H_2SO_4 + 5CaC_2O_4 \rightarrow K_2SO_4 + 2MnSO_4 + 5CaSO_4 + 8H_2O + 10CO_2$$

Potassium permanganate — Calcium oxalate — Potassium sulphate — Manganous sulphate — Calcium sulphate

Expt. 68   *To estimate the amount of calcium in soil sample*

MATERIALS REQUIRED

Laboratory glassware

Laboratory equipment

   Hot plate

   Titration assembly

REAGENTS

Solid ammonium chloride

Diluted ammonia

Diluted sulphuric acid

Diluted acetic acid

Potassium permanganate (0.1N)

Silver nitrate

Litmus paper

Filter paper No. 40

PROCEDURE

1. Take 50 ml of the hydrochloric extract in a 500-ml beaker.
2. With that, add 2 g of solid ammonium chloride and a piece of red litmus paper to indicate the pH of the beaker content.
3. To make the filtrate alkaline, gently add diluted ammonia to the flask.
4. This should be done till the filter paper turns blue.
5. Now add diluted acetic acid and check the litmus paper to turn red again.
6. Boil the contents of the flask; add 10 ml of saturated ammonium oxalate.
7. While adding, take care to stir the beaker continuously.

8. Again boil for 3 minutes. After this, leave the beaker undisturbed for another 5 minutes. Once the precipitate is formed, add little quantity of ammonium oxalate and observe for the formation of new precipitate (ammonium oxalate should be added using a pipette with its delivery end immersed in the beaker content).

9. Ammonium oxalate can be added until no more precipitate is formed.

10. After this, digest the content of the beaker in a sand bath for 30 minutes.

11. Ensure the granulation of the precipitate and then filter it through a filter paper (No. 40).

12. Flush lukewarm water over the precipitate till the filtrate runs free of chloride.

13. Finally, by piercing a hole in the filter paper flush the precipitate into the filtrate collecting flask.

14. For this, lukewarm water and slightly heated sulphuric acid should be used. Since the fine particles of the filtrate may adhere to the folds of the filter paper, spread its folds and wash it at least thrice.

15. To the filtrate collected in the conical flask, add 10 ml of dilute H_2SO_4 (along its sides) and heat in a hot plate at 70°C.

16. This can be performed till the first bubble comes.

17. Remove this from the hot plate, keep it for 3 minutes and then take the supernatant for titration against 0.1N $KMnO_4$.

18. The end point is the appearance of faint-pink colour.

19. For accurate results, transfer the filter paper into the flask and if the colour is retained record the value. Otherwise proceed with the titration and using the following calculation determine the calcium content.

CALCULATION

Amount of calcium in 100 g of soil on moisture free basis(%)

$$= 0.002 \times \frac{250}{50} \times \frac{500}{50} \times \frac{100}{W} \times \frac{100}{(100-M)}$$

where,

0.002 g of calcium is present in 1ml of 1 N $KMnO_4$

250/50 = 50 ml of iron oxide used, out of total volume of 250 ml

500/50 = 50 ml of HCl acid extract used, out of its total volume of 500 ml

W = Weight of soil taken

M = Moisture percent of soil

4.22 MAGNESIUM

4.22.1 Principle

In alkaline medium, addition of sodium phosphate precipitates magnesium as magnesium ammonium phosphate.

$$MgCl_2 + Na_2HPO_4 + NH_4OH \rightarrow Mg(NH_4)PO_4 + NaCl + H_2O$$

Magnesium chloride   Sodium phosphate   Ammonium hydroxide   Magnesium ammonium phosphate   Sodium chloride

$$2Mg(NH_4)PO_4 \rightarrow Mg_2P_2O_7 + 2NH_3 + H_2O$$

Magnesium ammonium phosphate   Magnesium phosphate

Expt. 69   *To estimate the percent of magnesium in soil sample*

MATERIALS REQUIRED

      Laboratory glassware

      Laboratory equipment

        Hot-air oven

    Litmus paper

    Filter paper no. 42

REAGENTS

Solid ammonium chloride

Diluted ammonium hydroxide (1:4;1:7)

Disodium phosphate

Silver nitrate

PROCEDURE

1. Take 50 ml hydrochloric acid extract in a 500-ml beaker.
2. Add 1 g of solid ammonium chloride and a piece of red litmus paper.
3. Add ammonium hydroxide solution (1:4) in small quantities till red litmus paper turns blue. After that, add in drops, 10 ml of disodium phosphate solution (10%).
4. While adding, ensure complete mixing by stirring the beaker with a glass rod (Do not swirl the beaker)
5. Keep this for about 12 hours and after complete precipitation, filter through Whatman No.42 filter paper.
6. Wash the precipitate with diluted ammonia several times till the filtrate runs devoid of chloride.
7. Take care not to flush the precipitate with the filtrate into the collecting flask.
8. Remove the precipitate with the filter paper. Place it in a silica crucible and dry it in a hot-air oven at 105°C.
9. Then ignite the dried precipitate in a red-hot flame with the white ash produced, heat the crucible for 30 minutes.

10. After sufficient cooling, weigh the crucible for final weight (If the ash is not white, add few drops of conc. nitric acid after cooling)

CALCULATION

Amount of magnesium in 100 g of soil on a moisture free basis(%)

$$= \frac{48}{222} \times (b - a) \times \frac{250}{50} \times \frac{500}{50} \times \frac{100}{W} \times \frac{100}{100 - W}$$

where,

48 g of Mg is present in one molecule of $Mg_2P_2O_7$ (222 g)

b = Weight of silica crucible + $Mg_2P_2O_7$ precipitate

a = Weight of silica

250 = Total amount of iron oxide filtrate prepared

50   = Amount of iron oxide used

500 = Total amount of hydrochloric extract prepared

50   = Amount of HCl extract used

M = Moisture content of soil

4.23 TOTAL NITROGEN

4.23.1 Principle

Nitrogen in the sample is converted into ammonium sulphate. It is then distilled with excess alkali to liberate ammonia. The liberated ammonia is then collected and quantified by back titration with standard alkali.

Expt. 70   *To determine the percent total nitrogen in soil sample*

MATERIALS REQUIRED

Laboratory glassware

Laboratory equipment

    Balance

    Kjeldahl apparatus

REAGENTS

a. Add 20 g of copper sulphate to 3 g of mercuric oxide and 1g of selenium powder. Grind the mixture to ensure thorough mixing. This can be stored and used when required. Catalyst mixture can be prepared by adding 2 g of this mixture with 40 g of sodium sulphate.

b. Concentrated sulphuric acid.

c. 40% NaOH

Dissolve 40 g of sodium hydroxide in 100 ml of distilled water.

d. Zinc granules

e. Boric acid cum indicator solution

    i. Take 100 ml of distilled water, warm it for few minutes and then add 4 g of boric acid.

ii.   Add 0.5 g of bromocresol green in 100 ml of ethyl alcohol.

iii.   Add 0.1 g of methyl red solution in 100 ml of ethyl alcohol.

To prepare indicator solution, mix solutions (ii) and (iii) in 2:1 ratio. From this, take 5 ml and add with 100 ml of solution (i).

Note   Blue colour of the solution indicates its alkaline nature. To neutralize, add 0.01 N HCl till it turns faint-pink to brown.

PROCEDURE

Estimation of total nitrogen involves two steps:

- Digestion
- Distillation

Digestion

1. Transfer 10 g of soil sample into a clean dry kjeldahl flask.
2. Add 20 g of catalyst mixture.
3. Mix the contents well and leave it for 20 minutes.
4. Add 35 ml of H_2SO_4 to the flask, mix well and leave it for another 15 minutes.
5. Now digest the content over the bunsen burner for about 2 hours.
6. After the expiry time, cool the content and add 100 ml of distilled water.
7. Shake it well and then carefully deliver the supernatant alone into a distillation flask (presence of even a small quantity of soil would interfere with the distillation process)
8. Add a little amount of distilled water to the digested content, shake well and transfer the supernatant to the distillation flask.
9. Do it several times until the complete transfer of the digested material to the distillation flask is ensured.

Distillation

1. Add 100 ml of NaOH solution and a few zinc granules to the distillation flask.
2. Now the content is highly alkaline (To ensure this, it can be tested with litmus paper and if not, still more NaOH can be added.)
3. Pipette out 25 ml of boric acid cum indicator solution in a 500-ml Erlenmeyer flask and keep it at its delivery end, i.e., below the condenser of the distillation flask.
4. Run the distillation process on a hot plate and continue till 150 ml of distillate is collected. Collection of distillate in the flask can be confirmed by the blue colour of the indicator solution.
5. Now remove the flask from the distillation assembly and titrate it against 0.1N HCl.
6. Colour change from blue to light brown to pink is the end point.

7.   Conduct a distilled water blank in similar manner.

8.   Using the following formula TN can be determined.

CALCULATION

Total nitrogen (mg/g) $= \dfrac{(V_1 - V_2) \times N \times 14}{S}$

Percent total nitrogen $= \dfrac{(V_1 - V_2) \times N \times 1.4}{S}$

where,

V_1 = Volume of titrant used against sample (ml)

V_2 = Volume of titrant used against blank (ml)

N = Normality of titrant (0.1)

S = Weight of soil used (g)

4.24 NITRATE NITROGEN

4.24.1 Principle

Soil extract for nitrate determination is obtained by the combined action of copper sulphate and magnesium carbonate. Under dry conditions, when nitrate comes in contact with sulphuric acid produces nitric acid which leads to the nitration of 6th position in 2,4-phenol disulphonic acid.

Expt. 71   *To estimate the amount of nitrate nitrogen in soil sample*

MATERIALS REQUIRED

Laboratory glassware

Laboratory equipment

Chemical balance

Spectrophotometer

Hot-water bath

REAGENTS

a.   Phenol disulphonic acid

Take 25 g of white phenol, dissolve in 150 ml concentrated sulphuric acid. After a while, add 85 ml conc. sulphuric acid. Heat the content in a hot water bath for about 2 hours. After cooling, store the solution in a dark bottle.

b.   Potassium hydroxide solution (12N)

Dissolve 336.5 g of potassium hydroxide in distilled water and make it up to 500 ml.

c.   Standard nitrate solution

To prepare 1 litre nitrate stock solution (100 μg NO_3N/l), about 0.722 g of anhydrous potassium nitrate is used. From this stock solution,

100 ml is pipetted out and diluted to 1000 ml using distilled water. This solution is kept as working solution (10 µg NO_3N/l).

    d.   Extraction reagent

        i.   Dissolve 12.5 g of copper sulphate in distilled water and make up to 100 ml.

        ii. Dissolve 0.6 g of silver sulphate in distilled water and make up to 100 ml.

        To prepare nitrate extraction reagent, add 20 ml of solution (i) with 100 ml of solution (ii) and then make up to 1000 ml.

    e.   Calcium hydroxide

    f.   Magnesium carbonate

Both (e) and (f) should be crushed and dried before use.

PROCEDURE

1. Take 50 g of soil sample in a 500-ml Erlenmeyer flask.
2. Add the following reagents in the same order to the soil sample
   - 250 ml extraction reagent
   - 0.4 g calcium hydroxide and
   - 3. 1 g of magnesium carbonate
3. Remember to shake the content of the flask every time after the addition of reagent.
4. Now filter the content using Whatman No. 50 filter paper, measure the volume and then following phenol disulphonic acid method, nitrate nitrogen can be determined.

CALCULATION

$$\text{Amount of nitrate nitrogen in the sample (mg/g)} = \frac{F \times V}{1000 \times W}$$

where,

$F = $ NO_3N in filtrate (mg/l)

$V = $ total volume of filtrate (ml)

$W = $ weight of soil used (g)

4.25 AVAILABLE NITROGEN

4.25.1 Principle

Potassium permanganate added in excess with a known quantity of soil favours the oxidation of organic matter present in the soil. Nascent oxygen liberated during oxidation combines with NaOH to release ammonia, which is subsequently absorbed by boric acid to form ammonium borate. Available nitrogen is quantified by titrating ammonium borate with standard acid solution.

$$2KMnO_4 \xrightarrow{\text{NaOH}} 3O + K_2O + 2MnO_2$$

$$R.NH_2 + H_2O \rightarrow R.OH + NH_3$$

Expt. 72   ***To determine the available nitrogen in the soil sample***

MATERIALS REQUIRED

Laboratory glassware

Laboratory equipment

Distillation apparatus

REAGENTS

a.   0.32% $KMnO_4$ solution

Dissolve 3.2 g of $KMnO_4$ in 1 litre of distilled water.

b.   2.5% NaOH solution.

Dissolve 25 g of NaOH in 1 litre of distilled water.

c.   Boric acid (2%)

Dissolve 20 g of boric acid in 1 litre distilled water.

d. $N/50\ H_2SO_4$

e.   Double indicator

Add 0.5 g bromocresol green and 0.1 g methyl red in 100 ml of ethyl alcohol.

PROCEDURE

1.   Weigh accurately 20 g of dried soil sample and transfer it into the distillation flask. Add 30 ml of distilled water and 1 ml of liquid paraffin to the soil sample.

2.   The purpose of adding distilled water is to moisten the soil and therefore add accordingly.

3.   Likewise wax avoids frothing of the soil.

4.   With this, add 100 ml each of 2.5% NaOH and 0.32% $KMnO_4$ to the soil in the distillation flask.

5.   To avoid bumping, add pieces of glass beads.

6.   Take 20 ml of 2% boric acid with few drops of double indicator in a 100-ml beaker.

7.   Keep this beaker at the outlet of the condenser.

8.   Now distil the contents of the flask till the distillation process liberates about 30 ml of distillate.

9.   Run the distillation process till the liberated distillate is devoid of ammonia.

10.   This can be confirmed by keeping the moist red litmus paper near the delivery end of the condenser.

11.   The litmus paper will appear blue as long as the ammonia is evolved.

CALCULATION

$$\text{N present in kg/ha} = \frac{0.00028 \times A \times 2 \times 10^{6}}{20}$$

where,

0.00028 = 1 ml of N/50 H_2SO_4

A = Volume of N/50 H_2SO_4 consumed

20 = Weight of the soil in grams

4.26 TOTAL PHOSPHORUS

4.26.1 Principle

Total phosphorous of the soil is precipitated as phosphomolybdate in nitric acid medium. When this combines with stannous chloride, it gives blue colour which can be measured spectrophotometrically at 690 nm.

Expt. 73   *To determine the total phosphorus content of the soil sample*

MATERIALS REQUIRED

Laboratory glassware

Laboratory equipment

Spectrophotometer

Whatnan No. 44 filter paper

REAGENTS

a. Ammonium molybdate solution

   i. Add slowly 62 ml of conc. sulphuric acid to 80 ml of distilled water and leave it for 20 minutes.

   ii. Dissolve 5 g of ammonium molybdate in 35 ml of distilled water. Mix (ii) with (i) and make up to 200 ml.

b. Stannous chloride solution

   Dissolve 0.5 g of stannous chloride in 2 ml of conc. hydrochloric acid. Using distilled water, dilute the acid concentrate to 20 ml. (Do not stock this solution, prepare fresh before use.)

c. Standard phosphate solution

   Dry anhydrous potassium hydrogen phosphate in a hot-air oven at 105°C for 30 minutes. Weigh accurately 4.388 g of this powder, dissolve in distilled water and make up to 1 l. From this solution, take 10 ml and further make up to 1 l. This stock solution contains 1 mg P/l and can be used to prepare standard phosphorus solutions from 0.1 to 1 mg P/l by appropriate dilutions with distilled water.

d. Concentrated nitric acid

e. Conc. perchloric acid

f.   Sulphuric acid solution

Dilute 5 ml of concentrated sulphuric acid with distilled water and make up to 100 ml.

PROCEDURE

1.   Grind the acid-dried soil sample to a fine powder and weigh accurately 0.5 g of it in a round-bottomed flask.
2.   Add 2 ml each of conc. nitric and perchloric acid to the soil sample.
3.   Leave sufficient time interval between the addition of reagents (15 minutes).
4.   Also add a few drops of distilled water.
5.   Now heat the flask in a hot plate till the content becomes dry.
6.   Now add 2 ml of H_2SO_4, heat vigorously for 15 minutes.
7.   Allow it to boil and the time limit should be counted from then on.
8.   After cooling filter the digested content through Whatman No. 44 filter paper.
9.   Make up the filtrate to 250 ml with distilled water.
10. Determine the phosphorus content of this filtrate following the method described for estimation of inorganic phosphorus in effluent in chapter 3.

CALCULATION

$$\text{Total phosphorus (mg/g)} = \frac{P_1 \times V}{1000 \times W}$$

where,

P_1 = PO_4P in digested content (mg/l)

V = Total volume in solution (ml)

W = Weight of soil sample (g)

Expt. 74   *To determine the phosphate content of the soil sample*

MATERIALS REQUIRED

Laboratory glassware

Laboratory equipment

Spectrophotometer

Sulphuric acid (0.002 N)

Dilute 2.8 ml of conc. H_2SO_4 to 1 litre by adding distilled water. Pipette out 20 ml of it   and again dilute it to 1 l with distilled water.

Whatman No.50 filter paper.

REAGENTS

Refer estimation of inorganic phosphorus in water.

PROCEDURE

1. Take 1 g of dried soil sample in a 500-ml flask.

2. Pipette out 200 ml of 0.002 N sulphuric acid and add with the soil sample.

3. Shake vigorously for 30 minutes and then filter the suspension in Whatman filter paper using the method given for inorganic phosphorus (chapter 3) and determine the phosphate content of the filtrate.

CALCULATION

$$PO_4 P \text{ in mg/l} = \frac{P_2 \times V}{1000 \times W}$$

where,

$P_2 = PO_4 \, P$ content in filtrate (mg/l)

V = Total volume of the filtrate (ml)

W = Weight of soil sample in g

4.27 AVAILABLE PHOSPHORUS (OLSEN'S METHOD)

4.27.1 Principle

Sodium bicarbonate extracts the phosphorus content in the soil. Addition of ascorbic acid added to the extracted phosphorus gives blue colour whose intensity can be measured in a calorimeter.

Expt. 75　　*To estimate the available phosphorus in soil sample by Olsen's method*

MATERIALS REQUIRED

Laboratory glassware

Laboratory equipment

　　Mechanical shaker

　　Photoelectric colorimeter

Filter paper (No. 40)

REAGENTS

a. $NaHCO_3$ (0.5 M): Dissolve 42 g of $NaHCO_3$ in distilled water and make up to 1 litre. Maintains the pH at 8.5.

b. Activated carbon (free of phosphorus)

c. Ascorbic acid

d. Reagent A

　　i. 12 g of ammonium molybdate dissolved in 250 ml of distilled water

　　ii. 0.291 g of antimony potassium tartrate in 100 ml distilled water

    iii. H_2SO_4 (5 N): Dilute 138 ml of conc. H_2SO_4 in distilled water and make up to 1000 ml

    Mix (i), (ii) and (iii) thoroughly and make up to 2000 ml.

e.   Reagent B

   1.056 g of ascorbic acid in 200 ml of reagent A (use fresh).

PROCEDURE

1. Take 5 g dried soil sample in a 250-ml conical flask.
2. Add 50 ml of 0.5N $NaHCO_3$ and a pinch of activated carbon to facilitate the extraction free of colour.
3. Now place the flask in a mechanical shaker.
4. Shake vigorously for about 30 minutes.
5. Then filter the content through Whatman No. 40 filter paper, in a volumetric flask.
6. Add 4 ml of freshly prepared reagent 'B' and make up to 25 ml with distilled water.
7. Mix well and wait for few minutes until the content develops blue colour.
8. Evaluate the colour intensity in photoelectric calorimeter at 660 nm.
9. For calibration, run a blank. (1 ml of distilled water with 4 ml of reagent B).
10. Plot out the standard graph and find out the phosphorus content in mg/l and the available phosphorus can be calculated using the following formula.

CALCULATION

Weight of soil taken   = 5 g

Volume of $NaHCO_3$ = 50 ml

Volume of extract used   = 50 ml

Amount of phosphorus present in the sample = x mg/ml = $x/10^6$ g/ml

Amount of phosphorus in 25 ml of sample   = $x/10^6 \times 25$ g
   (Prepared from 5 ml of extractant)

Amount of phosphorus in 50 ml of extractant = $x/10^6 \times 25 \times 50$ g

4.28 AVAILABLE PHOSPORUS (BRAY AND KURTZ NO.1 METHOD)

4.28.1 Principle

Addition of hydrochloric acid and ammonium fluoride to the sample extracts the acid-soluble phosporus bound to aluminium, iron and calcium. Extracted phosphate can be determined calorimetrically using phosphomolybdenum blue and ascorbic acid as reducing agents.

$$3NH_4F + 3HF + HIPO_4 \rightarrow H_3PO_4 + (NH_4)\,HIF_6$$

$$3NH_4F + 3HF + FePO_4 \rightarrow H_3PO_4 + (NH_4)\,FeF_6$$

Expt. 76   *To determine the amount of available phosphorus in soil sample by Bray and Kurtz No. 1 method.*

MATERIALS REQUIRED

Laboratory glassware

Laboratory equipment

   Spectrophotometer

   Reciprocal mechanical shaker

REAGENTS

a. NH_4F solution (1N)

   Dissolve 37 g of ammonium fluoride in one litre distilled water.

b.   Hydrochloric acid solution (0.5N)

   Dilute 20.2 ml of concentrated hydrochloric acid to 500 ml using distilled water.

c.   Bray No.1 extractant

   Add 15 ml of solution (a) to 25 ml of solution (b). Make up to 500 ml using distilled water.

d.   Reagent (B)

   Refer Olsen's method

e.   Phosphate standard solution

   Weigh approximately 0.5 g of potassium dihydrogen phosphate, air-dry at 40°C and from this amount, take accurately 0.2195 g and dissolve in 400 ml distilled water. To this add, 25 ml of 1N H_2SO_4 and make up to 1000 ml. One ml of this solution contains 50 mg of P.

f.   Phosphate working solution

   Transfer 100 ml of the stock solution into a 1000-ml volumetric flask. Make up to the mark using distilled water.

PROCEDURE

1.   Prepare several dilutions of phosphate standard solution into a series of beakers with their concentrations ranging from 0.1 ppm to 10 ppm at 0.1 ppm interval.

2.   To each beaker add 5 ml of Bray No.1 extractant and 4 ml of reagent B.

3.   Make up the volume of each beaker to 25 ml using distilled water.

4.   Similarly take 5 g of soil in a shaking bottle.

5.   To the bottle, add 50 ml of Bray No.1 extractant and then shake it in a reciprocatory mechanical shaker.

6.   After one minute, allow the contents of the bottle to pass through Whatman No. 40 filter paper.

7.   Siphon out 5 ml of the filtrate into a 25-ml volumetric flask.

8. To this add 5 ml of Olsen's reagent and 4 ml of Bray No.1 extractant and mix well.

9. Make up the volume to 25 ml using distilled water.

10. Measure the transmittance value of standards, sample and blank using calorimeter.

11. Draw a standard curve by plotting the concentration of standards against their transmittance value.

12. From the calibration curve measure the transmittance value of the extract.

13. Evaluate the phosphate content in the soil using the formula.

CALCULATION

Refer Olsen's method

4.29 AVAILABLE POTASSIUM

Available potassium refers to the readily exchangeable and soluble potassium. In general Indian soils constitute about 0.05–3.5% of potash out of which 2% is in available form.

4.29.1 Principle

With the help of ammonium acetate solution, available potassium can be extracted from the soil and then the amount can be determined by flame photometer.

Expt. 77 *To determine the amount of available potassium in soil sample*

MATERIALS REQUIRED

Laboratory glassware

Laboratory equipment

Centrifuge

Flame photometer

REAGENTS

a. Ammonium acetate solution (IN)

Weigh 154 g ammonium acetate, dissolve in 1.8 litre distilled water. Make up to 2 litres using distilled water. Adjust the pH to 7.0.

b. Standard potassium chloride solution

Weigh 2 g of potassium chloride, dry at 60°C for 1 hour and from this take 1.908 g and dissolve in one litre distilled water. One ml of this solution consists of 1 mg of KCl.

c. Potassium chloride working solution

Transfer 100 ml of the stock solution into a 1000-ml volumetric flask and make up to the marked level. One ml of this solution consists of 0.1 mg KCl.

PROCEDURE

Preparation of ammonium acetate extract

Method I

1. Transfer 5 g soil to an Erlenmeyer flask containing 25 ml of neutral normal ammonium acetate.
2. Place the flask on a rotatory shaker and rotate it for 5 minutes.
3. Allow the content of the flask to pass through Whatman filter paper No. 1 and collect the filtrate (25 ml).

Method II

1. Transfer 5 g soil into a centrifuge tube containing 25 ml of ammonium acetate solution.
2. Rotate it at 2000 rpm for 10 minutes.
3. Collect the supernatant in a 100-ml volumetric flask.
4. Repeat this procedure three times.
5. Every time use 25 ml ammonium acetate solution and collect the filtrate up to 100 ml.

Determination of available K in ammonium acetate extract solution

1. Go through the operation manual of the flame photometer. After setting the potassium filter, initiate the compressor and light the burner of the instrument. Set the air pressure at 5 lbs and gas feeder to produce sharp flame cones.
2. Using highest potassium standard solution, adjust the instrument to show full reading. Similarly using extract solution, set the zero reading.
3. Begin from the lowest to the highest standard and record their emission value.
4. Draw a standard curve by plotting the reading of standard against their concentration.
5. From the calibration curve, evaluate the concentration of potassium in the solution. Let this be a.

CALCULATION

Ammonium acetate extract prepared by Method I

$$\text{Amount of available K (kg/ha)} = \frac{a \times V}{W} \times 2.24$$

where,

a = concentration of K in the unknown sample read from calibration curve.

V = Volume of extractant (25 ml)

W = Weight of soil (g)

Ammonium acetate extract prepared by Method II

$$\text{Amount of available K (kg/ha)} = \frac{a \times V}{W} \times 2.24$$

Here, $V = 100$ ml

4.30 ORGANIC CARBON

4.30.1 Principle

Even in small quantities, organic carbon influences the physical and chemical properties of soil. It supplies energy and body-building constituents for soil microorganisms. In other words it gives "life" to the soil and determines its productivity. The source of organic matter in the soil is the living and dead organic materials of plant and animal origin. Living and dead organic content of the soil depends on factors like organic matter supply; physicochemical characteristics of soil; climatic condition and soil microbial activity. Chromic acid in combination with sulphuric acid favours the oxidation of organic matter in the sample. During the oxidation, potassium dichromate reacts with H_2SO_4 to give nascent oxygen, which ultimately reacts with carbon to form CO_2. Excess of chromic acid is quantified by back titration with ferrous ammonium sulphate using diphenylamine indicator.

$$2K_2Cr_2O_7 + 8H_2SO_4 \rightarrow 2K_2SO_4 + 2Cr_2(SO_4)_3 + 6(O)$$

| Potassium dichromate | Sulphuric acid | Potassium sulphate | Chromium sulphate | Nascent oxygen |

$$3C + 6(O) \rightarrow 3CO_2$$

| Carbon | Nascent oxygen | Carbon dioxide |

Expt. 78 *To estimate the organic carbon in the soil sample*

MATERIALS REQUIRED

Laboratory glassware

REAGENTS

Potassium dichromate (0.1N)

Diphenylamine indicator

Ferrous ammonium sulphate (FAS) (0.5N)

Concentrated H_2SO_4

Orthophosphoric acid (85%)

PROCEDURE

1. Take exactly 0.5 g of dried homogenized soil sample in a 500-ml conical flask.

2. Pour 10 ml of 1N $K_2Cr_2O_7$ into the flask.

3. Shake well and then add 20 ml of conc. H_2SO_4.

4. Allow the contents to mix well by swirling the flask gently for 3 minutes.

5. Place the flask in an aluminium tray filled with water to resist intense heat.

6. After 30 minutes, dilute the flask content with 200 ml distilled water.

7. With this, add 10 ml of phosphoric acid and 1 ml of diphenylamine indicator.

8. Now titrate the solution with 0.5N FAS. Colour change will be dull green at the initial stage, turbid blue in the middle and bright green colour at the end.

9. However the end point is very sharp.

10. Run a blank titration using distilled water.

11. Record the volume of 0.5N FAS consumed and determine the organic content of the soil.

CALCULATION

$$\text{Weight of the soil taken} = 0.5\,\text{g}$$
$$\text{Volume of potassium dichromate} = 10\,\text{ml}$$
$$\text{Volume of FAS used in blank} = a$$
$$\text{Volume of FAS used for sample} = b$$

a ml ferrous sulphate reduces 10 ml of potassium dichromate

Amount of potassium dichromate reduced by b ml of FAS $= b/a \times 10$

$$\left.\begin{array}{l}\text{Amount of potassium dichromate utilized}\\ \text{for oxidation of organic matter in 0.5 g soil}\end{array}\right\} = 10 - (10 - b/a)\,\text{ml}$$

$$= 10 - (10 \times b/a) \times 0.003\,\text{g}$$

0.003 g of carbon is present in 1 ml of 1N $K_2Cr_2O_7$.

$$\left.\begin{array}{l}\text{Amount of potassium dichromate}\\ \text{utilized for oxidation of organic}\\ \text{matter in 100 g soil}\end{array}\right\} = 10 - (10 \times b/a) \times 0.003 \times \dfrac{100}{0.5}$$

$$\text{Organic matter} = \text{organic carbon} \times 1.724$$

4.31 AVAILABLE MICRONUTRIENTS

4.31.1 Principle

Micronutrients in the soil such as copper, manganese, zinc and iron can be determined using atomic absorption spectrophotometer. Diethylene triamine pentaacetic acid acts as a chelating agent and favour the extraction of these micronutrients from the soil, and the extraction is performed under alkaline conditions. At pH 7.3, maximum amount of TEA buffer gets protonated to exchange Ca^{2+} and Mg^{2+} from soil. This elevates the concentration of Ca^{2+} ion and automatically inhibits the dissolution of calcium carbonate in calcareous soils.

Expt. 79   *To determine the amount of available micronutrients in soil sample*

MATERIALS REQUIRED

Laboratory glassware

Laboratory equipment

    Balance

    Reciprocating shaker

    AAS

    Hollow cathode lamps of the micronutrient to be analysed.

REAGENTS

a.   Extraction solution

Dissolve 1.47 g of calcium chloride in 500 ml double-distilled water. To this, add 13.3 ml of Triethanolamine (TEA) and 1.967 g of DTPA. Mix well. Using 1N hydrochloric acid adjust the pH to 7.3 and then make up the volume to 1000 ml.

b.   Standard solution—iron

Dissolve 0.4964 g of ferrous sulphate in 1000 ml distilled water. One ml of this solution contains of 0.1 mg of iron.

c.   Iron working solution

Prepare several dilutions of iron standard solution into a series of 100-ml volumetric flasks with their concentrations ranging from 1 ml to 10 ml at 2-ml interval. 1 ml of this solution contains 0.001 mg of iron. Make up the content of each flask to 100 ml using DTPA extracting solution.

d.   Standard solution—manganese

Dissolve 0.3076 g of manganese sulphate in 1000 ml distilled water. One ml of this solution contains 0.1 mg of manganese.

e.   Manganese working solution

Prepare several dilutions of Mn standard solution into a series of 100-ml volumetric flasks with their concentrations ranging from 1 ml to 10 ml at 2-ml interval. One ml of this solution contains 0.001 mg of manganese. Make up the content of each flask to 100 ml using DTPA extracting solution.

f.   Standard solution—copper

Dissolve 0.3928 g of copper sulphate in 1000 ml distilled water. One ml of this solution contains 0.1 mg of copper.

g.   Copper working solution

Prepare several dilutions of copper standard solution into a series of 100-ml volumetric flasks with their concentrations ranging from 1 ml to 10 ml at 2-ml interval. One ml of this solution contains 0.001 mg of copper. Make up the content of each flask to 100 ml using DTPA extracting solution.

h. Standard solution—zinc

Dissolve 0.4398 g of zinc sulphate in 1000 ml distilled water. One ml of this solution contains 0.1 mg zinc.

i. Zinc working solution

Prepare several dilutions of zinc standard solution into a series of 100-ml volumetric flasks with their concentration ranging from 1 ml to 10 ml at 2-ml interval. One ml of this solution contains 0.001 mg of zinc. Make up the content of each flask to 100 ml using DTPA extracting solution.

Procedure (Common for all the micronutrients)

1. Take a 100-ml conical flask, wash it thoroughly with acidified water and then with double-distilled water.

2. Transfer 10 g of air-dried soil into the conical flask.

3. To the flask, add 20 ml of DTPA extracting solution. Check whether the pH of the solution is exactly 7.3 and if not adjust the pH and then use.

4. Place the flask in a reciprocatory shaker for two hours. Set the temperature of the shaker to 25°C.

5. Following this, allow the content of the flask to pass through Whatman No.1 filter paper.

6. Use this filtrate to determine the amount of micronutrients.

7. Using blank, set the zero in the instrument.

8. Measure absorbance of standards of the element to be analysed and standardize the apparatus.

9. Now measure the absorbance of soil extract. Let this be (a).

10. Perform dilution if the concentration of the sample goes beyond the already set standardized range of the instrument.

11. Measure the absorbance of sample and standards in spectrophotometer at 42 nm.

12. Draw a standard graph by plotting the absorbance value of standards against their concentrations.

13. From the calibration curve, evaluate the amount of micronutrient in the sample. Let this value be a.

Calculation

Amount of micronutrient in the soil sample (mg/l) = $a \times 2$ mg/kg

where,

a = concentration of micronutrients read from the calibration curve.

2 = Total number of dilution.

4.32 AVAILABLE BORON

4.32.1 Principle

Available boron in the soil is water-soluble and therefore it can be easily extracted by hot-water extraction method. Boron can be analysed by several methods such as calorimetric methods, inductively coupled plasma and atomic emission spectrometry out of which calorimetric method is the most simple and reliable one.

Expt. 80 *To determine the amount of available boron in soil sample*

MATERIALS REQUIRED

Laboratory containers

Since available boron is water-soluble, borosilicate glass should be strictly avoided.

Laboratory equipment

Analytical balance

Calorimeter

REAGENTS

a. Buffer Solution

To 500 ml double-distilled water add 250 g ammonium acetate and stir well. Add slowly glacial acetic acid until the pH is adjusted to 10.5.

b. EDTA reagent (0.025M)

Dissolve 9.3 g EDTA in 1000 ml double-distilled water.

c. Azomethine-H

To 100 ml double-distilled water, add 0.45 g azomethine-H and 1.0 g ascorbic acid. Stir well. Heat the solution in a hot water bath (30°C) until the chemicals added are completely dissolved. Do not store the solution beyond 7 days.

d. Standard boric acid solution

To 1000 ml double-distilled water add 0.571 g of boric acid. Mix well. One ml of this stock solution contains 0.1 mg of boron. Transfer 5 ml of this stock solution into a 100-ml volumetric flask and make up to the mark using distilled water. This is the working boric acid solution. One ml of this solution consists of 0.005 mg B.

PROCEDURE

1. Transfer 25 g of soil into a beaker containing 50 ml of double-distilled water.

2. To this add, 0.4 g of activated charcoal. For complete dissolution, boil the content of the beaker for about 5 minutes.

3. Then allow the content to pass through a Whatman No. 42 filter paper and collect the filtrate.

4. Transfer 5 ml of this filtrate in a separate flask. To this add 2 ml each of buffer reagent, EDTA solution and azomethine H solution.

5. Allow the filtrate to stand for one hour.

6. Meanwhile prepare several dilutions of boric acid working solution into a series of beakers with their concentration ranging from 0.25 to 5 ml at 0.25-ml interval.

7. Similar to that of sample add 2 ml each of buffer reagent, EDTA solution and azomethine H solution to all dilutions.

8. Allow the beakers to stand for 30 minutes. Following this, make up each beaker content to 25 ml.

CALCULATION

Amount of boron in the soil $= a \times 10$ µg/g

where,

10 = Total no. of dilution

a = concentration of boron read from the calibration curve

4.33 AVAILABLE MOLYBDENUM

4.33.1 Principle

Added Brigg's reagent to the sample under suitable pH gets protonated to form oxalate ions. These ions can readily exchange with the molybdenum ions absorbed onto the colloids and clays of soil. Since the oxalate acid form a stable complex with molybdenum, reverse exchange is inhibited.

Expt. 81   *To determine the amount of available molybdenum in soil sample*

MATERIALS REQUIRED

Laboratory glassware

Laboratory equipment

Analytical balance

Electric Shaker

Multiple Furnace

Spectrophotometer

Water bath

REAGENTS

a. Conc. H_2SO_4

b. Thiourea (10%)

c. Ascorbic acid (5%)

d. Tartaric acid (50%)

e. H_2O_2 (30%)

f. Potassium iodide

g. Amyl acetate

i. Extracting solution

To 500 ml double-distilled water, 24.9 g ammonium oxalate and 12.6 g oxalic acid. Mix well. Slowly add oxalic acid to this mixture so that its pH is adjusted to 3.3. Finally make up the solution to 1 l.

j. Ferrous ammonium sulphate solution

Dissolve 63 g ferrous ammonium sulphate in 1 l double-distilled water.

k. Dithiol solution

Dissolve 1 g of analytical grade dithiol in 199 ml of 10% sodium hydroxide solution. Heat it in a hot plate (51°C) for about 15 minutes. To this add 1.8 ml of thioglycolic acid and mix well. Filter the solution if it is turbid.

l. Molybdenum stock solution

Dissolve 0.150 g of analytical grade molybdenum trioxide in 100 ml of 0.1N NaOH solution. Make up the solution to 900 ml using distilled water. Adjust the pH of the solution to acidic range by adding little amount of HCl solution. Finally make upto 1000 ml using distilled water. One ml of this solution contains of 0.1 mg of molybdenum.

m. Molybdenum working solution

Transfer 10 ml of this stock solution into a separate flask and make up to 1 litre using distilled water. One ml of this solution contains 0.001 mg Mo.

PROCEDURE

1. Prepare several dilutions of working molybdenum solution in a series of beakers with their concentration ranging from 0.5 to 10 ml at 2-ml interval.

2. Make up the content of each beaker to 50 ml using distilled water.

3. To each beaker, add 0.25 ml ferrous ammonium sulphate, 20 ml distilled water and 2 ml potassium iodide solution.

4. Liberated iodine produces colour in the solution. Clear out the colour by adding ascorbic acid in drops.

5. To each beaker, add 0.25 ml of 50% tartaric acid solution and 2-ml of 10% thiourea solution. Stir well the beaker content after every addition.

6. Then add five drops of dithiol solution, swirl the beakers to favour even mixing. Let them to stand for half an hour.

7. Following this, add 10 ml of isoamylacetate, stir well and let them to stand for another one hour.

8. After the layers have been separated, decant the upper aqueous phase and collect the lower organic phase into a series of centrifuge tubes.

9. Centrifuge at 2000 rpm for 15 minutes and then measure their absorbance at 680 nm.

Soil extraction

1. Transfer 25 g of soil in a conical flask containing 250 ml ammonium oxalate solution.

2. Place the flask in a shaker and run it for about 8 hours.

3. Following this, keep it undisturbed overnight.

4. Now allow the suspension to pass through Whatman No.42 filter paper and collect the filtrate in a 250-ml silica crucible.

5. Place the crucible containing 200 ml filtrate in a hot-water bath and evaporate to dryness.

6. To eliminate the oxalate and organic matter in the residue, ignite it in the muffle furnace for about 5 hours.

7. Keep it undisturbed overnight and then digest the crucible content using nitric acid and perchloric acid digestion method (4:1). Continue to digest till the content is evaporated to dryness. Then add 10 ml of sulphuric acid (1N) and 1 ml of hydrogen peroxide and continue to digest till it is evaporated to dryness.

8. To the residue formed, add 10 ml of hydrochloric acid and filter it through Whatman No. 42 filter paper. Repeat the filtration process by washing the filter paper with 0.1 N hydrochloric acid. Collect the filtrate in a separate beaker (100 ml).

9. To the filtrate, add 2 ml of KI and decolourise the solution by adding ascorbic acid in drops.

10. Then similar to that of standards add the following reagents to the beaker.

- 0.25 ml of 50% tartaric acid solution
- 2 ml of 10% thiourea solution
- 5 drops of dithiol solution
- 10 ml of isoamyl acetate

While adding the reagents, carefully follow the steps as given in the procedure for standards.

11. Measure the absorbance value in spectrophotometer at 680 nm.

12. Draw a standard graph by plotting the concentration of standards against their absorbance value.

13. From the calibration curve evaluate the amount of molybdenum in the soil extract. Let this be a.

CALCULATION

Amount of molybdenum in the soil (mg/kg) = $a \times 0.5$

where,

   0.5 = Total number of dilution

   a = Concentration of molybdenum as read from the standard graph.

4.34 HEAVY METALS

Heavy metals like cadmium, chromium, nickel and lead are the common soil pollutants. They reach soil environment from a variety of sources such as smelting activities, sewage sludge and other industrial wastes.

4.34.1 Principle

For the determination of total heavy metal content, aqua regia–hydrogen fluoride digestion method is highly preferable.

Expt. 82   *To evaluate the total heavy metal content of the soil sample*

MATERIALS REQUIRED

    Laboratory glassware

    Laboratory equipment

        Hot plate

        Atomic absorption spectrophotometer

REAGENTS

Nitric acid (1:1)

30% Hydrogen peroxide

Concentrated hydrochloric acid

PROCEDURE

1. Transfer 2 g of air-dried soil in a beaker containing 10 ml of 1:1 nitric acid solution.

2. Close the beaker with a lid, and reflux it at 95°C for 15 minutes in a hot plate.

3. Allow the beaker content to cool and then add 5 ml of concentrated nitric acid.

4. Reflux the content again at 95°C for 15 minutes.

5. Adjust the lid of the beaker to cover the beaker partially. This allows the fumes to escape and reduces the volume of the solution.

6. As soon as the solution volume is reduced to 5 ml, remove it from the hot plate and allow it to cool.

7. Now add 2 ml of double-distilled water and 3 ml of 30% hydrogen peroxide.

8. Again close the beaker and heat it till the effervescence comes out.

9. Then add 1 ml of 30% H_2O_2 and heat again. Continue to repeat this process until the effervescence stops.

10. Now heat it gently after adding 5 ml of conc. hydrochloric acid and 10 ml of double-distilled water.

11. After 15 minutes of heating, cool the sample and filter it through Whatman No. 42 filter paper. Collect the filtrate and make up to 50 ml using double-distilled water.

12. Following the instruction manual, operate the instrument and evaluate the total metal content in the sample.

4.35 AVAILABLE HEAVY METALS

4.35.1 Principle

It should be noted that not all pollutants present in the soil are available to plants and the evaluation of total metal content would be misinterpreted by the analyst. Therefore a procedure that could evaluate the available fraction of these metals is very essential and as per it, DTPA extraction method is highly preferable for cadmium and nickel. EDTA extraction method can be used for cadmium, copper, lead and zinc.

Expt. 83   *To evaluate the available heavy metal content of the soil sample*

MATERIALS REQUIRED

Laboratory glassware

Laboratory equipment

Rotatory shaker

AAS

REAGENTS

a. DTPA extraction solution

This solution is a mixture of DTPA, calcium chloride and triethanol amine. To one litre double-distilled water, add 74.6 g AR grade TEA, 9.835 g DTPA and 7.35 g calcium chloride. Mix well. Make up to 5 l using distilled water. Set the pH of the solution to 7.3 using 1:1 hydrochloric acid solution.

b. EDTA extraction solution

To one litre double-distilled water add 93.05 g of EDTA powder. Stir well and adjust the pH to 7.3 using ammonium hydroxide solution. Make up the solution to 5 l using distilled water.

PROCEDURE

DTPA EXTRACTION METHOD

1. Transfer 5 g of air-dried soil in a flask containing 25 ml DTPA extraction solution.

2. Place the flask in a rotatory shaker and rotate it for 2 hours at a rate of 120 cycles/min.

3. Following the shaking, remove the flask and allow its content to pass through Whatman No. 42 filter paper.

4. See to it that the filtrate is clear and if not refilter the filtrate.

5. Use this filtrate to determine the available metals in AAS.

EDTA EXTRACTION METHOD

1. Transfer 5 g of air-dried soil into a flask containing 25 ml of ethylene diamine tetra-acetate.

2. Place the flask in a rotatory shaker and shake it at a rate of 120 cycles/min for 2 hours.

3. Following this, remove the flask and allow its contents to pass through Whatman No. 42 filter paper.

4. Use this filtrate to determine the available metals in AAS.

BIOLOGICAL PARAMETERS

Soil provides a suitable habitat for a wide variety of microorganisms and it is the physical and chemical properties of soil that decide the type of microbial community in it. Therefore isolation and identification of these microorganisms give a clear idea about the soil status.

4.36 DETERMINATION OF SOIL MICROORGANISMS

Soil environment supports diverse variety of microorganisms such as bacteria, actinomycetes and fungi. To evaluate these different groups, a single technique is not sufficient because a single medium cannot meet the nutritional requirements of variety of microorganisms. Therefore a common method that determines relative number of bacteria, actinomycetes and fungi is adopted. However the media used to grow them vary.

Expt. 84   *To identify the microorganisms in the soil sample*

MATERIALS REQUIRED

Laboratory glassware

    Pipettes

    Petri plates

Glass marker

Laboratory equipment

    Autoclave

    Incubator

    Inoculating loop

    Bunsen burner

REAGENTS

a.   Nutrient Agar medium

Peptone	0.5 g
NaCl	0.5 g
Agar	1.5 g
Beef extract	0.25 g
Distilled water	100 ml

b.   Glucose Yeast Extract Agar

Glycerol	0.5 ml
Yeast extract	0.2 g
Dipotassium phosphate	0.1 g
Agar	1.5 g
Distilled water	100 ml

c.   Sabroud agar

Peptone	1.0 g
Dextrose	4.0 g
Agar	1.5 g

Add 10 mg/1 ml of aureomycin to the sterile cooled and molten medium.

PROCEDURE

1.   Sterilize 6 test tubes containing 9 ml of distilled water in each test tube.

2.   Take 1g of soil, dissolve it in 100 ml of sterile distilled water. This is designated as 10^{-2} dilution.

3.   Place the soil suspension in the water bath. Shake for about 2 hours.

4.   Carefully transfer 1ml of 10^{-2} dilution into test tube No. 1 using sterilized pipette (10^{-3}).

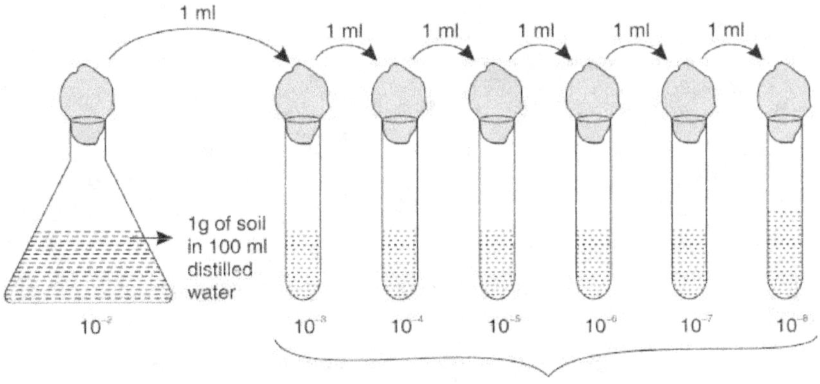

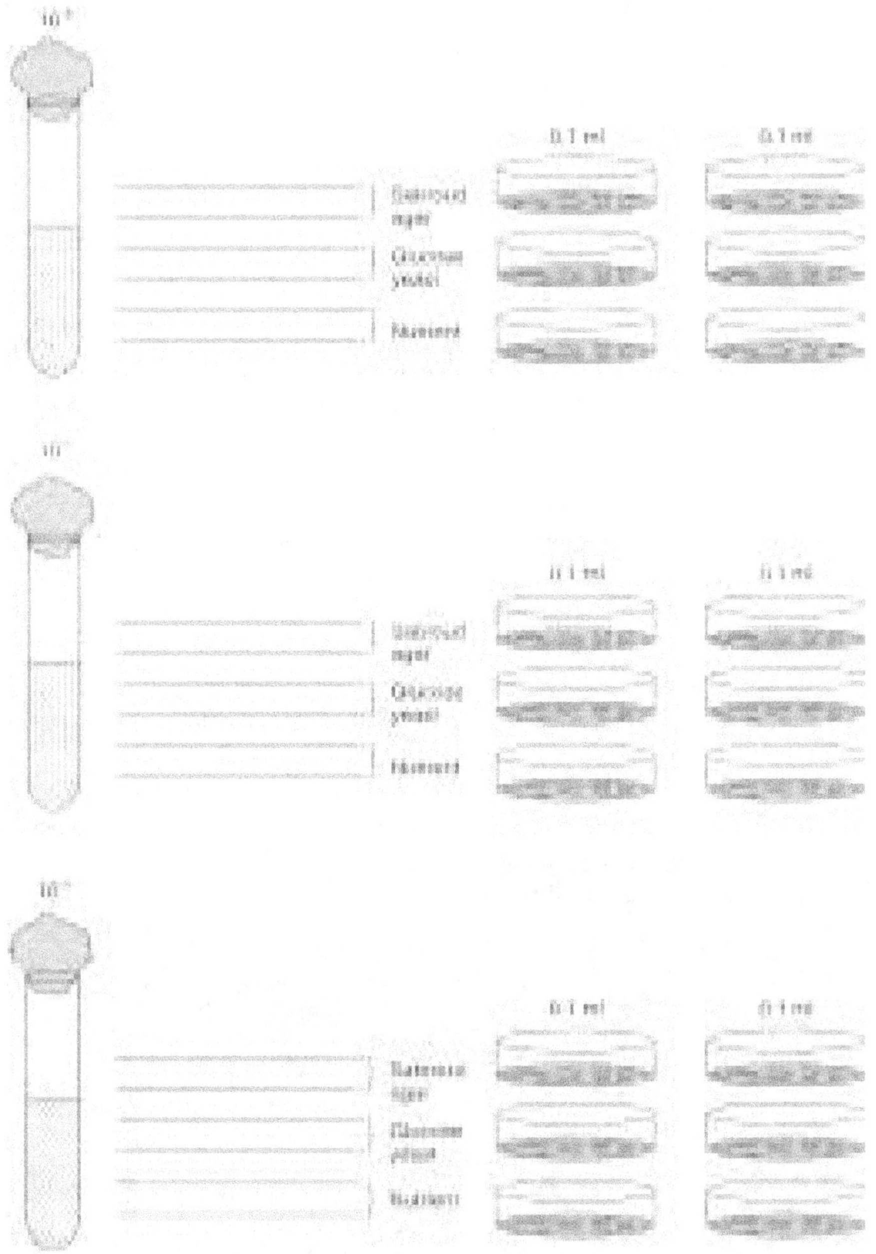

Figure 4.3 Isolation of soil microorganisms

5. Roll the test tube between the palms so that the soil is completely mixed in the medium.

6. Discard the pipette in the disinfectant tray.

7. Using sterile pipette, pipette out 1ml from the test tube No. 1 and transfer it into the test tube No. 2 (10^{-4})

8. Repeat step (5) and (6).

9. Following the procedure described above, transfer the sample from test tube No. 2 to test tube No.3. Continue to perform till the test tube No. 6 and ensure that the dilution in the test tube No. 6 is 10^{-8}.

10. For each test tube, take care to mix the suspension well.

11. Choose three dilutions (10^{-6}, 10^{-7}, 10^{-8}) to be transferred in petri plates.

12. For each dilution, use 6 petri plates, 2 petri plates each for bacteria, fungi and actinomycetes.

13. Use different pipettes for each dilution.

14. For convenience sake label the petri plates indicating the microbial group (bacteria, fungi and actinomycetes) and dilution.

15. Using the sterile pipettes, transfer required volume of the sample (0.1,1.0 ml) to the appropriate petri plates.

16. Rotate the petri plate in clockwise and anticlockwise direction.

17. Pour the molten nutrient medium into the labelled plates (Sabroud agar medium; glucose yeast agar medium; nutrient agar medium).

18. Do not pour the medium at temperatures greater than 45°C which would kill the microbes in the sample.

19. Allow it to solidify and keep in inverted position inside the incubator.

4.37 SCREENING OF ANTIBIOTIC-PRODUCING MICROORGANISMS FROM SOIL ENVIRONMENT

Antibiotics are the chemical substances secreted by specific microorganisms to inhibit the growth of wide range of organisms. *Streptomyces, Bacillus, Penicillium* and *Cephalosporium* are the four different microbial groups commonly cultured for antibiotic production. Till date intensive research is going on to screen the soil from different environments for the evaluation of new antibiotic-producing microorganisms. On the other hand, increasing the potency of already existing antibiotics by structural modification is being tried. The purpose behind is to maximize its efficiency to cover wide array of pathogens and to minimize the side effects in hosts.

Expt. 85   *To isolate antibiotic-producing microorganisms in soil sample*

MATERIALS REQUIRED

Laboratory glassware

Beaker

Test tubes

Pipettes

Petri plates

Glass marker pencil

Laboratory equipment

Autoclave

Incubator

Hot-air oven

Bunsen burner

Thermometer

Inoculating needle

REAGENTS

Trypticase soy agar

Trypticase	1.5 g
Phytane	0.5 g
Sodium chloride	0.5 g
Agar	1.5 g
Distilled water	100 ml

To prepare soil suspension take 0.1 g of finely sieved soil sample in 50 ml sterile distilled water. Shake well (1:500 dilution).

PROCEDURE

1. Take three test tubes 1, 2 and 3 with 5 ml of sterile distilled water.

2. Mix well the soil suspension for about 5 minutes.

3. Carefully transfer 5 ml of the 1:500 dilution to the test tube 1 with 5 ml distilled water (1:1000 dilution)

4. From test tube 1 pipette out 5 ml and transfer into the test tube 2 (1:2000 dilution)

5. Mix well and from test tube 2, pipette out 5 ml and transfer into test tube 3 (1:4000 dilution).

6. Use different pipettes for different test tubes.

7. Pipette 1 ml each from dilutions 1:1000; 1:2000; 1:4000 into the respective petri plates.

8. Rotate the petri plates in clockwise and anticlockwise direction. Then pour the molten trypticase soy agar into the petri plates.

9. While pouring, see to it that the temperature of the agar medium does not exceed 45°C.

10. Allow the plates to solidify.

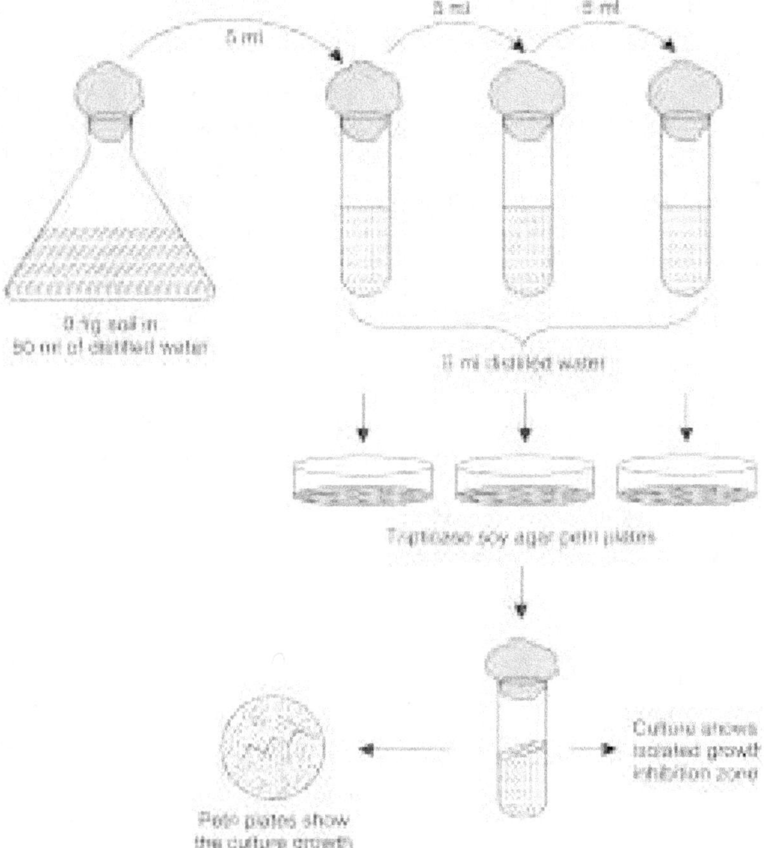

Figure 4.4   Screening of antibiotic-producing microorganisms

11. Incubate them at 25°C for four days.

12. Following the incubation period, isolate the colony with zone of growth, inhibition and streak into trypticase agar slant tubes.

13. Mark the test tubes accordingly.

14. Using this as stock cultures, determine their antimicrobial activity by streaking them onto trypticase soy agar plates.

15. A single line streak is sufficient. Incubate the plates for 5 days at 25°C.

16. Following the incubation period, streak the culture to be tested in perpendicular direction to the antibiotic-producing isolate.

17. While streaking, do not touch the antibiotic-producing isolate but see to it that the test cultures are quite close to the isolate.

18. Incubate at 37°C for 24 hours and check for growth inhibition zone.

4.38 AMMONIFICATION

4.38.1 Principle

Ammonification is the initial step in the nitrogen transformation process in which the nitrogenous compounds are broken down to release/liberate ammonia. Soil microorganisms belonging to the genera *Bacillus*, *Clostridium* and *Streptomyces* are involved in this process. They secrete extracellular proteolytic enzymes, which hydrolyse the proteins to amino acids. Removal of amino group from the amino acids (deamination) releases ammonia. In this experiment, peptone broth is supplemented with an organic nitrogen source, which can be utilized by the microorganisms. Occurrence of ammonification process can be verified by the addition of Nessler's reagent, which in the presence of ammonia turns yellow. The intensity of colour formation depends on the amount of ammonia produced.

Expt. 86   *To determine the ammonification process*

MATERIALS REQUIRED

Laboratory glassware

Bunsen burner

Inoculating loop

Glass marking pencil

Laboratory equipment

Autoclave

Incubator

REAGENTS

a. Peptone Broth

Distilled water	100 ml
Beef extract	0.25 g
Peptone	0.5 g
Sodium chloride	0.5 g

b. Nessler's reagent

To prepare Nessler's reagent, dissolve potassium iodide in 35 ml double-distilled water. Then add 400 ml of 50% aqueous potassium hydroxide. Dilute to 1000 ml with ammonia-free distilled water. Leave it undisturbed for seven days. Decant the supernatant and then store it in tightly capped amber bottles.

c. Potassium iodide 50 g

d. Double-distilled water 35 ml

PROCEDURE

1. Pour molten nutrient broth medium in previously sterilized test tubes.

2. Carefully transfer a loopful culture into the test tubes under aseptic conditions.

3.   Maintain a control tube.

4.   Incubate the tubes at 25°C for seven days.

5.   Following the incubation period, add few drops of Nessler's reagent.

6.   Check for the appearance of yellow colour.

4.39 NITRIFICATION

Under oxic conditions, ammonia liberated by ammonification process will be oxidized to nitrite (Ist step) and then to nitrates (2nd step). In the above-said process, two different genera are involved, *Nitrosomonas* spp. (Ist step) and *Nitrobacter* spp. (2nd step) and the total process is termed as nitrification. Nitrate being a soluble nitrogen form is readily utilized by plants and microorganisms for the biosynthesis of cellular proteins. The medium used in this experiment consists of ammonium sulphate as nitrogen source so that the efficiency of microorganisms to oxidize the ammonia to nitrate can be evaluated. Use of nitrite broth helps to identify the microorganisms with the potential to oxidize nitrite to nitrate.

Expt. 87   *To determine the process of nitrite production*

Materials Required

Laboratory glassware

Bunsen burner

Inoculating loop

Glass marking pencil

Laboratory equipment

Autoclave

Incubator

Reagents

a.   Ammonium sulphate broth

Ammonium sulphate	0.2 g
Magnesium sulphate	0.05 g
Ferric sulphate	0.003 g
Sodium chloride	1 g
Dipotassium carbonate	0.1 g
Distilled water	100 ml

b.   Nessler's regent

c.   Trommsdorf's reagent

Slowly add 100 ml of 20% aqueous zinc chloride solution to a mixture of 4 g of starch in water. Heat until the starch is dissolved as much as possible and the solution is almost clear. Dilute with water and add 2 g of potassium iodide. Again dilute it to 1000 ml, filter and store for further use.

PROCEDURE

1. Sterilize the test tubes containing ammonium sulphate broth.
2. Inoculate the loopful of culture into the test tubes.
3. Maintain a control tube.
4. Incubate the culture at 25°C for seven days.
5. Following the incubation period, add few drops of Nessler's reagent or Trommsdorf's reagent. If the Nessler's reagent turns yellow, it is a negative result indicating the accumulation of ammonium whereas absence of colour change confirms the conversion of ammonia to nitrate. If addition of Tromsdorff reagent and sulphuric acid to the culture brings about a colour change to blue-black, it confirms the conversion of ammonia to nitrate.

Expt. 88   *To determine the process of nitrate production*

MATERIALS REQUIRED

Laboratory glassware

Test tubes

Pipettes

Glass marking pencil

Laboratory equipment

Bunsen burner

Inoculating loop

Autoclave

Incubator

REAGENTS

Diphenylamine reagent

Slowly add 28.8 ml of distilled water in 60 ml of concentrated sulphuric acid. Dissolve 0.7 g of diphenylamine in this mixture. Leave it to stand for five minutes. After the mixture is cooled, add 11.3 ml of conc. hydrochloric acid. Again leave it to stand for 12 hours. Store it in amber bottles for further use.

PROCEDURE

1. Sterilize the test tubes containing nitrite broth.
2. Inoculate the loopful of culture into the test tubes.
3. Maintain a control tube.
4. Incubate the culture at 25°C for seven days.
5. Following the incubation period, add Trommsdorf reagent and sulphuric acid or diphenylamine reagent and sulphuric acid.

6. Appearance of blue-black colour with Trommsdorf reagent and sulphuric acid indicates the absence of nitrite whereas no colour change indicates the formation of nitrate. In the case of diphenylamine reagent and sulphuric acid, appearance of deep blue colour indicates presence of nitrate.

4.40 DENITRIFICATION PROCESS

4.40.1 Principle

In waterlogged soil, with minimum oxygen supply, nitrates are subsequently reduced to nitrites, ammonia and finally to nitrogen gas. The process is termed as denitrification and the microbial genera involved are *Pseudomonas, Bacillus* and *Micrococcus*. These organisms use nitrates as final electron acceptor and they are adapted to a fermentative type of metabolic pathway. This experiment consists of a nitrate substrate for microbial utilization. Nitrogen gas production can be identified by the bubble formation in inverted durham tubes.

Expt. 89   *To determine the process of denitrification*

MATERIALS REQUIRED

Laboratory glassware

Test tubes

Pipettes

Glass marking pencil

Laboratory equipment

Autoclave

Incubator

Bunsen burner

Inoculation loop

REAGENTS

Nitrate broth

Peptone   0.5 g

Beef extract   0.3 g

Distilled water 100 ml

PROCEDURE

1. Sterilize the test tubes containing nitrate broth.

2. Inoculate the culture to be analysed into the test tubes.

3. Insert the durham tubes into the test tubes in inverted position. See to it that the inner portion of the durham tubes are completely filled with the medium.

4. Do not entertain air bubbles. Incubate the tubes at 25°C for seven days.

5. Maintain a control tube.

6. Following the incubation period, check for the presence of bubble formation inside the durham tube.

4.41 NITROGEN FIXATION

4.41.1 Principle

Biological nitrogen fixation is the process by which the atmospheric nitrogen is converted to readily available nitrogen form such as ammonia or nitrate. Only the microorganisms with nitrogenase enzyme can carry out this process and they are called as nitrogen fixers. Nitrogen fixation process is mediated by two microbial groups.

1. Symbiotic
2. Nonsymbiotic

Symbiotic   They fix nitrogen in the root nodules of leguminous plants. Symbiotic form comprises the genus *Rhizobium* and soon after their infection, they penetrate into the root system and settle inside the tumour-like root nodules. The interaction between the microorganisms and plant is called mutualism (++) as microorganisms are nourished by the nutrients of the plants and the plants in turn receive the nitrogen fixed by the microbes for biosynthesis.

Nonsymbiotic forms   It consists of nonsymbiotic free-living forms and members of *Azotobacter, Beijerinckia, Clostridium* and *Cyanobacteria*. They fix the nitrogen outside the plant i.e. in soil. Under adverse environmental conditions, they cover themselves with a thick protective cell wall and establish themselves when suitable conditions revert. *Rhizobium* inside the plant are called bacteroids. They exhibit pleomorphism as they exist in different forms—X, Y, T, V and stellate.

Expt. 90   *To determine the process of nitrogen fixation*

MATERIALS REQUIRED

Laboratory glassware

Petri plates

Test tubes

Pipettes

Glass marking pencil

Laboratory equipment

Autoclave

Incubator

Bunsen burner

REAGENTS

a. Methylene blue

b. Methylene blue powder   0.3 g

c. Distilled water   100 ml

d.   Gram stain
     Solution A

Crystal violet	2 g
Ethyl alcohol	20.0 ml

   Solution B

Ammonium oxalate	0.8 g
Distilled water	80 ml

   Mix solution A and B.

e.   Grams iodine

Iodine	1 g
Potassium iodide	2 g
Distilled water	300 ml

f.   Ethyl alcohol

Ethyl alcohol	2 g
Distilled water	5 ml

g.   Safranin

Safranin 'O'	0.25 ml
Ethyl alcohol	10.0 ml
Distilled water	100 ml

h.   Nitrogen for mannitol agar

Mannitol	15 g
Dipotassium hydrogen phosphate	0.05 g
Magnesium sulphate	0.02 g
Calcium sulphate	0.01 g
Sodium chloride	0.02 g
Calcium carbonate	0.5 g
Agar	1.5 g
Distilled water	100 ml

PROCEDURE (FOR AZOTOBACTER ISOLATION)

1.   Take 1g of soil sample in previously sterilized mannitol-agar-containing test tubes.

2.   Mix the content well by vigorous shaking.

3.   Incubate the culture at 25°C for about seven days.

4.   Following the incubation period, check for the appearance of a thin film of microbial growth.

5.   Carefully transfer a loopful of this culture into nitrogen-free mannitol-agar-containing petri plates.

6. To isolate individual colonies, perform four way streak method.

7. Incubate the plates in inverted position for about 4 to 6 days at 25°C.

8. Individual colonies are subjected to Gram's iodine tests and the stained cells are examined under microscope.

9. Results are recorded based on their colony characteristics.

- Brown to black colonies which when examined under microscope appear as gram-negative, large-sized ovoid rods. They exist in pairs. Cysts may cover cells. — *Azotobacter chroococcum*

- Colourless colonies which when examined under microscope appear similar to that of *Azotobacter chroococcum*. — *Azotobacter vinelandii*

- Colourless colonies appearing single or in pairs. Cysts are absent. Cells cannot be identified as positive or negative. Shape may be ovoid or coccoid. — *Azotobacter macrocytogenes*

PROCEDURE (FOR RHIZOBIUM ISOLATION)

1. Carefully remove the root nodule from the leguminous plant.

2. Wash it thoroughly and homogenize the nodule between the two slides.

3. Spread the thin layer throughout the slides.

4. Allow it to air-dry and heat fix it.

5. Then flood the smear with methylene blue stain.

6. Allow it to stand for one minute.

7. Wash the slide under running water till the water dripping from the slide is colourless.

8. Allow it to air-dry.

9. Examine it under microscope.

EXERCISES

1. Water dissolves the ionic bonds binding the minerals when the ions are ______________.

2. Structural organization of soil in terms of organic and inorganic matter determines ______________ of soil.

3. Organic content of the soil depends on ______________ and ______________.

4. Nature of the parent material determines ______________ of soil.

5. Addition of Nessler's reagent turns ammonia to ______________.

6. Available potassium refers to ______________ and ______________ in soil.

7. Which layer of the soil act as the receiver of washed-out material?

8. What are the influencing factors that bring about major changes in the soil ecosystem?

9. List out the processes that favour the input and output of soil ecosystem.

10. Which is the deciding factor to initiate the leaching process in soil.

11. Explain the oxidation–reduction process of iron in gleying of soil.

12. Which layer is subjected to change in lessivage process and why?

13. Describe the formation of elution horizon.

14. Explain the sequential steps involved in podzolisation process.

15. In arid region, surface soil is saline, why?

16. What is the functional role of humus in soil?

17. Write down the forces that help to bind the humus with the mineral portion of soil.

18. Differentiate mull humus and mor humus.

19. What is the pH of the soil at the end of hydrolysis?

20. Soil microorganisms induce the carbonic and mediated acid hydrolysis, how?

21. Define clays.

22. What are phyllosilicates?

23. Explain the cation exchange capacity (CEC) with examples.

24. Which property of the soil indicates the nutrient availability of soil for plant growth?

25. Which constituent of soil is essential for its productivity?

26. List out the sources of CO_2 in soil.

27. What is a domain?

28. What is the relationship between bulk density and prodensity?

29. Mention the three forces that facilitate the soil water movement.

30. Write down the predominant ions in acid and alkaline soils.

31. Define respiratory quotient.

32. Which is the major influencing factor that favours the ingress of soil air?

33. What are the sources of carbon compounds in soil?

34. What are the mineral constituents in strongly weathered soil?

35. What is the main composition of secondary minerals?

36. What are the two end products when humus is subjected to chemical change?

37. What is the proportion of carbon, nitrogen, phosphorus and sulphur in humus?

38. Describe the favourable conditions for peat formation.

39. Specify the five major groups of soil microorganisms and their mode of nutrition.

40. Mention few microbial groups that are involved in ammonification process.

41. What is the chemical reaction behind the protein transformation to amino acids.

42. In which form can N_2 be readily utilized by plants and microorganisms?

43. What is the basic criterion for the nitrification process to occur?

44. In the mutual interaction between plant root and microorganism, what is the benefit acquired by the microorganisms from the plants?

45. What are bacteroids and in what way do they differ from normal forms?

46. What are the influencing factors that determine the moisture content of the soil?
47. Explain with chemical reaction the principle behind the quantification of calcium and magnesium.
48. In available N_2 estimation, what is the purpose of adding liquid paraffin to soil.
49. Discuss the role of soil enzyme activity in plant growth.
50. Categorize soil particles on the basis of its size.
51. What is a soil solution?
52. Name a few sources of organic matter in soil.
53. What happens when organic matter in soil is decomposed?
54. Write down the structural composition of humus.
55. Classify soil microorganisms on the basis of energy source.
56. Which temperature range is preferable for soil actinomycetes.
57. What are rhizomorphs?
58. Which microbial group predominate acid soils?
59. Define biological nitrogen fixation.
60. List out the microbial members involved in the nonsymbiotic nitrogen fixing bacteria.
61. Define antibiotics.
62. How will you estimate the soil moisture content?
63. Which constituent of the soil will be lost when it is ignited?
64. How will you estimate the specific gravity of the soil?
65. What are the two steps involved in soil nitrogen estimation?
66. Differentiate total nitrogen and available nitrogen.
67. How will you prepare Trommsdorf reagent?
68. Write down the principle of available molybdenum estimation in soil.
69. What are the reagents to be added to extract nitrate content from soil?
70. Which forms of sulphur are insoluble?
71. Which reagent acts as a chelating agent to extract micronutrients from soil?
72. Which method is preferable to estimate total metal content in soil?

(**5**)

MICROBIAL ANALYSIS OF WATER, WASTE WATER AND SOIL

5.1 INTRODUCTION

Microbiology is a branch of biology that deals with the study of microorganisms. Evolution of this subject began with the discovery of microorganisms in the 16th century. Thereafter, microorganisms have been subject to constant research and analysis. Precisely, microbial analysis plays a key role in the various fields of science. Successful results and accurate interpretation in these analyses depend on several factors such as thorough understanding of the objective and principle of the experiment to be carried out, strong theoretical background, refinement of procedures according to the need and careful performance. This chapter discusses each experimental procedure with a brief introductory part, principle and the methodology.

5.2 MICROBIOLOGICAL EQUIPMENTS

The commonly used microbiological equipments include culture tubes, petri plates, inoculation loops, inoculation needles, pipettes, incubators, shaking water bath and refrigerator.

5.2.1 Glassware

Microorganisms can be cultivated in test tubes and petri plates. To maintain sterile condition, test tubes are closed with caps made up of stainless steel, or heat-resistant plastics. Initially they were plugged with cotton but the major drawback is that they should be discarded after use. Petri plates consist of a medium-containing base and a slightly loose cover. Petri plates of various sizes are available but petri plates of 15 cm diameter are commonly used. After sterilization of petri plates, 15 to 20 ml of sterile agar medium is transferred. The medium is allowed to solidify by cooling. Then the entire culture to be grown is inoculated into the petri plates. Then the petri plates are kept in an inverted position, otherwise the water vapour formed during solidification may drip into the hardened agar and contaminate them.

5.2.2 Instruments Used for Transfer of Cultures

Transfer of culture from one medium to the other is called subculturing and the transfer is done under aseptic condition. For this, inoculation loops, inoculation needles and pipettes are used.

Inoculation loops and needles   Inoculation loops and needles are made up of nichrome or platinum wire inserted into the metal handle. Sterilization of this equipment is done by heating the loop or needle to red-hot in the bunsen burner.

Pipette   Sterile pipettes are used to transfer cultures using sucking pressure at the mouthpiece which draws up the liquids. They are made up of glass or plastic with calibrations on the surface. Different volume of pipettes are available (10 ml, 1 ml and 0.1 ml). However sucking microbial cultures by mouth is strictly prohibited.

The steps to be taken during culture transfer are the following.

- Once sterilized, do not place the loop needle on the table.
- Before taking the loopful of culture, allow the loop needle to cool for 10 to 20 seconds.
- Do not place caps/cotton plug on the laboratory bench.
- Immediately after cooling, touch the inner sterile wall of the petri plate tube for another few seconds and then take the inoculum.
- Use loops or needles appropriately, i.e., loops are used for broth and solid cultures whereas needles are used for agar deep tube.
- Gentle touch of the agar slant is enough. In the case of broth, immerse the needle to half its length. Do not allow the metal shaft to enter into the broth.
- To take the sample from the agar deep tube, insert the needle straight to the bottom and withdraw it along the inserted line.
- Flame the needle or loop red-hot to completely kill the inoculum in them.

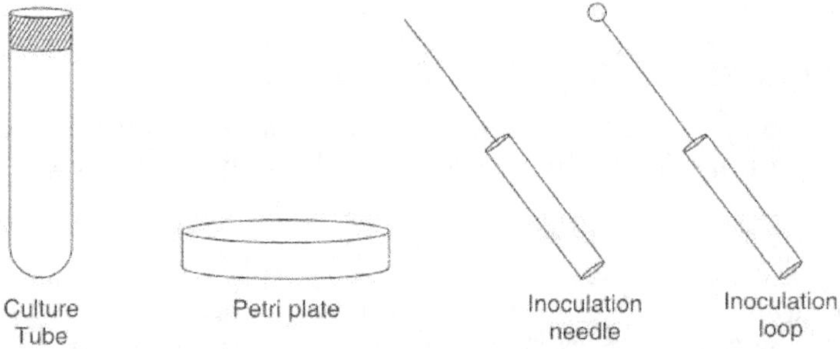

Figure 5.1   Instruments used for transfer of cultures

5.2.3 Incubators

Incubators with thermostats operate like hot-air ovens. The required temperature is adjusted and then the cultures are incubated for about 24 hours.

5.2.4 Water Bath Shakers

Cultures can be grown in a shaking water bath. The advantages of this instrument is

    i.   Uniform heat distribution

    ii.   Continuous aeration by agitation

However cultures in solid medium cannot be grown here.

5.3 BASIC REQUIREMENTS OF MICROORGANISMS

Growth and survival of microorganisms depends on several factors like nutrient and oxygen availability, pH, temperature and enzyme activity. Microorganisms cannot utilize complex nutrients and therefore need specific enzymes to break them into simpler substances. To overcome this difficultly, along with the favourable growth environment, the microorganisms should be provided with nutrients in available form. A solution that is able to meet the growth requirements of microbes is called as culture medium. They may exist in different forms, solid, semi-solid and liquid. Liquid medium is termed as broth and the broth supplemented with the solidifying agent agar is called solid medium or semi-solid medium. Agar does not possess any specific nutritive value. It exists in solid form at 40°C and liquid form at 100°C. The concentration required for a solid medium is about 1.5 to 1.8% and a concentration less than this (1.0 g) forms a semi-solid medium. In solid medium the microorganisms are equally distributed and therefore individual colonies can be isolated. Furthermore, the number of microorganisms per gram of sample inoculated can be evaluated. Liquid medium is maintained in test tubes whereas solid medium is maintained both in test tubes and petri plates. In the test tubes, immediately after sterilization they are allowed to cool in a slanted position so that the agar slants are produced. If the agar is solidified in an upright position, they are called agar tubes. Agar slants can be used to grow pure cultures and the culture is inoculated as depicted in the Figure 5.2.

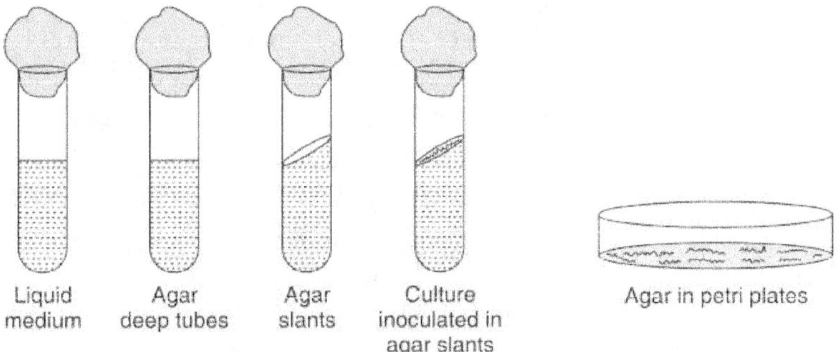

Figure 5.2　Inoculation of cultures

5.3.1 Nutritional Needs

The basic nutritional requirements of microorganisms are a) carbon, b) nitrogen, c) nonmetallic elements (sulphur and phosphorus), d) metallic elements (calcium, zinc, sodium, potassium, copper, manganese, magnesium and iron), e) vitamins, f) water and g) energy.

Carbon It is the most important constituent of all cell structures and therefore the microorganisms need them most. Based on the carbon utility, two different types are recognized.

1. Autotrophs (Use only inorganic compounds)
2. Heterotrophs (Use organic compounds)

Nitrogen It is the important constituent of all proteins and nucleic acids. Proteins serve as bodybuilders and are responsible for the metabolic activities of the cell. Nucleic acids form the genetic basis of the cell. Microbes utilize different forms of nitrogen.

1. Atmospheric nitrogen (nitrogen fixers)
2. Inorganic nitrogen (soil microbes)
3. Organic nitrogen (decomposers)

Nonmetallic elements Sulphur and phosphorus are the two nonmetallic elements of prime importance.

Sulphur It is an important component of certain amino acids. Microorganisms utilize them either from organic (sulphur-containing amino acids) or from inorganic forms (sulphur, sulphate).

Phosphorus They are present in DNA, RNA and in the energy-yielding compound ATP (adenosine triphosphate). Microorganisms utilize them as phosphates.

Metallic elements These are otherwise called as micronutrients as they are needed in trace amounts. Metallic elements play an important role in maintaining normal cellular constituents such as osmoregulation, regulation of enzyme action and electron transport in the oxidation of organic substance.

Water Water is required for the maintenance of osmoregulation.

Energy Based on the energy requirement, microorganisms are classified into two types a) Phototrophs (use solar energy as energy source) and b) chemotrophs (use either organic molecules such as glucose or inorganic molecules such as H_2S or $NaNO_2$ as energy source).

The media used for cultivation of microorganisms can be categorized into two types:

1. Chemically defined media
2. Artificial media

Chemically defined media Here the composition of organic and inorganic compounds is exactly defined. It is used to cultivate the microorganisms with specific needs. Typical examples are inorganic and glucose both.

Inorganic broth	Glucose broth
Sodium chloride – 0.5 g	Sodium chloride – 0.5 g
Magnesium sulphate – 0.02 g	Magnesium sulphate – 0.02 g
Ammonium dihydrogen phosphate – 0.1 g	Ammonium dihydrogen phosphate – 0.1 g
Dipotassium hydrogen phosphate – 0.1 g	Dipotassium hydrogen phosphate – 0.1 g
Distilled water – 100 ml	Distilled water – 100 ml

Artificial media   This media consists of complex substances whose composition is not exactly known. It is used to grow the heterotrophs. Typical examples are nutrient broth and yeast extract broth.

Nutrient broth	Yeast extract broth
Peptone – 0.5 g	Peptone – 0.5 g
Beef extract – 0.25 g	Beef extract – 0.25 g
NaCl – 0.5 g	NaCl – 0.5 g
Distilled water – 100 ml	Yeast extract – 0.5 g
	Distilled water – 100 ml

Apart from this, two different types of media are recognized, a) differential media and b) selective media. It is used for a variety of applications. A few applications are listed below.

i.   To isolate specific microbial groups from mixed cultures.

ii.   To isolate a specific microbial group from its closely related species.

iii.   To enumerate the bacteria in water, sewage, food and dairy products.

iv.   To access the microbial products such as antibiotic vitamins, enzymes.

v.   To characterize the microbial groups based on their metabolic activities.

Given here are examples of a few media:

a) Blood agar

Infusion from beef heart	50 g
Tryptose	1.0 g
Sodium chloride	0.5 g
Agar	1.5 g
Distilled water	100 ml

b) EMB agar medium

Peptone	1.0 g
Lactose	0.5 g
Dipotassium phosphate	0.2 g
Eosin	0.04 g
Methylene blue	.0065 g
Distilled water	100 ml

c) Mc Conkey agar

Bactopeptone	1.7 g
Proteose peptone	0.3 g
Lactose	1.0 g
Bile salt mixture	0.15 g
Sodium chloride	0.5 g
Agar	1.5 g
Neutral red	0.003 g
Crystal violet	0.0001 g
Distilled water	100 ml

d) Mannitol salt agar

Beef extract	0.1 g
Peptone	1.0 g
Sodium chloride	7.5 g
d-Mannitol	1.0 g
Agar	1.5 g
Phenol red	0.0025 g
Distilled water	100 ml

5.3.2 Temperature

Temperature changes influence the enzyme activity which in turn affect the growth of microorganisms. With increasing temperature, enzymatic activity is accelerated however, with continuous rise in temperature, enzyme structure gets denatured. Also, a drop in temperature inactivates the enzyme action thus blocking the total metabolic activity. Therefore every microbial group has optimum temperature at which maximum growth is exhibited.

Based on the temperature requirements, bacteria are classified into three types namely psychrophiles, mesophiles and thermophiles. Psychrophiles grow at low temperature whereas mesophiles and thermophiles prefer medium and high temperatures respectively.

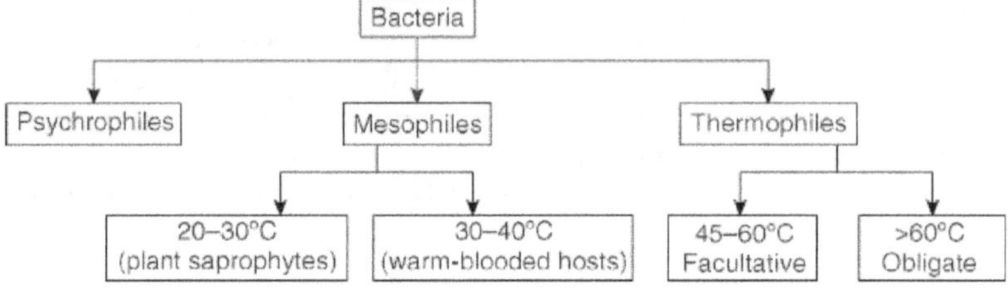

5.3.3 pH

pH requirements differ between microbial groups and the pH range at which the microorganisms grow well reflects the ability of the microorganism to adapt itself to the environments. In general, the pH range for bacteria is between 4 and 9 and the optimum pH range is between 6.5 and 7.5 whereas it is between 4 and 6 for fungi, moulds and yeasts. However for the cultivation of microorganisms in the laboratory, a neutral pH is maintained as majority of microorganisms can be grown in that range. Metabolic activities of microorganisms may result in the production of acid or alkali thus shifting the pH to acidic or alkaline range. This can be used as an indicator for the presence of that specific organism. But for the continuous culturing of microorganisms, the pH should be maintained in the neutral range. Therefore suitable buffer reagents like K_2HPO_4 (weak base) and KH_2PO_4 (weak acid) should be added to the medium. Weak acid combines with excess OH to form weak base while weak base combines with excess H to form weak acid.

5.3.4 Oxygen Availability

Based on the oxygen utility for cellular respiration, microorganisms are classified into five types. They include aerobes, microaerophiles, aerotolerant anaerobes, facultative anaerobes and anaerobes.

Aerobes   Here O_2 acts as final electron acceptor and therefore for the oxidation of organic compounds, O_2 is required.

Microaerophiles   They require O_2 for growth. But in excess amount, their oxidative enzymes get deactivated and finally they die.

Aerotolerant anaerobes   They are fermenting microbes and therefore do not need O_2 as final electron acceptors. They produce superoxide dismutase and therefore can tolerate small amounts of oxygen.

Facultative anaerobes   They use O_2 as final electron acceptors. However, they can manage small amounts of oxygen and even in the absence of oxygen they can survive utilizing nitrates and sulphates as final electron acceptors.

Anaerobes　These microorganisms have low oxidation–reduction potentials and therefore instead of oxygen, use nitrate or sulphate as final electron acceptors.

Growth distribution of microorganisms reveals their oxygen requirement. For example, aerobes occupy the surface while the anaerobes restrict themselves to the bottom of the tube. Facultative anaerobes, based on the oxygen requirement, distribute themselves throughout the medium.

5.4 BACTERIAL GROWTH

Microorganisms, when provided with favourable environmental conditions, multiply rapidly and show a steady rate of increase in their population. Microbial growth is evaluated by plotting the cell number against the incubation period and the growth curve drawn (Figure 5.3) shows four different stages such as lag phase, log phase, stationary phase and death phase.

5.4.1 Lag Phase

a.　Growth, i.e., increase in cell size is observed.

b.　Microbial cells take time to adapt themselves to the new environment.

c.　Metabolic activity is speeded up.

d.　Biosynthesis of enzymes takes place.

e.　No cell division and therefore no increase in microbial numbers is observed.

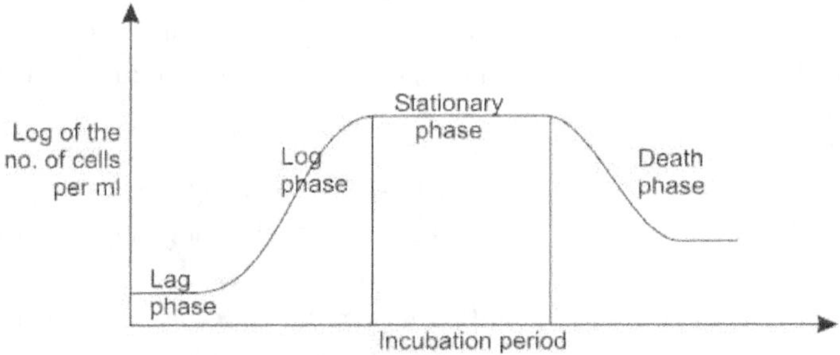

Figure 5.3　Curve showing bacterial growth

5.4.2 Log Phase

a.　Reproduction occurs at a steady and rapid rate by binary fission.

b.　This results in the logarithmic increase in population which doubles at a uniform speed.

c.　The time taken by the population to double itself is called generation time.

d.　It varies between microbial groups.

e.　Also it depends on the nutrient availability in the medium.

f.　Average incubation period for the log phase is between 6 and 12 hours.

5.4.3 Stationary Phase

a.　The microbial population shows a gradual death rate and therefore the microbial proliferation rate is equal to that of death rate.

b. No increase in population is recorded.

c. The population growth remains constant for a specific period and this peak is called as saturation point.

d. Depletion of essential metabolics and accumulation of toxic compounds are the two factors that inhibit further growth.

5.4.4 Death Phase

a. After a period of time, the death rate outranges the growth rate.

b. Lack of metabolites and continuous accumulation of toxic products bring growth to an end.

c. Except a few resistant microbial groups, others die.

5.5 CHARACTERISTICS OF MICROORGANISMS

Microorganisms when grown in different media exhibit different growth characteristics. These are called cultural characteristics. Based on the culture characteristics and growth pattern, they can be classified into different taxonomic groups and therefore, evaluation of cultural characteristics plays a key role in the identification of microbial species.

5.5.1 Nutrient Agar Slants

a. Growth appearance   none, slight, large
b. Pigmentation   chromogenic, non-chromogenic
c. Light transmittance   opaque, translucent, transparent
d. Growth form   filiform, echinilate, beaded, effuse, arborescent, rhizoid.

5.5.2 Nutrient Agar Plates

a. Size   pinpoint, small, medium, large
b. Pigmentation Chromogenic/nonchromogenic
c. Growth form   Circular, irregular, rhizoid
d. Margin   Entire, lobate, undulate, serrate, filamentous
e. Elevation   Flat, raised, convex, umbonate

5.5.3 Nutrient Broth Cultures

Growth pattern   a. Fine particles dispersed uniformly
                      b. Floc dispersed uniformly
                      c. Pellicle on the surface
                      d. Sediment on the bottom

5.5.4 Nutrient Gelatin

1. Crateriform   Saucer-shaped liquefaction
2. Napiform   Bulb-shaped liquefaction
3. Infundibuliform   Funnel-shaped
4. Saccate   Elongated, tubular
5. Stratiform   Liquefaction from the top to the middle

5.6 SAFETY MEASURES

- Before adopting a new method for the determination of a specific microorganism the microbiologist should go through the procedure thoroughly. This prevents the wastage of chemicals and microbial contamination, and helps him to perform it in a efficient manner.
- Take care not to spread the laboratory manual or laboratory note books on the working table.
- At the end and beginning of each day's work, disinfect the lab area. Start the work after it is dried.
- Do not place any chemicals especially acids in racks near the burner.
- Check whether all the reagents are of high grade as the accuracy of results depends on the reagents used.
- Use quality distilled water to prepare solutions and frequently check the pH of the distilled water. Service the distillation unit every six months. Use double-distilled water whenever the procedure warrants.
- Do not lick the fumed labels to paste it in the culture tubes.
- Inside the laboratory, make sure that your hands do not touch your face.
- Keep the working table clean and tidy. Never keep the pen or pencils in the mouth.
- Do not exchange the laboratory furniture with the one used for other purposes (office rooms, library, classrooms).
- Do not pipette by mouth. Discard the contaminated glassware in closed receptacles.
- Disinfect the used microbial slides before discarding them in sinks.
- Laboratory coats should be laundered at least twice a week.
- Wash hands thoroughly with a germicidal soap and wipe it dry. See to it that you remove your coats before washing.

5.7 ISOLATION OF PURE CULTURES

5.7.1 Principle

To study the cultural, morphological and biochemical properties of each species of microorganisms, they should be available in individual colonies. Culture of single microbial species is called pure culture. Therefore for any microbial characterization, a pure culture is essential. Under natural environmental conditions, they exist as mixed forms. By using suitable techniques they should be isolated. For isolation, they should be allowed to grow in discrete colonies. For this the concentration of inoculum should be progressively reduced. Gradual decrease in inoculant size ensures the development of discrete colonies at the end.

Expt. 91   *To isolate pure cultures*

MATERIALS REQUIRED

    Laboratory glassware

        Petri plates

    Laboratory equipment

        Inoculation loops

        Bunsen burner

        Incubator

Nutrient agar medium

Mixed culture

PROCEDURE

1. Isolation of pure culture is accomplished by four-way quadrant method in petri plates.

2. As an initial step, sterilize the petri plates and pour 15–20 ml of sterile molten agar into them.

3. Allow the medium to solidify.

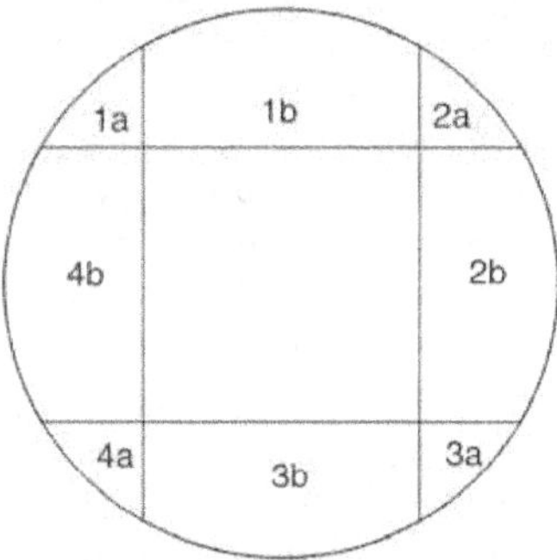

Figure 5.4   Division of agar surface of petri plates

4. Now divide the agar surface of the petri plates into four equal regions. For convenience sake, they are represented as 1 (a, b), 2 (a, b), 3 (a, b) and 4 (a, b).

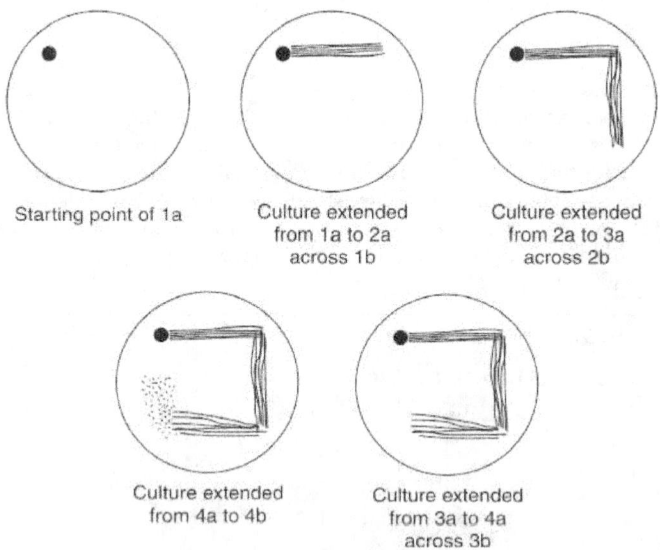

Figure 5.5   Streaking of culture

5. Carefully take a loopful of mixed culture and place it on the area 1 of the agar surface.

6. Flame the loop to red-hot and cool it for few seconds.

7. Then from the starting point of area 1, drag the loop several times till area 2a across 1b.

8. Flame the loop to red-hot and cool it for a few seconds.

9. Keeping area 2a as the starting point, drag the loop towards area 3a across 2b.

10. Repeat step (6).

11. Now from 3a drag the culture towards 4a across 3b.

12. Repeat step (6).

13. Keeping 4a as the initial point drag the culture several times up to 4b.

14. Do not reach 1a and see to it that there exists considerable distance between 4b and 1a.

15. Flame the loop to red-hot, close the petri plate and incubate it for 24 hours at 37°C.

16. After the incubation period, examine for the development of discrete colonies.

5.8 SERIAL DILUTION TECHNIQUE

5.8.1 Principle

This ensures the quantitative enumeration of microorganisms in a variety of samples like food, water, milk, sewage, soil and even air. The number of cells grown can be counted by devices like Petroff hauser chamber, colony counter and Neubaur chamber. On the other hand, serially diluted cultures can be allowed to grow by pour plate method. The advantage of this technique is that this type of enumeration takes into consideration only the viable cells and the colonies grow discrete. Also they can be subcultured for further analysis. It too has its own drawback i.e., it requires 24 hours incubation period, needs more glassware and at times it gives erroneous results due to dilution errors.

Expt. 92 *To perform the serial dilution technique*

MATERIALS REQUIRED

Laboratory glassware

Petri plates

Test tubes

Beakers

Marker pencil

Test tube rack

Laboratory equipment

Hot plate

Water bath

Thermometer

Neubaur chamber

Incubator

Nutrient agar medium

PROCEDURE

1. Sterilize 6 test tubes, each containing 9 ml of distilled water.

2. Take 1 g/ml of the sample; dissolve it in 100 ml of sterile distilled water. This is designated as 10^{-2} dilution.

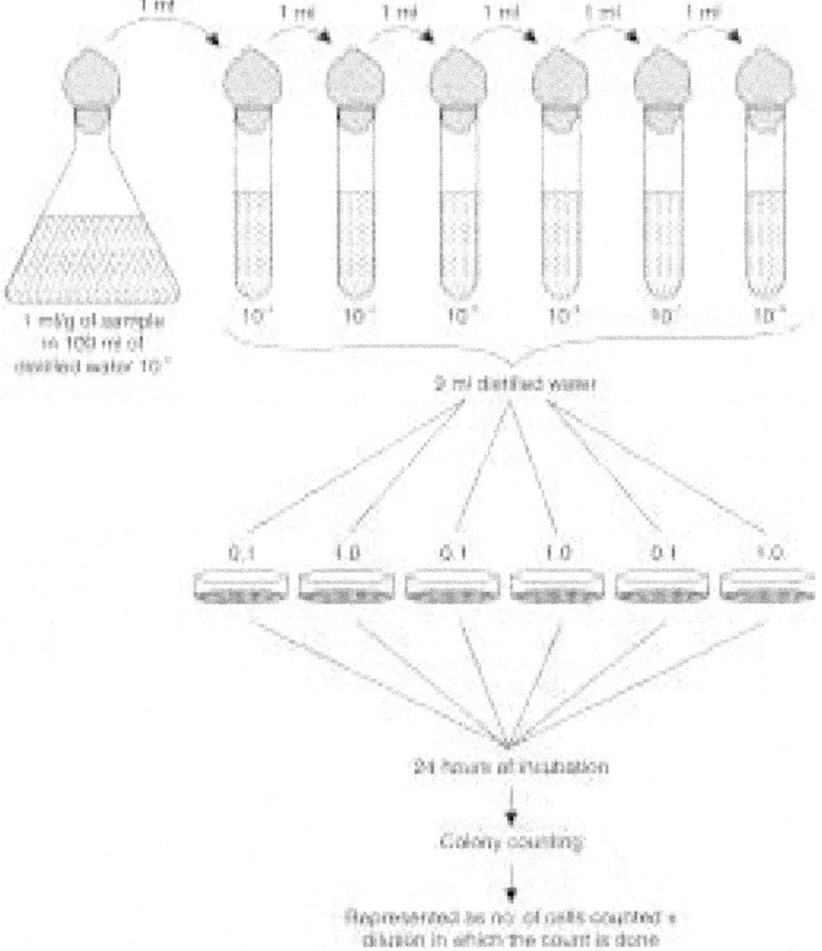

Figure 5.6   Steps involved in serial dilution technique

3. Carefully transfer 1 ml of 10^{-2} dilution into the test tube No. 1 using a sterilized pipette (10^{-3}).

4. Place the sample suspension in the water bath shaker for about one hour.

5. Roll the test tube between the palms so that the cells are equally dispersed in the medium.

6. Discard the pipette in the disinfectant tray.

7. From the test tube No. 1 pipette out 1 ml using a sterile pipette and transfer it into test tube No. 2 (10^{-4}).

8. Repeat steps (5) and (6).

9. Follow the procedure described above for all the six test tubes and ensure that the dilution in the test tube No. 6 is 10^{-8}.

10. Do not forget to mix each test tube before taking the sample.

11. Choose 3 dilutions to be transferred to the petri plate for colony counting.

12. For each dilution use two petri plates one for 0.1 ml and another one for the 1.0 ml of the sample.

13. Use separate pipettes for different dilutions.

14. Label the petri plates as 10^{-4} (a), 10^{-4} (b); 10^{-5} (a), 10^{-5} (b) and 10^{-6} (a), 10^{-6} (b).

15. Using the sterile pipettes, transfer required volume of the sample (0.1, 1.0 ml) to the appropriate petri plates.

16. Rotate the petri plate in clockwise and anticlockwise direction.

17. Pour the molten agar medium into the plates, rotate it again as described in step (16).

18. Do not pour the agar medium at greater then 45°C because it would kill the microbes in the sample.

19. Allow it to solidify and keep them in inverted position inside the incubator.

5.9 STAINING METHODS

The minute size of microorganisms is not a limiting factor for their identification, as microscopes can easily solve the problem. However their transparency and their generally colourless structures make their visualization most difficult. Therefore to study the microbial properties and to classify them into specific groups, staining is very essential. Basically a stain is an organic compound consisting of a benzene ring, a chromophore and an auxochrome group. Benzene is an organic colourless solvent (chromogen). It is coloured when combined with a chromophore and the function of the auxochrome is to induce the ionization property of chromogen that helps them to bind with living cells.

$$\text{organic compound with benzene ring} + \text{coloured chemical group} \rightarrow \text{chromogen (coloured organic compound)}$$

$$\text{chromogen} + \text{auxochrome} \rightarrow \text{stain}$$

Based on the charge on the surface of the chromogen and the microbial component, the stain may bind either to the nucleic acid or to the proteins. Accordingly two different stains are identified—acidic dyes and basic dyes.

5.9.1 Acidic Dyes

These dyes exhibit negative charge on the chromogen portion (anionic) and hence bind with the positive constituent of microbial cells to be stained. Since proteins are positively charged they acquire the colour of negatively charged chromogen, e.g., picric acid.

5.9.2 Basic Dyes

These dyes exhibit positive charge on the chromogen portion (cationic) and therefore bind with the negative constituents, i.e., nucleic acids of the microbial cells. As a result the cells acquire the colour of positively charged chromogen, e.g. methylene blue.

Staining is adopted for diagnostic purposes, and according to the purpose, different techniques are used.

A few examples are given below.

a.   Morphological studies—simple staining

b.   Group identification—gram staining, acid-fast staining

c.   Identification of specific structures—flagella stain, capsule stain, spore stain, nuclear stain

Except simple staining, other staining techniques involve 2 contrast stains. Staining procedures are relatively easy to perform. However for accurate results, great care should be taken. Steps to be taken while doing the staining are as follows

a.   Always use clean slides for smear preparation. Give a thorough washing with soap powder. Rinse it well and immerse the slides in 95% alcohol for about 15 minutes. Dry and place them on tissue papers or clean lab towels. While handling the slides, take care not to touch the central portion of the slide. As a safety measure lift them by placing the index finger and thumb on either rims of the slide.

Figure 5.7   Handling of a slide

b.   Smear preparations should not be thick or dense. Smears appearing as a thin film or as a white transparent layer after drying give better results. Broth cultures should be spread evenly on the slide using sterile inoculation loop. For solid cultures, placing a loop of water on the slide and then diluting the culture can ensure equal distribution. While transferring a culture, care should be taken to touch the culture with the tip of the inoculation needle. Now by exhibiting a slow and steady circular motion on the drop of water, the cells can be evenly spread. After drying, check for the appearance of semitransparent white layer. Allow the smear to air-dry.

c. To prevent the bacterial cultures from being washed away, heat-fix it. While heating, the protein constituents of the microbial cells are coagulated and fixed on the glass surface. Do not show the glass slide directly on the heat. Instead, pass the slide rapidly over the flame 3 to 4 times.

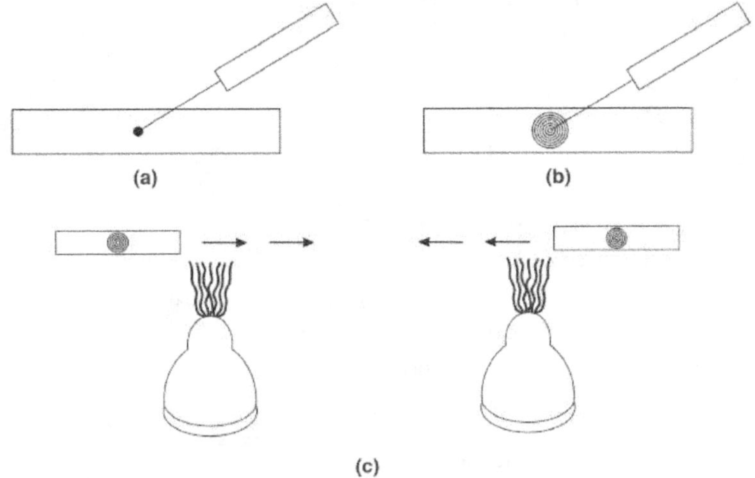

Figure 5.8   **(a) Solid cultures placed on the drop of water, (b) circular motion exhibited to spread the solid cultures on the drop of water, (c) Heat-fixing the air-dried cultures**

5.10 SIMPLE STAINING

5.10.1 Principle

In this technique, positively charged basic stains that can stain the negative constituents of the microbial cells are used. It is adopted to evaluate the shape and arrangement of microbial cells. Carbol fuchsin, methylene blue and crystal violet are commonly used. The experimental procedure is the same for these stains except for the time of exposure.

Stains	Exposure time
methylene blue	1–2 minutes
crystal violet	20–60 seconds
carbol fuchsin	15–20 seconds

Expt. 93   *To perform the simple staining technique*

MATERIALS REQUIRED

Laboratory glassware

Test tubes

Glass slides

Laboratory equipment

    Bunsen burner

    Microscope

    Inoculation loop

Cultures

Staining tray

REAGENTS

Methylene blue

Crystal violet

Carbol fuchsin

PROCEDURE

1. Clean the glass slides, air-dry and prepare bacterial smears on it.
2. Air-dry the smears and heat-fix them.
3. As per the exposure time indicated, flood the smear with the stains in a staining tray.
4. Keep the glass slide under running tap water so that the excess stain is washed away.
5. While washing, do not allow direct contact of the culture with tap water. Instead hold it in such a way that the tap water flows on one corner of the slide.
6. Blot-dry the smear.
7. Examine under the microscope

5.11 NEGATIVE STAINING

5.11.1 Principle

In this technique, acidic stains that can bind with the positive constituents of the microbial cells are used. These stains cannot penetrate the bacterial cells bearing negative charge on their surface and therefore even after saffranin application, these cells remain colourless. This technique is useful for studying the bacterial cultures.

Expt. 94   *To perform the negative staining technique*

MATERIALS REQUIRED

Laboratory glassware

    Test tubes

    Glass slides

Laboratory equipment

    Bunsen burner

    Microscope

    Inoculation loop

Staining tray

REAGENTS

Nigrosin stain

PROCEDURE

1. Transfer a loopful of culture to a drop of nigrosin stain placed at one end of a clean glass slide.

2. Take another glass slide. Place it on the first slide in such way that its edge is at an angle of 30° to the bacterial culture of the first slide.

3. Now carefully push the edge of the second slide over the culture and move it to form a thin smear.

4. Allow the slide to air-dry

5. Do not heat-fix.

6. Identify the culture under oil immersion.

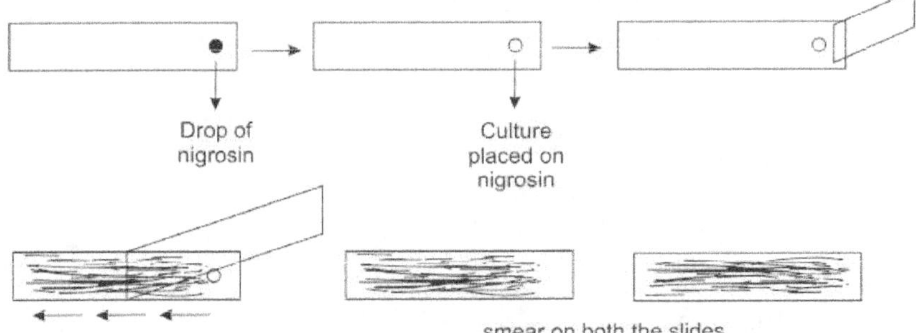

Figure 5.9   Negative staining

5.12 ACID-FAST STAINING

5.12.1 Principle

Microorganisms that come under the genus *Mycobacterium* possess a thick lipid layer. In general, these microbial groups are pathogens and hence their identification for diagnostic purposes is very important. For this purpose, a powerful staining solution containing phenol with the application of heat is used. This stain after penetrating through the lipid layer will not be decolourised and this gives a characteristic yellowish brown colour to the cell. However acid-fast bacteria that have been decolourised, take up the second stain. Hence the technique involves the usage of three different stains.

1. Carbol fuchsin (primary stains)
2. Decolourising agent (acid alcohol)
3. Methylene blue (counterstain)

5.12.2 Carbol Fuchsin

Carbol fuchsin is a red coloured phenolic stain that readily solubilizes the lipid cell wall of the *Mycobacterium* species. After penetration, they are retained inside the cell permanently. Heat fixation favours further penetration into the cytoplasm and this causes the cell to appear red.

5.12.3 Acid Alcohol

It is a mixture of 3% hydrochloric acid and 95% ethanol. Before the application of discolouring agent, cool the smear so that the lipid layer is hardened. This will prevent the entrance of acid alcohol into the cell. Therefore even the excess amount of acid alcohol cannot decolourise the primary stain. In the case of non-acid-fast cells that do not possess lipid cell wall, acid alcohol removes the primary stain.

5.12.4 Counterstain

Methylene blue is used as the counterstain. In the absence of primary stain in non-acid-fast cells methylene blue adhere to the cell components and stains them blue whereas acid-fast cells remain red.

Expt. 95　　*To perform the acid-fast staining technique*

MATERIALS REQUIRED

Laboratory glassware

　Glass slides

　Lens paper

Laboratory equipment

　Bunsen burner

　Microscope

Staining tray

REAGENTS

Carbol fuchsin

Acid-alcohol

Methylene blue

PROCEDURE

1. Prepare bacterial smear on the clean glass slides.
2. Allow them to air-dry and heat-fix it.
3. Flood it with carbol fuchsin in a staining tray.
4. Heat fix it once again in a hot plate.
5. Allow the stain to evaporate but take care not to boil the culture.
6. Cool it immediately.
7. Flush the stain with tap water; hold the glass slide under tap water as described in procedure for simple staining.
8. Now add the decolourising agent to the culture in drops.
9. Continue to add till the added drops cease to wash away the stain.
10. 'Counterstain the slide with methylene blue. Keep it for another 2 minutes.
11. Wash it with tap water.
12. Blot it dry and examine under microscope.

5.13 SPORE STAINING

5.13.1 Principle

This technique is used to identify the spore formation in mixed cultures of certain aerobic (*Bacillus*) and anaerobic genera (*Clostridium* and *Desulfotomaculum*) when they exist as inactive spores due to unfavorable conditions. These spores have an inner endospore covered by an impervious spore coat. The endospore structure is protected from high and low temperatures, radiation, desiccation, chemical agents and even stains. Under suitable environmental conditions, these endospores regain their original structure by the process of germination. Two different reagents are used in this technique.

1.   Malachite green (primary stain)
2.   Safranin (counterstain)

5.13.2 Malachite Green

It is a primary stain. Presence of impervious cell wall prevents the staining of microbial cells by malachite green. Therefore after the application of primary stain, the slide is heat-fixed. This facilitates the stain penetration and in this step both the spores and regulative cells are stained green.

5.13.3 Decolourising Agent

Tap water is used to flush the excess primary stain from the cellular components. Since the spores are deeply stained, they remain green whereas the stain from the vegetative cells is completely washed away making them colourless.

5.13.4 Safranin (Counterstain)

Colourless vegetative cells take the colour of the safranin to appear red and spore cells continue to remain green.

Expt. 96   *To perform the spore staining technique*

MATERIALS REQUIRED

Laboratory glassware

Glass slides

Lens paper

Laboratory equipment

Bunsen burner

Microscope

Staining tray

REAGENTS

Malachite green

Decolourising agent

Saffranin

Procedure

1. Prepare the smear from the culture to be identified, on a clean glass slide.
2. Allow it to air-dry and heat-fix it.
3. Flood the smear with malachite green in a staining tray.
4. Heat it in a hot plate for about 3 minutes.
5. Do not boil the stain, instead, warm it gently.
6. Cool it immediately.
7. Hold the slide under running tap water.
8. Flood the smear with counterstain. Leave it undisturbed for another 30 seconds.
9. Wash it under running tap water.
10. Blot it dry and examine it under the microscope.

5.14 CAPSULE STAINING

5.14.1 Principle

Virulent bacteria protect themselves from immune cells of the host by a gelatinous outer layer called capsule. It is made up of polysaccharides or polypeptides. Identification of microorganisms with capsule is useful in preliminary investigation of pathogens, which may be confirmed by further tests. However, capsule staining is difficult as capsule materials are water-soluble and therefore will be removed while washing. Also, the smears should not be heat fixed, because cells without capsule may shrink on heating and form a clear zone around the cell. This, when examined under microscope looks like a capsule. Stains used to identify capsule are

1. Crystal violet (1%)
2. Copper sulphate (20%)

5.14.2 Crystal Violet

Application of primary stain will render purple colour to the microbial cells to be analysed. Since the capsule is non-ionic, the primary stain will adhere to it rather than get absorbed or penetrated into the cell.

5.14.3 Copper Sulphate

Copper sulphate is used both as a decolourising agent and as a counterstain. Since the primary stain can be easily washed away, copper sulphate from cells with capsule gives light blue colour to these cells.

Expt. 97   *To perform the capsule staining technique*

Materials Required

Laboratory glassware

Glass slides

Laboratory equipment

Bunsen burner

Microscope

Staining tray

REAGENTS

Crystal violet

Copper sulphate solution

PROCEDURE

1. Prepare bacterial smear on clean glass slides.
2. Allow it to air-dry and heat-fix it.
3. Flood the stain with 1% w/v aqueous solution of crystal violet.
4. After two minutes, wash the stain with 20% copper sulphate solution.
5. Drain the solution and blot it dry.
6. Then examine the smear under the microscope.

5.15 GRAM STAINING

5.15.1 Principle

This method involves the use of four chemical reagents

1. Primary stain
2. Mordant
3. Decolourising agent
4. Counterstain

Crystal violet is used as primary stain which stains the microbial cells to appear purple blue. Gram's iodine acts as a mordant, a substance that forms an insoluble complex with crystal violet (CV-I complex) thus intensifying the colour of the primary stain. Now the cells appear purple-black in colour. Being a lipid solvent and protein dehydrating agent, ethyl alcohol plays a dual role. Lipid concentration in the cell wall of microorganisms determine its action in gram-positive cells, the lipid concentration is low and the primary stain is retained. Smaller the lipid content in microbial cell wall, lesser will be the dissolution to form minute cell wall pores. Soon, the pores will be clogged by the dehydration of cell wall proteins which in turn retain the primary stain. In contrast, large amount of lipids in gram-negative cells will lead to large-sized pores for which the dehydrating effect of ethyl alcohol in sufficient. This results in rapid release of CV-I complex thus making the cells colourless. Application of counterstain (safranin) at this stage will stain the cells pink.

Expt. 98   *To perform the gram staining technique*

MATERIALS REQUIRED

Laboratory glassware

Glass slides

Laboratory equipment

Bunsen burner

Microscope

Staining tray

REAGENTS

Crystal violet

Gram's iodine

PROCEDURE

1. Place a drop of water on a clean glass slide.

2. Aseptically transfer a loopful of culture on the slide and prepare a smear (this can be done by exhibiting circular movement by the inoculating needle).

3. Air-dry the smear with crystal violet and allow it to stand for 1 minute.

4. See to it that the smear is submerged in the stain.

5. Wash with tap water. Continue the washing till water dripping from the slide appears colourless.

6. Do not expose the smear directly to running water.

7. Flood the smear with Gram's iodine and allow it to stand for one minute. Repeat the washing procedure as described above.

8. Wash it with ethyl alcohol. Now counterstain the smear with safranin and allow it to stand for 45 seconds.

9. Again wash it with tap water and air-dry. Observe for microbial growth. Presence of gram-negative non-spore-forming rod-shaped bacteria indicates faecal contamination of the analysed sample.

5.16 LACTOPHENOL BLUE STAINING

This stain is used in the identification of fungi as the lactophenol blue gives colour to the fungal cytoplasm. Thus in a light blue background hyphal walls can be clearly examined.

Expt. 99   *To perform the lactophenol blue staining technique*

MATERIALS REQUIRED

Laboratory glassware

Laboratory equipment

Bunsen burner

Microscope

REAGENTS

Lactophenol blue stain

Phenol crystals   20 g

Lactic acid	20 ml
Glycerol	40 ml
Cotton blue/methyl blue	0.05 g
Distilled water	1 l

PROCEDURE

1. Place a drop of lactophenol blue stain in a clean dry slide.
2. Carefully transfer a loopful of mould over the lactophenol stain drop.
3. Using sterile needles, tease the hyphae and spread it evenly.
4. Now cover the spread culture with a cover slip. Take care not to entertain air bubbles inside the coverslip.
5. Examine the slide under the microscope and check out for the presence of light blue fungal cytoplasm covered by the unstained hyphal cell well against a light blue background.

5.17 STANDARD PLATE COUNT (SPC)

Enumeration of total bacteria in the sample taken, say for example water, directly reflects its quality. The presence of aerobic and facultative anaerobic hydrophilic bacteria in water can be determined by the standard plate count method. However, the result obtained may not be very accurate. This is because microorganisms with varied growth requirements cannot be expected to grow under similar environmental conditions and single growth medium. In spite of these disadvantages, this method is universally accepted as it represents the predominant microorganism, which in turn is dependent on the treatment processes. This method is employed in the inlet and outlet so as to evaluate the efficiency of the treatment performance.

5.17.1 Sample Collection

Composite samples of water are analysed following the procedures described in chapter 1. The sample analysis is performed immediately after sample collection. If this not possible, it is done before 8 hours. Beyond this time, the collected samples should be stored at temperatures greater than 10°C. In this case, the analysis should be done within 30 hours from the time of collection. Outside the refrigerator, exercise care to maintain the temperature of the bottled samples between, 20 and 25°C. Prior to analysis, shake the bottle up and down, and back and forth to facilitate complete mixing.

Expt. 100 *To perform the standard plate count technique*

MATERIALS REQUIRED

Laboratory glassware
> Pipettes
> Petri dishes

Laboratory equipment
> Incubator
> Water bath
> Colony counter

Dilution water

Take 34 g of potassium dihydrogen phosphate (KH_2PO_4) in 50 ml distilled water. Carefully add 1N NaOH in drops until the pH of the solution reaches 7.2. Make up to 1 l. To this add 1.25 ml of phosphate buffer stock solution and 5.0 ml of $MgSO_4$ (Prepare this by dissolving 50 g of $MgSO_4$ in 1 l distilled water). Autoclave the solution at 121°C for 15 minutes. See to it that it is freshly prepared before use.

Tryptone glucose yeast agar

Tryptone 5.0 g

Yeast extract 2.5 g

Glucose 1.0 g

Distilled water 1 l

Maintain the pH at 7.0

PROCEDURE

1. Prepare different dilutions based on the quality of water kept for analysis. However in common, two dilutions 1 : 0 and 1 : 100 are prepared.

2. Pipette out 0.1 ml and 1 ml of sample to get 1 : 0 and 1 : 10 dilution states respectively.

3. Similarly to acquire 1 : 100 and 1 : 1000 dilutions, dilute 0.1 and 1 ml of the sample with 100 ml of diluent water.

4. Heat the contents of tryptone glucose yeast agar till it melts completely.

5. Shake well intermittently to avoid pellet formation.

6. Then sterilize the medium in the autoclave at 121°C for 15 minutes.

7. Allow it to cool for 10 minutes.

8. Meanwhile, pour 0.5 ml from each of the four dilutions into sterilized petri plates and rotate in clockwise and anticlockwise direction.

9. Add 15–20 ml of tryptone glucose extract agar into each petri plate.

10. To ensure mixing of medium and sample, repeat the rotation procedure.

11. Take care to maintain the temperature of the medium at 44– 46°C .

12. While pouring the medium keep the petri plates undisturbed for about 10 minutes. After complete solidification, incubate it at 37°C ± 0.5°C for 48 hours. In the case of waste water, this can be extended up to 72 hours.

13. Using the colony counter, count the number of colonies and express it as CFU/ ml of water. 30 and 300 are the two extremities below and beyond which the results cannot be considered and therefore should be neglected as too numerous to count (TNTC) for colonies greater than 300 and too low to count (TLTC) for colonies less than 30. If all the plates exhibit values above 300, consider the number which is nearer to 300.

Expt. 101   ***To perform the standard plate count technique for faecal streptococci***

MATERIALS REQUIRED

Laboratory glassware
  Pipettes
  Petri plates
Laboratory equipment
  Inoculation loops
  Autoclave
  Incubators

MEDIUM

KF streptococcus agar used in membrane filter technique.

PROCEDURE

1. Pour molten KF streptococcus agar medium in previously sterilized petri plates.
2. Allow it to solidify.
3. Depending on the pollution status of the water to be analysed, dilute it. From the diluted sample pipette out 1 ml or 0.01 ml into the petri plates.
4. Rotate the petri plates in clockwise and anticlockwise direction so as to distribute the sample equally.
5. Allow it to solidify and incubate it at $35 \pm 0.5°C$ for 48 hours.
6. Following the incubation period, observe for the dark red to pink colonies.
7. Consider both the surface and the submerged colonies with complete edges.
8. Express the result as number of faecal streptococci/ml of sample used.

5.18 MOST PROBABLE NUMBER (MPN) METHOD

Total coliforms involve the *E. coli* and *Enterobacter* spp. The only source of the former is faecal matter whereas the latter may enter the water source from the soil. Since only *E. coli* can utilize the lactose broth at higher temperature, i.e., at $44.5 \pm 0.2°C$, by performing the same test at this temperature of incubation for 24 hours, coliforms of faecal source can be confirmed.

Expt. 102   ***To perform the MPN technique for faecal coliforms***

MATERIALS REQUIRED

Laboratory glassware
  Petri plates
  Pipettes
  Durham tubes

 Lab equipment

  Incubator

  Autoclave

REAGENTS

 EC medium

Tryptone/trypticase	20.0 g
Lactose	5.0 g
Bile salts	1.5 g
K_2HPO_4	4.0 g
KH_2PO_4	1.5 g
Sodium chloride	5.0 g
Distilled water	1 l

 pH 6.9

PROCEDURE

1. The procedure is similar to that of standard total coliform MPN tests except for the medium used in presumptive and completed tests and the temperature to be maintained during incubation period.

2. Instead of BGLB medium, inoculate the water samples in EC medium, a differential medium for *E. coli*.

3. Likewise set the temperature at $44.5 \pm 0.2°C$ during incubation and follow the same procedure as described in coliforms.

5.19 MEMBRANE FILTER TECHNIQUE

5.19.1 Principle

This is used to determine the quality of water sample. The results obtained are highly reliable and this method has several advantages over the conventional multiple tube method. This technique is widely accepted. However the adoption of this method is not preferable for turbid samples, as suspended particles may block the pores not allowing sufficient water to pass through, thus misinterpreting the results. Similarly presence of toxic metals in the sample may show low coliform count in fixed sample volume.

Expt. 103 *To perform the membrane filter technique*

MATERIALS REQUIRED

 Laboratory glassware

  Pipettes

 Forceps

 Absorption pads

Laboratory equipment

 Filtration apparatus

 Filter membranes (0.45 mm)

 Incubator

 Autoclave

MEDIUM

a. Endo agar medium

b. *E. coli* medium

c. KF streptococcus agar

Proteose peptone (No. 3)	0 g
Yeast extract	10 g
Sodium chloride	5 g
Sodium glycerophosphate	10 g
Maltose	20 g
Lactose	1 g
Sodium azide	0.4 g
Agar	20 g
Distilled water	1 l

PROCEDURE

1. Sterilize the whole membrane filter unit and assemble the apparatus as follows.

2. Carefully insert the sintex glass filter base of the apparatus into the neck of suction flask.

3. Now place the sterilized bacteria retaining the membrane filter disc ($<$ 45 mM) into the perforated receptacle of the sintex glass platform.

4. Take care to fix it properly. Connect the rubber tube from the side arm of the suction flask to the vacuum pump.

5. As given below, take appropriate volume of the sample and filter it using vacuum pressure.

Potable water	100 ml
Running water	0.001 to 1 ml
Treated sewage	0.01 ml to 1 ml
Sewage	0.0001 to 0.01 ml
Stagnant water	1 ml to 100 ml

6. After filtration, flush the inside of the apparatus with 30 ml sterilized diluent water and filter the same.

7. Now carefully remove the membrane filter disc with sterile forceps and place it above the absorption pad of petri plates.

8. For a specific sample volume, use three sterilized petri plates and label them as TCC (total coliform count), FCC (faecal coliform count) and FSC (faecal streptococcal count).

9. Under strict aseptic conditions, insert the sterile absorbent pad into the petri plates. Dip the forceps in 95% alcohol; heat it over red-hot flame before use.

10. Add 2 ml each of endo broth, EC broth and KF broth into the absorbent pad of petri plates labeled as TCC, FCC and FSC respectively.

11. Incubate the TCC and FCC petri plates for 24 hours at 37°C.

12. Seal the FSC petri plates with waterproof tape, wrap it in waterproof plastic bag and incubate in the water bath at 44.5°C for 24 hours.

13. To keep the plastic bag submerged in water, tie it with a weight, following the incubation period.

14. Take the filter disc out of the petri plates and allow it to dry on an absorbent paper for 45 minutes.

15. Now examine it for coliform growth. Colonies with a golden metallic sheen in M–Endo broth, blue colonies in m–FC broth and pink-red colonies in KF agar confirms the presence of coliforms.

16. For all the three mediums, petri plates with less than 20 colonies should be discarded as "too few to count" (TFTC). But the number of colonies designated as "too numerous to count" (TNTC) varies with the medium and it is 50 for M–Endo broth; 60 for M–FC broth and 100 for KF broth. Values in between this range can be expressed as

$$\frac{\text{No. of colonies} \times \text{dilution factor} \times 1000}{\text{ml of sample used}}$$

Expt. 104 *To perform the membrane filter technique for faecal streptococci*

MATERIALS REQUIRED

Laboratory glassware

    Pipettes

    Culture dish

Membrane filter

Laboratory equipment

    Autoclave

    Incubator

    Inoculation loop

MEDIUM

Streptococcus agar medium

Proteose peptone	1 g
Yeast extract	1 g
Sodium chloride	0.5 g

Sodium glycerol phosphate	1.0 g
Maltose	2.0 g
Lactose	0.1 g
Sodium azide	0.04 g
Agar	2 g
Distilled water	100 ml

PROCEDURE

1. Pour molten streptococcus agar medium into the culture dish.

2. Allow it to solidify.

3. Meanwhile, filter the sample to be analysed in a sterile membrane filter. See to it that the sample size is sufficient to give 20–100 colonies on the membrane surface. Reduce the size of the sample if it is highly polluted.

4. Now using a sterile forceps carefully transfer the filter to the agar medium in a petri dish.

5. Do not entertain air bubbles while transferring the culture.

6. After it is properly placed, incubate the plates at $35 \pm 0.5°$ C for 48 hours.

7. Following the incubation period, observe for the growth of dark red to pink colonies.

8. Express the result as faecal streptococci density/amount of the sample used for filtration.

5.20 MULTIPLE TUBE FERMENTATION TECHNIQUE FOR COLIFORMS

This technique evaluates the most probable number of coliforms (MPN) present in the sample. It involves three tests conducted in series for each sample. They are

1. Presumptive test
2. Confirmed test
3. Completed test

Expt. 105   *To perform the multiple tube fermentation technique for coliforms*

MATERIALS REQUIRED

Laboratory glassware

    Petri plates

    Durham tubes

    Pipettes

    Glass slides

Laboratory equipment

    Autoclave

    Incubator

    Inoculating needle

    Bunsen burner

MEDIUM

a.   Brilliant green lactose bile medium (BGLB)

Peptone	10 g
Lactose	10 g
Oxgall	20 g
Brilliant green	0.0133 g
Distilled water	1 l

pH 7.2

b.   Endo agar medium

Peptone	10 g
Lactose	10 g
K_2HPO_4	3.5 g
Agar	15 g
Sodium sulphite	2.5 g
Basic fuchsin	0.5 g
Distilled water	1 l

pH 7.4

c.   EMB agar medium

Peptone	10 g
Lactose	5 g
Dipotassium phosphate	2 g
Agar	13.5 g
Eosin Y	0.4 g
Methylene blue	0.065 g
Distilled water	1 l

d.   Crystal violet

e.   Gram's iodine

f.   Ethyl alcohol (95%)

g.   Saffranin

PROCEDURE

Presumptive Test

This technique being specific for coliforms detects their presence in water samples. Out of several enteric organisms, only coliforms can utilize lactose in brilliant green lactose bile (BGLB) as carbon source and hence can be identified. The presence of surface tension depressant and bile salts prevent the growth of other bacteria. All the factors facilitate the growth of coliforms whose growth is confirmed by the gas and acid production.

1. Arrange 3 sets of three sterile fermentation tubes each, label the water source and the quantity of sample to be taken as given in Figure 5.10.

2. As per the dilution procedure illustrated in figure, fill the test tubes with BGLB medium.

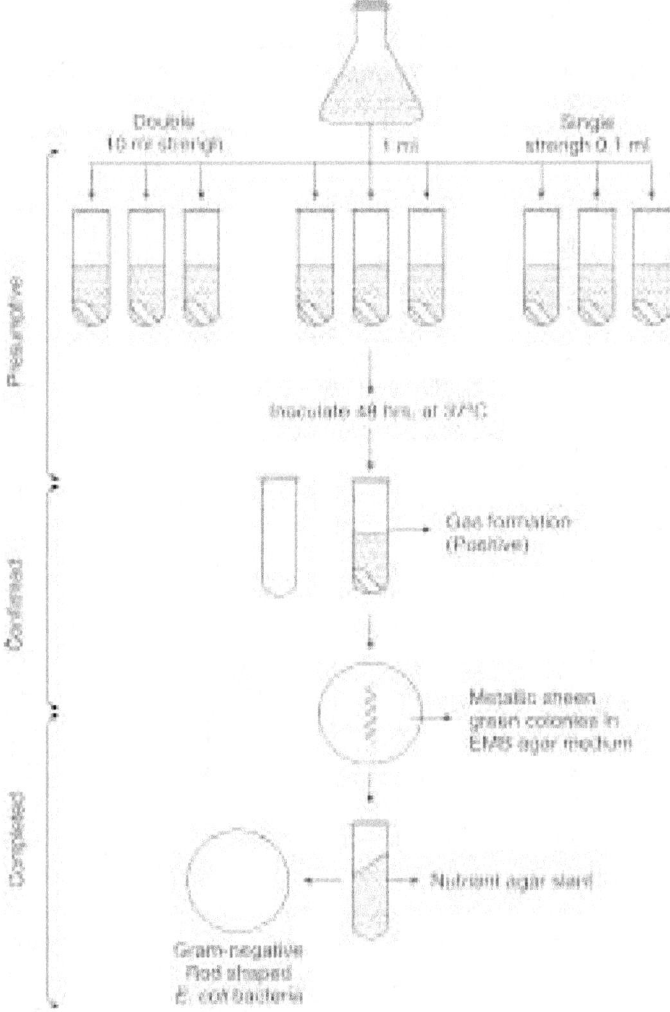

Figure 5.10   Multiple tube fermentation technique for coliforms

3. Cotton plug the test tubes and autoclave at 121°C for sewage water sample aliquots into the appropriate fermentation tubes.

4. Use separate pipettes for different volumes (0.1, 0.01, and 0.001 ml).

5. Perform these steps near the flame and flame the rim of both the sampling and fermentation tubes before and after transfer of sample.

6. Repeat the same procedure for running and stagnant water samples and incubate the inoculated fermentation tubes for 48 hours at 35–37°C.

Table 5.1 MPN Index table (Determination of MPN using 3 samples of 10 ml, 1 ml and 0.1 ml)

Combination	MPN/100ml	Combination	MPN/100ml
0.0.0	>2	430	28
0.0.1	2	431	33
0.1.0	2	440	34
0.2.0	4	500	23
1.0.0	2	501	31
1.1.0	4	502	43
1.1.1	4	510	33
1.2.0	6	511	46
2.0.0	6	512	63
2.0.1	5	520	69
2.1.0	7	521	70
2.1.1	7	522	94
2.2.0	9	530	79
2.3.0	9	531	110
3.0.0	12	532	140
3.0.1	8	533	180
3.1.0	11	540	130
3.1.1	11	541	170
3.2.0	14	542	220
3.2.1	14	543	280
4.1.0	17	544	350
4.0.1	13	550	240
4.1.0	17	551	350
4.1.1	17	552	840
4.1.2	21	553	920
4.2.0	26	554	1600
4.2.1	22	555	72400

7. After this period, observe for the gas production and record it as positive whereas its absence is considered as negative. Positive tubes showing gas production can be taken to be the confirmatory test.

8. To evaluate the most probable number of coliforms, refer the MPN index table (table 6.1).

9. Carefully introduce the sterile durham tubes in the inverted position into the fermentation tubes so that in each tube it reaches the bottom, with the inside of the tube completely filled with BGLB medium.

10. See to it that there exist no bubbles inside the inverted tube.

A schematic representation showing the procedure for total coliforms is shown below.

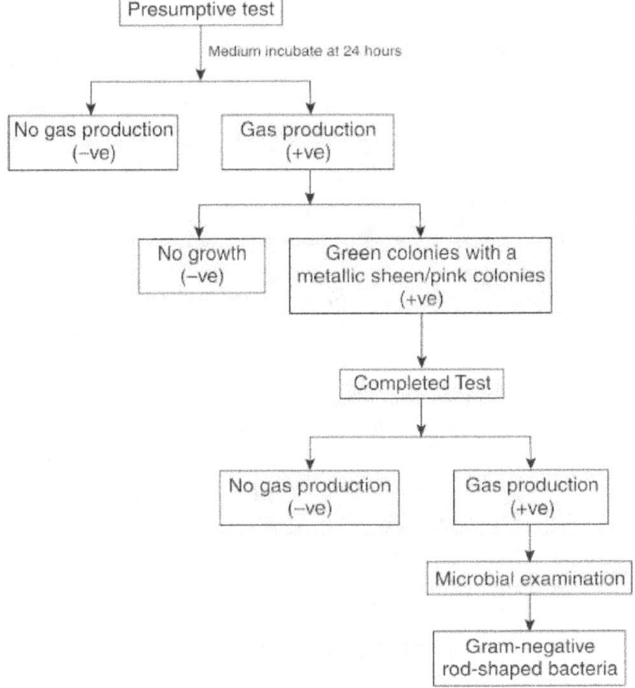

Confirmed Test

The result acquired in the presumptive test involve both gram-positive and gram-negative enteric organisms. Confirmation of result is necessary and for this, selective and differential media such as EMB or endo agar are used.

EMB agar

The principle is that the eosine methylene blue in EMB agar suppresses the growth of gram-positive microorganisms. Also it forms a complex with the acid produced by *E. coli* to form-dark centred colonies with a green metallic sheen.

Endo agar

In acidic environment, fuchsin dye in the endo agar medium turns into dark pink colour. Obviously, the characteristic acid-producing ability of *E. coli* colonies will exhibit pink colour in this medium.

1. Take 2 sets of three sterile petri plates each, first set for EMB and second for endo agar respectively.

2. Pour the sterilized EMB and endo agar medium into the respective plates and allow it to solidify.

3. From the positive BGLB tubes, take a loop of inoculum, streak into the petri plates. Take care to flame the rim of petri plates and sample tubes before and after transfer. To acquire discrete colonies, use four-way streak plate method. Keep the petri plates in an inverted position in the incubator for 48 hours at 35–37°C and record the result. Presence of nucleated, fat, dark-centered colonies with a metallic sheen or atypical growth with pink, non-nucleated, mucoid colonies confirms the coliform growth.

Completed Test

BGLB tubes showing positive results in confirmed test are used for completed tests. It is the final analysis in which the colonies which appeared in the confirmed test are examined. A loopful of isolated colony is transferred to BGLB medium and then incubated for 24 hours at 35–37°C. Test tubes showing positive result are further confirmed by microscopic examination.

1. Transfer a loopful of culture showing positive results in EMB/Endo agar petri plates to sterilized BGLB medium-containing fermentation tubes.

2. Incubate it for 24 hours, and following incubation period, observe for gas production.

3. From these test tubes, take a loopful and perform gram-staining.

5.21 MULTIPLE TUBE FERMENTATION TECHNIQUE FOR FAECAL STREPTOCOCCI

Faecal streptococci refer to a variety of species such as *S. faecalis*; *S. faecalis*, var *liquifaciens* , *S. faecalis* var *zymogenes*, *S faecium*, *S. faecium* var *durans*, *S. bovis* and *S. equines*. These organisms reside in the intestine of animals and humans and their presence in water therefore indicates faecal contamination. *S. bovis* and *S. equines* have limited host range and cannot survive outside for a long time. Hence the status of the water body with their presence can be categorized as a "recently polluted". In contrast, *S. faecalis* var *liquifaciens* has a wide range of host habitats (intestine of man and animals, vegetation, insects and soil) and therefore cannot be taken into consideration.

Expt. 106   *To perform the multiple tube fermentation technique for faecal streptococci*

MATERIALS REQUIRED

Laboratory glassware

    Pipettes

    Fermentation tubes

Laboratory equipment

    Autoclave

    Inoculation loop

    Incubator

MEDIUM

a. Azide dextrose broth

Beef extract	4.5 g
Tryptone/polypeptone	15 g
Glucose	7.5 g
Sodium chloride	7.5 g
Sodium azide	0.2 g
Distilled water	1 l

pH 7.2

b. Ethyl violet azide broth

Tryptone	20 g
Glucose	5.0 g
Sodium chloride	5.0 g
K_2HPO_4	2.7 g
KH_2PO_4	0.4 g
Ethyl violet	0.00083 g
Distilled water	1 l

pH 7.0

PROCEDURE

This method involves two steps

1. Presumptive test

2. Confirmed test

Presumptive Test

1. Take 2 sets of 5 test tubes each, add 10 ml of single-strength azide dextrose broth to the first set and 10 ml of double-strength azide dextrose broth to the second set respectively.

2. Sterilize it in the autoclave at 121°C for 15 minutes.

3. After cooling, incubate 10 ml and 1 ml of the sample in the first and second set and incubate both the sets at 35 ± 0.5°C for 24 ± 2 hours.

4. After incubation period, observe for the turbidity in the test tubes. In the absence of turbidity, the incubation period can be extended up to 48 ± 3 hours. Interpret the result with the aid of MNP index table.

Confirmed Test

1. From the positive tubes of azide dextrose broth, transfer three loopsful of culture into the test tubes containing ethyl violet azide broth.

2. Incubate the tubes at $35 \pm 0.5°C$ for 24 hours.

3. Following the incubation period, observe for the dense turbid growth or the formation of purple precipitate at the bottom of the tube.

4. In the absence of growth, re-inoculate three loopsful of culture and inoculate them for another 24 hours.

5. Observe the results and consider this as final.

Expt. 107 *To isolate Rhizobium from root nodules*

MATERIALS REQUIRED

Laboratory glassware

Laboratory equipment

    Autoclave

    Incubator

REAGENTS

a. Mercuric chloride (0.1%)

b. Ethyl alcohol (70%)

c. Congo red yeast extract mannitol agar

Mannitol	1.0 g
K_2HPO_4	0.05 g
$MgSO_4$	0.02 g
NaCl	0.01 g
Yeast extract	0.1 g
$CaCO_3$	0.3 g
Agar	1.5 g
Distilled water	100 ml
Congo red	1.5 ml

For preparation of congo red, dissolve 1 g of congo red powder in 400 ml distilled water.

PROCEDURE

1. Carefully uproot a 6–8 week-old healthy leguminous plant. To acquire the whole plant, remove the plant with a soil block of 25 cm diameter and 10–15 cm depth.

2. Gently hold the plant under running water so that the soil particles adhering to the roots are completely removed.

3. Observe for the pink, multilobed nodules on the taproot.

4. Always select large-sized nodules and carefully separate nodules from the root. While doing this, take care not to damage or cut the rootlets attached to the surface of the nodule.

5. Transfer the nodule into a test tube and wash it several times with water.

6. Again transfer the nodule into a sterile test tube containing 0.1% mercuric chloride solution. Let the nodule be inside the tube for 3 minutes.

7. Discard the solution and add a little amount of 70% ethyl alcohol solution. Let the nodule be inside the tube for another one minute. Shake the tube well.

8. Following the exposure of root nodule to sterilants, wash it with sterile distilled water at least ten times. This will completely eliminate the sterilants from the nodule.

9. Now add a few drops of sterile distilled water into the test tube and smash the nodule using the flat end of a sterile glass rod. Perform this action without damaging the test tube.

10. Wash the glass rod with a few ml of sterile distilled water by keeping it inside the test tube itself.

11. Now the suspension is rich in rhizobium strain. Take 3 test tubes (a), (b) and (c) containing 5, 6 and 7 ml of sterile distilled water respectively.

12. Transfer a loopful of suspension to the test tube a, shake well and from a transfer a few drops to b. After thorough mixing transfer few drops from tube b to c.

13. Pour molten Congo red yeast extract mannitol agar into three petri plates labeled as a, b and c.

14. Transfer a drop of suspension from the three test tubes a, b and c into the respective petri plates a, b and c.

15. Incubate the plates at 28°C and observe for the development of small round colourless colonies. However *Agrobacterium* species may resemble Rhizobium and any of the tests given below can be done to confirm the species.

Test I

Reagents

Hoffer alkaline medium

1. Thymol blue solution (1.6%)

2. Yeast extract mannitol agar medium

    For every 100 ml of yeast extract mannitol agar medium, add 0.1 ml of 1.6% thymol blue solution. Add 2.8 ml of in 1N NaOH and adjust the pH to 11.

Procedure

1. Prepare slants using the Hoffer alkaline medium.

2. Streak the colonies developed from the different dilutions of nodule suspension on the slants.

3. Incubate the slant at 28°C for 15 days.

4. Absence of growth/absence of colour change is the positive result confirming the colony to be *Rhizobium*.

Test II

Reagents

Glucose peptone agar medium

Glucose	1.0 g
Peptone	2.0 g
NaCl	0.5 g
Agar	1.5 g
Bromocresol purple (1.6%)	0.1 ml
Distilled water	100 ml

pH 7.1

Procedure

1. Prepare slants using glucose peptone agar medium.

2. Streak the colonies developed from different dilutions of nodule suspension on the slants.

3. Incubate the slant at 28°C for 15 days.

4. Rhizobium colonies exhibit very little or no growth.

Test III

Reagents

1. Yeast extract lactose agar

    Yeast extract lactose agar preparation is similar to that of yeast extract mannitol agar except that mannitol is replaced by lactose agar.

2. Benedict's reagent

    Solution A Dissolve 173 g sodium citrate and 100 g sodium carbonate in 600 ml distilled water.

    Solution B Dissolve 17.3 g of copper sulphate in 100 ml distilled water.

    Add slowly, solution B to A. Stir well. If the solution remains turbid, filter it.

Procedure

1. Pour molten yeast extract lactose medium into a series of petri plates.

2. Allow it to solidify and streak the colonies developed from the different dilutions of *Rhizobium* suspension.

3.   Incubate the plates and observe for the growth formation.

4.   Now flood the colonies with Benedict's reagent.

5.   Since *Agrobacterium* species are able to produce 3 ketolactose in lactose medium, they exhibit yellow ring of cuprous oxide around their growth.

6.   Lack of such ring formation confirms the presence of *Rhizobium* strain.

5.22 BACTERIAL IDENTIFICATION BY BIOCHEMICAL TESTS

Cellular metabolism of any organism is governed by biological catalysts called enzymes. Even the simplest microorganisms are no exception. Here too, enzymes determine the biochemical pathways by which the microorganisms acquires energy (ATP), synthesize new protoplasmic requirements and get the ability to degrade a particular substance. Since these biochemical activities vary between species, evaluation of biochemical pathways is a typical way for the microbiologists to identify and characterize them.

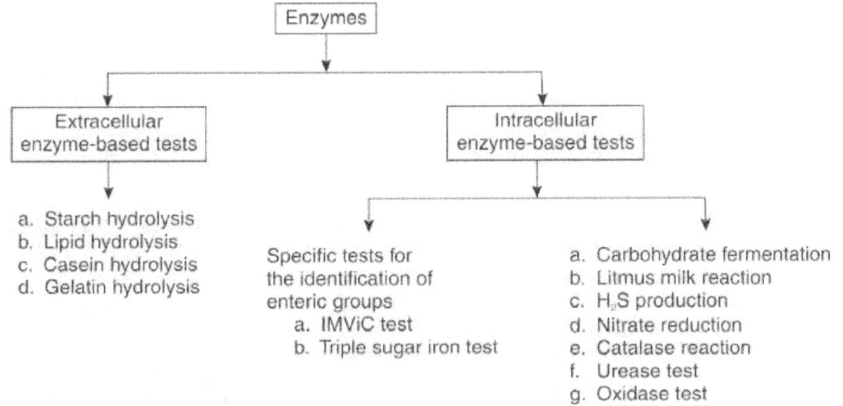

As described above, enzymes take in the control of whole cell. They act both inside (endoenzymes) and outside (exoenzymes) the cell. Exoenzymes due to their hydrolytic nature break down the high molecular weight complex substances into simpler ones which is then absorbed into the cell for further breakdown and assimilation. ATP production from the assimilated substances is facilitated by endoenzymes. This energy is utilized for daily cellular activities and synthesis of new cellular components. The ultimate result of this biochemical transformation is the excretion of metabolic products by the cell into the environment. Evaluation of the end products helps to identify the enzyme and thereby the identification of the bacterial strain. In general bacterial identification serves in finding out the infectious microorganisms and in screening out the microorganisms for commercial purpose.

5.23 STARCH HYDROLYSIS

Starch is composed of several molecules of glucose linked together by glycosidic bonds. It is a mixture of amylose (linear form) and amylopectin (highly branched form). Being complex in nature, starch needs the amylase enzymatic activity for hydrolysis. Amylase acts on 1–4 glycosidic linkage of starch and hydrolyses them into fragments of dextrins and then to maltose molecules. Further hydrolysis to soluble glucose requires the presence of another enzyme maltase.

$$\text{Starch} \xrightarrow{\text{I}} \text{Soluble starch} \xrightarrow{\text{II}} \text{Amylodextrin} + \text{Maltose} \xrightarrow{\text{III}} \text{Erythrodextrin} + \text{Maltose}$$

$$\xrightarrow{\text{IV}} \text{Achrodextrin} + \text{Maltose} \xrightarrow{\text{V}} \text{Maltose} \xrightarrow{\text{VI}} \text{Glucose}$$

It should be noted that the concentration of maltose increases at every stage. In this experiment, starch agar is used as supplement for carbon source. Only the microorganisms with amylase activity can utilize the substrate and grow. Following the incubation period, when the petri plate is flooded with iodine, appearance of blue-black colour denotes the presence of insoluble starch whereas the formation of clear zone indicates microbial growth, hydrolysing the starch by starch-splitting enzymes.

Expt. 108 *To demonstrate starch hydrolysis*

MATERIALS REQUIRED

Laboratory glassware

Petri plates

Pipette

Laboratory equipment

Autoclave

Insulator

MEDIUM

a. Starch agar medium

Peptone	5 g
Beef extract	3 g
Soluble starch	2 g
Agar	15 g

pH 7.0

b. Iodine solution

Iodine	1 g
Potassium iodide	2 g
Distilled water	300 ml

Crush the iodine and potassium iodide in a mortar and pestle. Dissolve the mixed powder in 300 ml water and stir vigorously. Store in amber bottles for further use.

PROCEDURE

1. Heat the starch agar till the medium content melts.

2. Autoclave it at 121°C for 15 minutes.

3. After 10 minutes pour the medium into the sterilized petri plates (15–20 ml).

4. Allow it to solidify.

5. Then under aseptic conditions, transfer a loopful of growth from sample tube and perform single-line streaking across the centre of the petri plate.

6. Incubate the plates at 37°C for 24 hours.

7. Following the incubation period, flood the plates with iodine solution.

8. Appearance of clear zone around the microbial growth indicates positive results.

5.24 LIPID HYDROLYSIS

5.24.1 Principle

Lipids are the complex mixture of fatty acids and glycerol linked together by ester bonds. They possess high molecular weight and also act as reservoirs of large amount of energy. Their degradation therefore requires the enzymatic activity of lipase without which their entry through the microbial cell membrane would be impossible. However the presence of lipid-splitting enzymes enables their entry after which they undergo a series of metabolic activities, either to acquire, provide or to synthesize new protoplasmic components. In this experimental procedure tributyrin agar serves as a supplement for lipid source. In combination with agar, tributyrin forms an emulsion. Following the incubation period, a clear zone around the microbial growth confirms the presence of lipase.

Expt. 109   *To demonstrate lipid hydrolysis*

MATERIALS REQUIRED

Laboratory glassware

Petri plates

Test tubes

Pipettes

Laboratory equipment

Autoclave

Incubators

MEDIUM

Peptone	0.5 g
Beef extract	0.3 g
Agar	1.5 g
Tributyrin	1.0 g
Distilled water	100 ml
pH 7.2	

PROCEDURE

1. Prepare the medium as mentioned above.

2. Heat the medium well till the content of the flask melts.

3.   Shake well to avoid pellet formation.

4.   Then autoclave the medium at 121°C for 15 minutes.

5.   Under aseptic conditions pour 15–20 ml of the molten medium into the petri plates.

6.   Allow it to solidify.

7.   Then transfer a loopful of growth from the sample culture tube into the petri plates.

8.   Perform single-line streaking across the centre of the plate.

9.   Observe for the formation of clear zone around the microbial growth.

5.25 CASEIN HYDROLYSIS

5.25.1 Principle

Casein is a high molecular weight milk protein made up of amino acids, i.e., several amino acids are linked together by peptide bonds to form this complex structure. The enzyme protease facilitates its hydrolysis at the peptide bond by adding water molecules. Amino acids formed as a result of hydrolysis can be easily transported into the cell membrane where it is used for the synthesis of structural and functional cellular proteins.

In this experimental procedure, skimmed milk agar in supplemented for protein source. In combination with agar, it forms a colloidal suspension to give a milky white colour and opaque nature to the medium. The ability of the microorganism to undergo proteolysis is indicated by the formation of a clear zone and loss of opacity around the microbial growth whereas no change in the colour and nature of the medium indicates negative result.

Expt. 110   *To demonstrate casein hydrolysis*

MATERIALS REQUIRED

Laboratory glassware

    Petri plates

    Pipettes

Laboratory equipment

    Autoclave

    Incubator

MEDIUM

Skimmed milk agar medium

Skimmed milk powder	100 g
Peptone	5 g
Agar	15 g
Distilled water	1 l

pH 7.2

PROCEDURE

1. Prepare the medium as mentioned above.
2. Heat the medium well till the contents melt and shake well to avoid pellet formation.
3. Then autoclave the medium at 121°C for 15 minutes.
4. Pour 15–20 ml of medium aseptically into the sterilized petri plates.
5. Allow it to solidify.
6. Then transfer a loopful of growth from the sample culture tube into the petri plates.
7. Perform single-line streaking across the centre of the petri plate.
8. Observe for the clear zone formation around the microbial growth.

5.26 GELATIN HYDROLYSIS

5.26.1 Principle

Gelatin gains more significance in bacterial identification. It is an incomplete protein existing in two forms with change in temperature, i.e., at temperatures less than 25°C, it exists in gel state whereas at temperatures greater than 25°C it is in liquid state. Microorganisms possessing gelatinase enzyme can hydrolyse gelatin to low molecular weight soluble amino acids. Under these conditions even at a very low temperature (4°C), the gelatin cannot maintain its gel state and it is this liquefaction property that indicates the presence of gelatinase enzyme.

Expt. 111   *To demonstrate gelatin hydrolysis*

MATERIAL REQUIRED

Laboratory glassware

    Petri plates

    Pipettes

Laboratory equipment

    Autoclave

    Incubator

MEDIUM

Gelatin medium

Peptone	5 g
Beef extract	3 g
Gelatin	120 g
Distilled water	1 l

pH 6.8

PROCEDURE

1. Prepare the medium as mentioned above.
2. Heat the medium well till the content melts and shake well to avoid pelletformation.
3. Then autoclave the medium at 121°C for 15 minutes.
4. Allow the test tubes to cool.
5. Under aseptic conditions transfer a loopful of growth from the sample culture tube into the gelatin tubes.
6. Incubate it at 37°C for 24 hours.
7. Following the incubation period, keep the tubes inside the refrigerator at 4°C for at least 4 hours.
8. Observe for the liquefaction of gelatin.

5.27 OXIDASE TEST

In aerobic and facultative anaerobic microorganisms, oxidase enzymes catalyse the bio-oxidation of organic substrates. The proton transferred through a series of components in the electron transport chain finally combines with molecular oxygen to form H_2O_2 or H_2O. Cytochrome oxidase catalyses this step and this test helps to differentiate oxidase-positive genera (*Neisseria* and *Pseudomonas*) from oxidase-negative genera Enterobacteriaceae. Microorganisms producing cytochrome oxidase can be identified by allowing the culture to be grown in a medium supplemented with tetramethyl p-phenyl-diamine dihydrochloride. This reagent acts as an electron donor from an external source such that the microbial colonies with oxidase enzyme get oxidized to show a characteristic colour change from pink to maroon and finally to black. No colour change is observed in the case of oxidase activity.

Expt. 112 *To perform oxidase test*

MATERIALS REQUIRED

Laboratory glassware

    Test tubes

    Pipettes

Laboratory equipment

    Autoclave

    Incubator

MEDIUM

a. Trypticase soy agar medium

Trypticase	15 g
Phytane	5 g
Sodium chloride	5 g
Agar	15 g
Distilled water	1 l

pH 7.3

    b.   Tetramethyl p-phenyl-diamine dihydrochloride reagent (1%)

        Dissolve 1 g of this reagent in 100 ml distilled water.

PROCEDURE

1. Sterilize the medium in the autoclave at 121°C for 15 minutes.
2. Before sterilizing, ensure complete mixing of medium contents.
3. This can be accomplished by gentle heating.
4. Under aseptic conditions pour the medium into sterilized petri plates.
5. Transfer a loopful of growth from the sample culture tube to the petri plates and perform single streaking across the centre of the plate.
6. Incubate it at 37°C for 24 hours.
7. Following incubation period, add the reagent solution in drops over the microbial growth.
8. Observe for the colour change from pink to maroon and pure black at the end.

5.28 CATALASE TEST

Aerobic and facultative anaerobic microorganisms, during the energy-yielding bio-oxidation process, produce H_2O_2 and at times superoxide. Since superoxide is highly toxic, it should be immediately degraded by suitable enzymes. The enzymes involved in the degradation process are catalase or peroxidase. They act on the hydrogen peroxide to give water and free oxygen. In the absence of catalase or peroxidase, they possess superoxide dismutase. Those organisms that lack these enzymes possess nitrates or sulphate as final electron acceptor instead of oxygen. Presence of catalase enzyme can be determined by adding hydrogen peroxide to the substrate so that the O_2 evolved is identified by bubble formation.

Expt. 113   *To perform catalase test*

MATERIALS REQUIRED

Laboratory glassware

    Petri plates

    Pipettes

Laboratory equipment

    Autoclave

    Incubator

MEDIUM

a. Trypticase soy agar medium
b. $3\% H_2O_2$

    Dissolve 3 ml of hydrogen peroxide in 100 ml distilled water.

PROCEDURE

1. Prepare trypticase soy agar slant tubes.

2. Transfer a loopful of growth from the sample culture tube to the TSA slant tubes.

3. Incubate the inoculated test tubes at 37°C for 24 hours.

4. At the end of incubation period add in drops, 3% H_2O_2 over the slant surface and observe for the bubble or foam formation.

5.29 NITRATE REDUCTION TEST

Presence of nitrate or sulphate as final electron acceptor is the characteristic property of anaerobes. Instead of O_2, NO_3 acts as the final hydrogen acceptor and gets reduced to nitrite.

$$NO_3 + 2e^- + H^+ \rightarrow NO_2^- + H_2O$$

Some microorganisms still continue to reduce either to ammonia (NH_3^+) or to molecular nitrogen (N_2). Nitrate reduction process can be determined by supplementing 0.1% potassium nitrate on a semi-solid agar medium. Since the denitrification process requires anoxic condition, O_2 penetration is prevented by semi-solid medium. After incubation period, addition of sulphanilic acid and alpha naphthyl amine gives a characteristic cherry red colour, a positive result for nitrate reduction. However, the colour change indicates the conversion of nitrate to nitrite and the colourless cultures therefore have the possibility of getting reduced to ammonia and sometimes even to molecular nitrogen. To confirm this, zinc powder is added. The presence of red colour indicates the conversion of nitrates to nitrites, a negative result. In the absence of colour change it can be confirmed that nitrates were reduced to ammonia or to nitrogen gas.

Expt. 114   *To perform nitrate reduction test*

MATERIALS REQUIRED

Laboratory glassware

Test tubes

Pipettes

Laboratory equipment

Autoclave

Incubator

MEDIUM

Trypticase soy nitrate broth

Nitrite-free potassium nitrate   0.2 g

Peptone   5 g

Distilled water   1 l

REAGENTS

Solution A   8.0 g of sulphanilic acid dissolved in 1 l of acetic acid

Solution B   5.0 g of 2-napthalamine dissolved in 1 l of acetic acid

Store both solutions in separate bottles and mix them in equal proportions before use.

PROCEDURE

1.   Prepare nitrate broth tubes.

2.   Under aseptic conditions, transfer a loopful of culture into the nitrate broth tubes.

3.   Incubate the tubes for 96 hours at 35°C.

4.   Following the incubation period, add 0.1 ml of the reagent and immediately observe for the sharp colour change from orange to red.

5.30 LITMUS MILK REACTION

Milk sugar lactose, milk protein casein, lactoalbumin and lactoglobulin are major milk substrates susceptible to microbial transformation. Based on their enzymatic action, microorganisms undergo different biochemical changes such as lactose fermentation, gas production, litmus red action, curd formation, proteolysis and alkaline reaction. This helps to identify them. Microorganisms with β glycosidase act on lactose substrate to give lactic acid as the end product. Using litmus indicator, the acid production can be identified, i.e., the colour change from purple to pink indicates the pH shift from neutral to acid range. Fermentation-induced CO_2 and H_2 production can be identified by the curd separation. This is because, the gas evolved may cause the pressure developed in the upper layer to form cracks or fissures. The fermentation process occurs in anoxic condition and the process removes hydrogen from the lactose which is subsequently accepted by litmus indicator. Litmus under reduced condition turns from pink to milky white. Oxidized lactose produces end products such as lactic acid, butyric acid, CO_2 and H_2. Two types of curd formation are observed—acid curd and rennet curd.

5.30.1 Acid Curd

Acid end products, lactic acid or organic acid formed by the biochemical oxidation of milk substrates precipitate the milk protein casein to form calcium caseinate. It exists in the form of a hard insoluble clot. This can be identified by inverting the tube where the clot remains in its original position.

5.30.2 Rennet Curd

Organisms with rennin enzyme convert the casein to paracasein which combines with calcium ions to form calcium paracaseinate. It is semi-solid in nature and exhibits mobility even to slight tilting. Microorganisms that lack the efficiency to utilize lactose, use proteins as their energy source. With the presence of proteolytic enzymes, they hydrolyse milk proteins into amino acids. The end product is ammonia that creates alkaline condition in the medium. Following the incubation period, the positive result is exhibited by the deep purple upper portion and translucent brown whey-like middle portion where the protein has been completely hydrolysed.

5.30.3 Alkaline Reaction

Incomplete breakdown of casein into shorter polypeptide chains results in the production of alkaline end products. This ultimately reflects in the colour change of litmus from purple to deep blue.

Expt. 115   *To demonstrate litmus milk reaction*

MATERIALS REQUIRED

Laboratory glassware

Test tubes

Pipettes

Laboratory equipment

Autoclave

Incubator

MEDIUM

Skimmed milk agar

Skimmed milk powder	100 g
Litmus	0.075 g

pH 6.8

PROCEDURE

1. Prepare skimmed milk medium as described above.
2. Heat the contents of the flask and ensure complete dissolution of the medium.
3. Autoclave the medium at 121°C for 15 minutes.
4. Pour the molten agar medium into the test tubes.
5. Streak and stab the sample culture into the solidified medium.
6. Incubate the tubes at 37°C for 24 hours.
7. Observe for the change in colour and nature of the medium.

5.31 UREASE TEST

The test is specific for the identification of protein species which produce urease enzyme. These microorganisms can utilize urea by the hydrolytic action of urease enzyme and the enzyme breaks down the nitrogen and carbon bond of the substrate to give ammonia. This ultimately shifts the pH of the medium to the alkaline range. Presence of a suitable indicator (Phenol red) reveals the hydrolysis by a characteristic colour change from phenol red to light pink. This confirms the positive reaction.

$$\text{Urea} \xrightarrow{\text{Urease}} CO_2 + \text{Ammonia} + \text{Water}$$

Expt. 116 *To perform urease test*

MATERIALS REQUIRED

Laboratory glassware

Test tubes

Pipettes

Laboratory equipment

Autoclave

Incubator

MEDIUM

Urea broth

Urea	90 ml
Sterile distilled water	10 ml

PROCEDURE

1. Prepare urea broth tubes and under aseptic conditions, transfer a loopful of growth from sample culture tube to urea broth tubes.

2. Incubate it for 24–48 hours at 37°C.

3. Following the incubation period, observe for the colour change from phenol red to deep pink.

5.32 HYDROGEN SULPHIDE TEST

Microorganisms can utilize the sulphur in both organic and inorganic forms.

5.32.1 Utilization of Organic Sulphur

Breakdown of proteins by proteolytic microorganisms leads to the release of amino acids. Degradation of sulphur-containing amino acids requires the presence of an enzyme called cysteine desulphurase. Those microorganisms that produce this enzyme can only utilize the sulphur-containing amino acids. These enzymes act on the sulphur component and reduce it to sulphide.

$$\text{Cysteine} \xrightarrow{\text{Cysteine desulphurase}} \text{Pyruvic acid} + \text{Hydrogen sulphide} + \text{Ammonia}$$

5.32.2 Utilization of Inorganic Sulphur

Certain microorganisms can utilize inorganic sulphur compounds such as thiosulphate (S_2O_3) and sulphates (SO_4^-), and reduce them to sulphites (SO_3) and hydrogen sulphide. The process involved is oxidation and the sulphur atom of the inorganic compound acts as hydrogen acceptor.

$$\text{Thiosulphate/sulphate} \xrightarrow{\text{Thiosulphate reductase}} \text{Sulphite} + H_2S \text{ gas}$$

Since the process requires anoxic condition, semi-solid agar medium is used. The medium is supplemented with peptone, sodium thiosulphate and ferrous ammonium sulphate (indicator). The black precipitate formed along the growth of culture reveals the evolution of H₂S. The precipitate (coloured) is obtained by the complex reaction between hydrogen sulphide and ferrous ammonium sulphate.

Expt. 117 *To perform hydrogen sulphide test*

MATERIALS REQUIRED

Laboratory glassware

Test tubes

Pipettes

Laboratory equipment

Autoclave

Incubator

MEDIUM

Sim agar

Peptone	3 g
Beef extract	0.3 g
Ferrous ammonium sulphate	0.02 g
Sodium thiosulphate	0.0025 g
Agar	0.3 g
Distilled water	100 ml

pH 7.3

PROCEDURE

1. Prepare sim agar medium as described above.

2. Heat the contents of the flask and ensure complete dissolution of the medium.

3. Autoclave the medium at 121°C for 15 minutes.

4. Pour the molten agar medium into the test tubes.

5. Streak and stab the sample culture into the solidified medium.

6. Incubate the tubes at 37°C for 24 hours.

7. Observe for the growth pattern of microorganisms.

8. Black precipitate formed along the growth reveals the evolution of hydrogen sulphide.

5.33 IMViC TEST

This test is specific for the identification of enteric bacilli. The presence of these microorganisms in the environment is a clear indication of faecal contamination and hence evaluation of this microbial group in soil water is of public health significance. Enteric bacilli belong to the family Enterobacteriaceae recognized as the common inhabitants of the intestine of humans and lower mammals. Structurally they are short, gram-negative and non-spore-forming. Three classes of microorganisms come under this category—pathogens (*Salmonella, Shigella*), occasional pathogens (*Proteus, Klebsiella*) and normal intestinal flora (*Escherichia* and *Enterobacter*). These groups can be easily identified by characterizing their enzyme action on different substrates. Based on this, four different tests are adopted.

1. Indole production test
2. Methyl red test
3. Voges Proskauer test
4. Citrate utilization test

5.33.1 Indole Production Test

This test differentiates the microorganisms that can utilize the amino acid tryptophan using tryptophanase enzyme. Tryptophan is an essential amino acid which on hydrolysis yields indole and only specific microorganisms can metabolize it. The medium used is SIM agar with substrate tryptophan. After incubation period, Kovac's reagent is used as an indicator to evaluate indole production. Positive result is indicated by the formation of cherry red layer.

$$\text{Tryptophan} \xrightarrow{\text{Tryptophanase}} \text{Indole} + \text{Pyruvic acid} + \text{Ammonia}$$

Expt. 118   *To perform indole production test*

MATERIALS REQUIRED

Laboratory glassware

   Pipettes

   Test tubes

Laboratory equipment

   Bunsen burner

   Inoculating needle

   Autoclave

   Incubator

MEDIUM

SIM agar

Peptone	3.0 g
Beef extract	0.3 g
Ferrous ammonium sulphate	0.02 g
Sodium thiosulphate	0.0025 g
Agar	0.3 g

REAGENTS

Kovac's reagent

PROCEDURE

1. Sterilize the two semi-solid agar-containing test tubes.
2. Inoculate the culture to be analysed by stab inoculation method.

3. Maintain a control tube.

4. Incubate it for 24 hours at 37°C.

5. Following the incubation period, add 2 to 3 drops of Kovac's reagent.

6. Observe for the formation of cherry-red ring layer.

5.33.2 Methyl Red Test

Glucose is the main energy-yielding substrate for all enteric organisms. However the end products vary with microorganisms with different metabolic pathways. One such example is *Escherichia coli* and *Enterobacter aerogenes.* Even though both the species produce acid end products, with an increased incubation period the latter produces non-acid products. This increases the pH of the medium, which can be used as an indicator to detect the change, whereas *E. coli* stabilizes the pH (<4) by accumulating the acid products. If methyl red indicator added after incubation, turns red the presence of *E. coli* is confirmed. If it turns yellow indicating the alkaline pH of the medium, the presence of *E. aerogenes* is confirmed.

Expt. 119 *To perform methyl red test*

MATERIALS REQUIRED

Laboratory glassware

Test tubes

Pipettes

Laboratory equipment

Autoclave

Incubator

REAGENTS

a. MR – VP Broth

Peptone	0.7 g
Dextrose	0.5 g
Potassium phosphate	0.5 g
Distilled water	100 ml

b. Methyl red indicator

PROCEDURE

1. Sterilize the two semi-solid agar-containing test tubes.

2. Inoculate the culture to be analysed by stab inoculation method.

3. Maintain a control tube.

4. Incubate it for 24 hours at 37°C.

5. Following the incubation period, add 2–3 drops of methyl red indicator.

6. Observe for the formation of red colour, which is a positive result.

5.33.3 Voges Proskauer Test

This test evaluates the ability of microorganisms to convert the acid end product of glucose oxidation into non-acid products such as acetyl/methyl carbinol. The medium used to grow the microorganisms consists of alcoholic, α-naphthol and 40% KOH solution and on oxidation they yield diacetyl compound. A guanidine group in the peptone and α-naphthol catalyst induces the process. On addition of Barritt reagent, a rose colour end point is developed indicating the production of non-acidic diacetyl compound. This favours the differentiation of enteric groups such as *Enterobacter aerogenes*, *Escherichia coli* and *Klebsiella pneumoniae*.

Expt. 120   *To perform Voges Proskauer test*

MATERIALS REQUIRED

Laboratory glassware

Pipettes

Test tubes

Laboratory equipment

Incubator

Autoclave

REAGENTS

Barritt's reagent

PROCEDURE

1. Sterilize the two semi-solid agar-containing test tubes.

2. Inoculate the culture to be analysed by stab inoculation method.

3. Maintain a control tube.

4. Incubate it for 24 hours at 37°C.

5. Observe the colour change.

5.33.4 Citrate Utilization Test

This test is used to differentiate the microorganisms which have the ability to utilize citrate as carbon source. The presence of citrate permease enzyme favours the citrate utilization. In the fermentative pathway of these microorganisms, citrate is produced by the condensation of acetyl compound with oxaloacetic acid and by the activity of citrate permease they are broken down to oxaloacetic acid and acetate. By a series of enzymatic reactions, they get converted into pyruvic acid and CO_2. The liberated CO_2 combines with the sodium and water in the medium to form sodium carbonate. This raises the pH of the medium and changes the bromothymol blue from green to blue.

Expt. 121   *To perform citrate utilization test*

MATERIALS REQUIRED

Laboratory glassware

Test tubes

Pipettes

Laboratory equipment

Incubator

Autoclave

MEDIUM

Simmon citrate agar

Ammonium dihydrogen phosphate	0.1 g
Dipotassium phosphate	0.1 g
Sodium chloride	0.5 g
Sodium citrate	0.2 g
Magnesium sulphate	0.02 g
Agar	1.5 g
Bromothymol blue	0.008 g

PROCEDURE

1. Sterilize the two test tubes containing simmon citrate agar medium.
2. Inoculate the culture to be analysed by stab inoculation method.
3. Maintain a control tube.
4. Incubate it for 24 hours at 37°C.
5. Observe the colour change.

5.34 TRIPLE SUGAR IRON AGAR TEST

All the intestinal bacilli including those of the family Enterobacteriaceae can ferment glucose with acid end product. However the pattern of fermentation process and the ability to produce hydrogen sulphide vary between microbial growth. To evaluate the difference between the different enteric bacilli the medium is supplemented with different carbohydrate forms such as glucose, lactose and sucrose. Utilization of these substrates is indicated by the colour change of the phenol red indicator from red to yellow. Since the culture is streaked and stabbed, change induced by microbial activity has to be observed on both streak and slant. Also the medium has sodium thiosulphate and ferrous sulphate. Sodium thiosulphate acts as a substitute for H_2S production and therefore evolution of gas can be detected by the blackening of the butt, formed by the complexing of H_2S with FeS. Three different changes are noticed

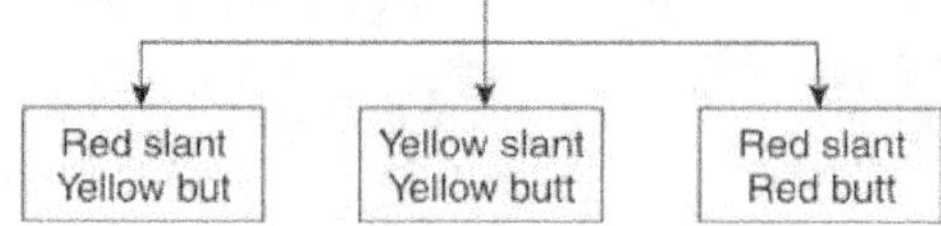

5.34.1 Red Slant and Yellow Butt

It indicates the acid condition of the butt and alkaline condition of the slant. This shows that the microorganisms have utilized only the glucose substrate and minimal amount of acid reduced could not acidify the slant. Moreover, the microbes have aerobically broken down the peptones in the slant to produce ammonia. This was not possible in the butt due to the aerobic condition, thus allowing it to remain acidic.

5.34.2 Yellow slant and Yellow Butt

Utilization of lactose and sucrose favours the fermentation process to occur continuously, thus maintaining the acidic condition in the slant and butt.

5.34.3 Red slant and Red Butt

Instead of carbohydrate, peptone has been the substrate for active utilization. This produces ammonia and the alkaline condition in the slant and butt confirms the aerobic degradation of protein in the slant and anaerobic degradation of protein on the butt.

Expt. 122 *To perform triple sugar iron agar test*

MATERIALS REQUIRED

Laboratory glassware

Test tubes

Pipettes

Laboratory equipment

Incubator

Autoclave

MEDIUM

TSI Agar

Beef extract	0.3 g
Yeast extract	0.3 g
Peptone	1.5 g
Protease peptone	0.5 g
Lactose	1.0 g
Saccharose	1.0 g
Dextrose	0.1 g
Ferrous sulphate	0.02 g
Sodium chloride	0.5 g
Sodium thiosulphate	0.03 g
Phenol red	0.0024 g
Agar	1.2 g
Distilled water	100 ml

PROCEDURE

1. Prepare TSI agar slants and inoculate the cultures using stab and streak method.

2. With an inoculating needle, transfer the culture and insert straight to the bottom of the slant.

3. Immediately withdraw the needle along the line of insertion and streak along the butt.

4. Incubate the culture at 37°C for 24 hours.

5. Following the incubation period, check for the colour change in butt and slant.

5.35 CARBOHYDRATE FERMENTATION

For most of the microorganisms, glucose is the major constituent for the energy-yielding bio-oxidation process. Based on their enzymatic action, they utilize glucose either aerobically or anaerobically. Some facultative anaerobes are efficient enough to utilize both aerobic and anaerobic pathways. Microbial utilization of carbohydrate in anaerobic condition is called fermentation and the organisms involved are fermenters of carbohydrates. End products of this process are lactic, formic or acetic acid.

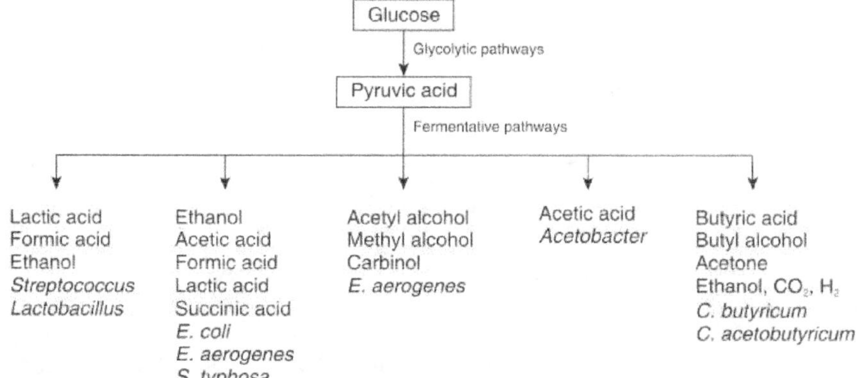

In this experiment, carbohydrate acts as substrate for fermenters. Conversion of carbohydrate into acid products turns phenol red indicator to yellow and the CO_2 accompanied with the acid is visible as gas and the gas is trapped in inverted Durham tubes. Certain microorganisms utilize peptones to release amino acids. Under several enzyme actions, oxidative deamination is favoured thereby converting amino acids to keto amino acids.

Expt. 123　*To demonstrate carbohydrate fermentation*

MATERIALS REQUIRED

Laboratory glassware

Test tubes

Pipettes

Laboratory equipment

Autoclave

Incubator

MEDIUM

Trypticase soy broth culture

PROCEDURE

1. Pour molten trypticase soy broth in fermentation tubes previously sterilized.
2. Carefully place the durham tubes in inverted position inside the test tubes.
3. See to it that the durham tubes (inside portion) are completely filled with medium and that it does is not contain any air bubble.
4. Transfer a loopful of culture into the test tubes.
5. Maintain a control tube.
6. Incubate it at 37°C for 48 hours. Do not extend the incubation period.
7. Check for the presence of yellow or red colour.

EXERCISES

1. Proteins are responsible for _____________ cell.
2. _____________ media is used to grow heterotrophs.
3. _____________ or _____________ act as final electron acceptor in anaerobes.
4. Urease test is specific for _____________ species.
5. Microorganisms possessing _____________ enzyme can utilize inorganic sulphur compounds.
6. IMViC test is specific for _____________.
7. Total coliforms involve _____________ and _____________ species.
8. Lipids are the complex mixture of _____________ and _____________.
9. _____________ enzyme facilitates the hydrolysis of casein.
10. Presence of nitrate or sulphate which act as final electron acceptors is the characteristic property of _____________.
11. Lactophenol blue stain is used to differentiate _____________.
12. What are the basic constituents of a typical stain?
13. What is the functional role of auxochrome in a stain?
14. Explain the principle behind the staining of microorganisms.
15. Why should the bacterial cultures be heat-fixed on a slide?
16. Why do bacterial cultures after negative staining appear colourless?
17. How will you identify microorganisms with thick lipid outer layer?
18. List out the microbial species that can be identified by spore staining method.
19. In capsule staining, the microbial cultures should not be heat-fixed. Why?
20. What is the function of crystal violet in capsule staining?
21. What is a culture medium?
22. What is the advantage of solid medium over liquid medium?
23. What is the difference between agar slants and agar tubes?
24. List out the basic nutritional requirements of microorganisms.

25. Write down the carbon source of autotroph and heterotroph.
26. Why do microorganisms need sulphur?
27. Explain the relation between temperature and enzyme.
28. Classify microorganisms on the basis of temperature.
29. pH can be used as an indicator to identify specific microorganisms. How?
30. Differentiate aerotolerant anaerobes from microaerophiles.
31. Microbial numbers do not show any increase in lag phase. Why?
32. Explain the saturation point in stationary phase of microbial growth curve.
33. Cotton plugs maintain sterilized condition inside the tube. Justify.
34. Why should the petri plates be kept in an inverted position while incubating them?
35. What is the role of agar in culture medium?
36. Explain subculture.
37. Pipetting by mouth is strictly prohibited. Why?
38. Define pure culture.
39. What are the different types of chambers in which the microbial cells can be counted?
40. What is the function of oxidase enzymes in aerobes?
41. Mention two microorganisms possessing oxidase activity?
42. Why do microorganisms degrade H_2O_2 produced during bio-oxidation process?
43. Why do anaerobes die in the presence of O_2?
44. Which process of nitrogen transformation demands anoxic conditions?
45. What is the role of zinc powder in nitrate reduction test?
46. What is the principle behind urease test. Explain with the chemical reaction.
47. Only certain microorganisms can utilize organic sulphur compounds. Why?
48. Evolution of H_2S is the positive result for inorganic sulphur-utilizing bacteria. Justify.
49. Enteric bacilli are associated with faecal pollution. Justify.
50. What are occasional pathogens?
51. In spite of common energy-yielding substrate, how do enteric bacilli differ?
52. Explain the principle behind Voges Proskauer test.
53. Describe the fermentative pathway of citrate-utilizing bacteria.
54. How will you identify the different microbial classes of enteric bacilli.
55. Describe the dilution procedure of MPN method for different water sources.
56. What are the four different stains used in gram-staining method?
57. What is the function of mordant in gram-staining method?
58. How will you estimate faecal coliforms in water?
59. Write down the two streptococcus species that have limited host range.
60. Discuss the function of enzymes in microorganisms.
61. In what way do biochemical tests serve for bacterial identification.
62. Write the structural composition of starch.
63. In lipid hydrolysis test, which agar acts as a supplement for lipid source?
64. In nitrate reduction test, which type of medium is preferred?

65. Name a few milk substrates that are susceptible to microbial transformation.
66. Name two agar medium used to isolate carbon-transforming microorganisms.
67. Which medium is used to isolate phosphate-solubilizing microorganisms?
68. What is the basic difference between acidic dyes and basic dyes?
69. How will you prepare a smear?
70. What is the sample volume of sewage in MPN method?

TECHNIQUES IN BIOTECHNOLOGY

6.1 INTRODUCTION

Biotechnology has long been practised as an art. As these techniques were food-based (production of bread, cheese, wine and beer, etc.) they became a part of daily life activities. Our ancestors performed these techniques without knowing the molecular background, but by mere experience it worked out well.

Later on with the discovery of microorganisms and their biochemical activities, biotechnology started to gear up as a 'Science'. Actually, revolution in the field of biotechnology has made it possible to alter even the genetic constitution of living organisms by way of "Genetic Engineering". This paved the way for new gene combinations thus crossing the boundaries of traditional breeding practices. Then with the development of fermentation technology to produce antibiotics, vaccines and monoclonal antibodies on one side and new molecular innovations such as gene therapy on other side, biotechnology set its imprint in all disciplines of science such as microbiology, plant and animal anatomy, biochemistry, immunology, cell biology, plant and animal physiology, morphogenesis, ecology and genetics.

However it should be remembered that the methodologies for these techniques have emerged by a series of trial and error over the last two decades. When compared to the other techniques, gene techniques are most sensitive and the successful results requires skilled workforce, understanding of the technique such as its potential, limitation, and accuracy, patience and careful planning. In this chapter several experiments ranging from the basics to the advanced techniques have been presented in a simple form and the students are requested to go through the procedures thoroughly and then perform the experiment.

In this chapter we have tried to cover advanced methods in biotechnology, immunological techniques, enzyme analysis and a few biochemical methods.

ENZYME ANALYSIS

Living cells produce enzymes to conduct biological reactions in an orderly and well defined manner. These enzymes with a high degree of specificity and efficiency speed up the reaction rate. The most characteristic feature of enzymes is that they are never used up by the substrate and product as well. Even a small quantity of enzyme is adequate to carry out chemical reactions liberating numerous metabolic end products. Based on their role in metabolic pathways, they are located in different components of cells, tissues and body fluids. Some of the significant properties of enzymes are

- catalytic property
- proteinaceous in nature
- specificity of isoelectric points
- denaturation by changes in pH or temperature
- substrate specificity

Basically they are classified into six major groups. They are

1. Oxidoreductases
2. Transferases
3. Hydrolases
4. Lyases
5. Isomerases
6. Ligases

Oxidoreductases   They are the group of enzymes that catalyse the transfer of hydrogen atoms/electrons from one substrate to another and thus favour the oxidation–reduction reaction between two substances.

Transferases   These enzymes transfer a group from one substrate to the other. The transferred group may be an amino group (transaminases), an acyl/acetyl group (transacylases) or peptides (transpeptidases).

Hydrolases   These enzymes (e.g. carbohydrases, esterases and proteases) act on substrates by addition of water to the substrate and their subsequent cleavage.

Lyases   Lyases are involved in the removal of a chemical compound from the substrate in free state. The liberated chemical group may be a molecule of water (dehydratases); H_2S (desulphydrases) or CO_2 (decarboxylases).

Isomerases   They catalyse the interconversion of isomers. For example, epimerase enzyme catalyses the conversion of L. alanine to D. alanine.

Ligases   These enzymes combine with ATP/GTP and unite two compounds.

6.2 ENZYME ACTION

An enzyme as a whole cannot take part in a reaction. They have specific sites onto which the substrate can fit and react. Therefore these sites are reactive sites and any substance that can undergo enzymatic reaction can be called as substrate. It should be remembered that the union between the enzyme and substrate is temporary and the combination is referred to as Enzyme Substrate Complex (ES Complex) or Michaelis complex. At this stage, the substrate gets activated and the ES complex is called activated complex. This weakens the substrate structure and finally it breaks down. Obviously, this results in the liberation of end products thus regenerating the enzyme molecule. The enzyme refreshes itself and continues to combine with a new substrate.

6.2.1 Coenzymes

In order to be activated, enzymes need the help of certain substances which, when attached to enzymes, initiate them to catalyse chemical reactions. They are low-molecular weight non-protein compounds and based on their firm attachment to enzymes they are termed as coenzymes or prosthetic groups i.e., if these substances are incorporated into the enzyme protein, they are called as prosthetic group whereas coenzymes get attached to the enzyme only during the reaction period. The union of enzyme and coenzyme as a whole is called as holoenzyme in which the protein part is called as apoenzyme. Coenzymes differ from enzymes in their stable and dialysable property.

Mechanism of coenzyme action   Similar to that of enzymes, coenzymes have reaction sites. During the cleavage of the substrate, the reaction site of coenzymes accepts one of the cleavage products. Now the liberation of cleavage products from the apoenzyme part becomes easier and the enzyme is also made free to go in for another substrate attachment. Only then does the dissociation of cleavage products and coenzyme take place. This process is highly advantageous in that one enzyme can act on many substrate molecules, e.g., thiamine pyrophosphate, flavin mononucleotide and pyridoxal phosphate.

6.2.2 Proenzymes

If enzymes are secreted in inactive form (zymogenes) they need to be activated by certain agents. These agents are called as activators or kinases.

6.2.3 Factors Influencing Enzyme Action

Factors that influence enzyme action are temperature, hydrogen ion concentration, enzyme concentration, substrate concentration, oxidation of enzyme, light and radiation, and inhibitors and activators.

Temperature   Enzyme action is directly proportional to the temperature till 40°C, beyond which it becomes inversely proportional. This is due to the fact that enzymes being proteinaceous in nature get denatured at high temperatures. The optimum temperature is different for different enzymes.

Hydrogen ion concentration   Similarly enzyme action depends on the pH. However beyond the optimum pH, the enzyme activity slows down.

Concentration of enzymes   Unlike other factors, enzyme concentration beyond a limit does not influence its enzymatic activity. In general, increase in enzyme concentration hastens the enzyme action. As enzyme–substrate complex essential for the reaction process depends on the equal proportion of enzyme and substrate, a higher concentration of either the substrate or enzyme will do nothing good to favour the enzymatic activity.

Substrate concentration   With the increase in substrate concentration, the enzymatic activity gears up. However, this can be achieved only till a saturation point. Beyond this, further progress is blocked due to the lack of sufficient enzyme to initiate the substrate.

Oxidation of enzyme   Enzymes with sulphydryl groups (SH) when exposed to oxidizing agents like O_2/air get oxidized to disulphide groups (S—S). This inactivates them and their regeneration is achieved by combination with compounds having SH groups like cysteine.

Light and radiation   Light rays influence the enzymatic activity. They may accelerate or retard the reaction rate. For instance, red and blue rays speed up the enzyme activity. In the case of radiation, they have negative effect on enzymes as they denature or inactivate them.

Inhibitors and inactivators   Heavy metals (Hg, Ag, Au, Cu, Zn, Pb, and Cd), oxidizing agents (I, HNO_3, H_2O_2), iodo acetic acid, sodium arsenite, cyanides and carbon monoxide are reported to block enzyme action.

6.3 STEPS TO BE FOLLOWED WHILE PERFORMING ENZYME ANALYSIS

- To maintain the stability of enzymes, store them in their original form. Follow the instructions given in enzyme kits and perform the experiment accordingly.
- Water used for the preparation of enzyme solution and buffer solution should be of high quality. It is best to use double-distilled water.
- Since enzymes are temperature-sensitive, prepare enzyme solutions in ice-cold buffer/water. Also store them in ice buckets.
- Dilution process makes the enzymes highly unstable. Therefore prepare the solution a few hours before use. Do not store it for a long period. If the storage period extends beyond this, take care to check its activity at regular intervals. If the activity is found to be slowed down, immediately discard the solution and start up the procedure from the beginning.
- In the case of crystalline suspension, even mild shaking would denature the enzyme. Instead, for uniform suspension, swirl the bottle or roll it gently on the desk.
- Freezing enzymes, denatures them and leads to loss of activity.
- Before opening, lyophilized cultures should be acclimatized to room temperature. This is because moisture will condense on the powder form which will ultimately reduce its activity.
- Store the enzymes in small aliquots. Do not subject the diluted enzymes or lyophilizates to repeated freezing and thawing.
- While using preservatives or detergents, exercise great care as they denature the enzymes.
- Transfer the enzymes from the parent vial in small aliquots. Every time, when you draw it, use fresh pipettes. Once drawn, the enzymes should not be returned back to the vial. Use automated filters to draw the aliquots.
- Check the pH of the enzyme solution after adding the buffer solution. This is due to the fact that even a slight fluctuation in pH will affect the enzyme activity.

6.4 AMYLASE

This enzyme is secreted mainly by the pancreas. A small amount of amylase is secreted by the parotid glands. Under normal conditions, it exists in low levels in plasma. However, acute pancreatitis disease associated with the severe damage of pancreatic tissues allows the amylase to enter into the bloodstream. Also complications like obstruction of pancreatic duct and paratitis enhance the serum amylase level.

6.4.1 Principle

Amylase favours the hydrolysis of the internal alpha 1, 4-glucon links in polysaccharides to give an end product containing maltose and glucose. Its enzymatic activity varies with different tissues even in the same species. Reduction of 3,5-dinitrosalicylic acid liberates maltose from the soluble starch and one unit of maltose liberated from one micromole of reducing groups at 25°C and pH 6.9 is calculated.

Expt. 124   *To analyse the presence of enzyme amylase*

MATERIALS REQUIRED

Laboratory glassware

Laboratory equipments

Spectrophotometer

Enzyme

Dilute the enzyme to a concentration of 1–10 µg/ml of distilled water. Prepare at least five concentrations in this range.

REAGENTS

a. 0.02M sodium phosphate buffer (pH 6.9) using 0.006M sodium chloride

b. 2N sodium hydroxide

c. Dinitrosalicylic acid colour reagent

Dissolve 1 g of 3, 5-dinitrosalicylic acid in 20 ml of 2N NaOH. To this, slowly add 30 g of sodium potassium tartrate tetrahydrate. Make up to 100 ml with distilled water. Protect it from carbon dioxide.

d. Starch (1%)

Dissolve 1 g of soluble starch in 100 ml of 0.02M sodium phosphate buffer with 0.006 sodium chloride. Adjust the pH to 6.9. Gently boil and ensure the complete dissolution of starch. Incubate at 25°C for 5 minutes prior to assay.

PROCEDURE

1. Prepare several dilutions of amylase enzyme preferably from the range of 2 µg to 10 µg at 2-µg interval.

2. Make each dilution to 5 ml using distilled water.

3. From each dilution, pipette 0.5 ml into a series of test tubes.

4. To accomplish temperature equilibrium, incubate all test tubes at 25°C for about 5 minutes. At this temperature to each of the test tubes add 0.5 ml of starch solution.

5. Allow it to stand for 3 minutes.

6. Exactly after this period add 1 ml of dinitrosalicylic acid colour reagent.

7. Exercise great care while adding these reagents to all the test tubes at specific time interval.

8. Boil the test tubes in a water bath for 5 minutes.

9. After cooling, add 10 ml distilled water.

10. Shake well and read the absorbance value in the spectrophotometer at 540 nm at 25°C.

11. Run a blank using 0.5 ml distilled water following the same procedure described above.

6.5 PHOSPHATASES

Hydrolysing enzymes involved in digestion and intermediary metabolism are categorized under a group phosphatases. They exhibit enzymatic activity at a wide range of pH. Based on this, they are classified into two types.

1. Acid phosphatases
2. Alkaline phosphatases

6.6 ACID PHOSPHATASE

Acid phosphatase with its broad esterase activity acts on pyrophosphate to yield alcohol and phosphoric acid.

$$\text{Orthophosphoric monoester} + H_2O \rightarrow \text{alcohol} + \text{phosphoric acid}$$

It is commercially available as lipase and acid phosphatase. Since it is highly concentrated in human prostate gland, the serum enzyme level is used as an index for prostate cancer. It is active at acid pH (4.9). The main source of this enzyme is prostate gland. Soon after its secretion from matured prostatic epithelial cells, it enters into prostatic fluid. That is why, adult prostatic secretion has acid phosphatase concentration 100 times more than that of any other tissues. Evaluation of enzyme content in blood is an index of prostate cancer in particular at its metastatic stage where the enzyme finds its way into the bloodstream.

6.6.1 Principle

Breakdown of o-carboxyphenyl phosphate liberates salicylic acid which in turn increases the absorbance at 300 nm thus determining the reaction mixture. Under suitable conditions of temperature (25°C) and pH (5), one unit hydrolyses one micromole of o-carboxy phenyl phosphate per minute.

Expt. 125   *To estimate the presence of the enzyme acid phosphatase*

MATERIALS REQUIRED

Laboratory glassware

Laboratory equipment

Spectrophotometer

Enzyme

Dissolve the enzyme to a concentration of 0.1 mg/ml of distilled water.

REAGENTS

a. 0.15 M sodium acetate buffer (pH – 5.0)

b. 3.65 mM o-carboxy phenyl phosphate

PROCEDURE

1. To the cuvette add 2.0 ml of 0.15M sodium acetate buffer and 0.5 ml of carboxyphenyl phosphate.

2. Shake well and record the blank rate by measuring the absorbance value in spectrophotometer at 300 nm.

3. Record the absorbance value.

4. To this add 10 µl of diluted enzyme solution and record the sample rate.

5. From the linear portion of the curve, calculate the OD value/min.

CALCULATION

$$\text{Amount of acid phosphate present in the sample (µg/mg)} = \frac{OD \text{ value/min}}{3500 \times E}$$

where,

E = enzyme concentration in the sample

6.7 ALKALINE PHOSPHATASE

A group of nonspecific phosphomonoesterases is collectively termed as alkaline phosphatase. It is active at alkaline pH and the following reaction takes place.

$$\text{Orthophosphoric monoester} + H_2O \rightarrow \text{alcohol} + \text{phosphoric acid}$$

It is active in alkaline pH (9) and its main origin is bone, though a small amount is formed in the liver. Skeletal diseases like osteomalacia, rickets, hyperparathyroidism and Paget's disease formed due to improper mineralization shows high level of acid phosphatase in serum. Also, impairment of liver allows the enzyme to enter into the lymphatic system and eventually into the bloodstream. In general the function of acid phosphatase is to favour the mobilization of Ca^{+++} and PO_4^- from the bloodstream to bone so as to strengthen its basic constituent, calcium phosphate.

6.7.1 Principle

The reaction rate is measured by the increase in absorbance at 405 nm which is the outcome of p-nitrophenyl phosphate hydrolysis. Under suitable conditions of temperature (30°C) and pH (10.2), one unit liberates one micromole of p-nitrophenol per minute.

Expt. 126   *To estimate the presence of the enzyme alkaline phosphatase*

METHOD I

MATERIALS REQUIRED

Laboratory glassware

Laboratory equipments

Spectrophotometer

Enzyme

Dissolve the enzyme to a concentration of 1 mg in 1 ml of 0.1M tris buffer. Adjust the pH to 8.5.

REAGENTS

    a.   0.1M tris, (pH 8.5)

    b.   0.2M glycine (pH 8.8)

    c.   0.05M magnesium chloride

    d.   3.65 mM o-carboxyphenyl phosphate

PROCEDURE

1. To 0.1 ml enzyme diluent, add 2.9 ml reaction mixture, mix well and transfer the content into a cuvette.

2. Measure the increasing absorbance for about 8 minutes in spectrophotometer at 405 nm.

3. Draw a standard curve and from its linear portion calculate the absorbance/minute.

METHOD II (CHICK ENZYME)

MATERIALS REQUIRED

Laboratory glassware

Laboratory equipments

    Spectrophotometer

Enzyme

    Dilute the enzyme to a concentration of 0.02–0.1 units/ml of distilled water.

REAGENTS

    a.   1.5 M tris HCl buffer (pH 8.0)

    b.   0.003 M p-nitrophenyl phosphate

PROCEDURE

1. Take 0.1 ml of the diluted enzyme solution in a clean dry test tube.

2. Place it in a water bath maintained at 25°C for 30 minutes. This enables the enzyme to get activated.

3. Transfer the enzyme solution into a cuvette and to this add 2 ml glycine, 1 ml OCPP and 0.5 ml magnesium chloride.

4. Mix well and measure the increasing absorbance for 4 minutes in the spectrophotometer at 300 nm.

5. Draw a standard curve and from its linear portion, evaluate the absorbance/minute.

6.8 PEROXIDASE

Peroxidase is a haemoprotein. It act as a catalysing agent to oxidize substrates like ascorbate, ferrocyanide, cytochrome C and leucoform of many dyes in the presence of hydrogen peroxide.

$$\text{Peroxidase} + H_2O_2 \rightarrow B \text{ (intermediate compound)}$$

$$B + AH_2 \text{ (substrate to be oxidized)} \rightarrow 2H_2O + A \text{ (end product)}$$

Two important characteristics of this enzyme, stability and ability to yield chromogenic products makes it suitable for the preparation of enzyme-conjugated antibodies. It is used for a variety of applications and they are as follows.

a. They are used as immunohistology probes for demonstration of tissue antigens.

b. They are used in enzyme-amplified immunoassay system for the quantitative determination of soluble and insoluble antigens.

c. They are useful for assay systems producing hydrogen peroxide.

6.8.1 Principle

The reaction rate is measured by the increase in absorbance at 510 nm, which in turn is dependent on the decomposition of H_2O_2 at 25°C, and pH 7.0. One micromole of H_2O_2 decomposed/minute results in an increase of one unit.

Expt. 127   *To estimate the presence of the enzyme peroxidase*

MATERIALS REQUIRED

Laboratory glassware

Laboratory equipments

Spectrophotometer

Enzyme

Dissolve enzyme to a concentration of 1 mg/ml of distilled water. Just before use, dilute it further to a concentration of 0.05–0.25 units/ml of distilled water.

REAGENTS

a. 0.2M potassium phosphate buffer, (pH 7.0)

b. 0.0017M H_2O_2

Dilute 1 ml of 30% H_2O_2 to 100 ml of distilled water. Take 1 ml from this solution and make up to 50 ml with 0.2M potassium phosphate buffer. Adjust the pH to 7.0. Prepare fresh.

c. 0.0025M 4-amino antipyrine

Dissolve 810 mg phenol in 40 ml distilled water. To this add 25 mg of 4-amino antipyrine and make up to 50 ml using distilled water.

PROCEDURE

1. To the cuvette, add 1.4 ml of amino antipyrine solution and 1.5 ml H_2O_2.

2. Mix well and to establish the blank rate measure the absorbance value in the spectrophotometer at 510 nm for 3–4 minutes.

3. This value is considered as blank.

4. Now to the cuvette, add 0.1 ml of diluted enzyme and measure the increase in absorbance for about 5 minutes.

5. Draw a standard curve by plotting the increase in absorbance value against time. From the linear portion of curve, calculate the OD value/minute.

CALCULATION

$$\text{Amount of peroxidase (unit/mg)} = \frac{\text{OD value/minu te}}{6.58 \times E}$$

where,

E = mg of enzyme/ml of reaction mixture

6.9 CARBONIC ANHYDRASE

It is present commonly in plants, animals and certain bacteria. However they exhibit different functions in these organisms. For instance, during respiration process of animals it helps in transporting CO_2 from the tissues to the lungs. This enzyme is also involved in the transfer and accumulation of H^+ and HCO_3^-. In plants, it resides in the chloroplast and favours photosynthetic fixation of CO_2. In animals, it exists in several forms differing in enzymatic properties, amino acid sequence and inhibitor binding. In human erythrocytes, it exists in 3 forms, carbonic anhydrase *a*, carbonic anhydrase *b* and carbonic anhydrase *c*.

6.9.1 Principle

The time period required for a saturated CO_2 solution to drop the pH of tris HCl buffer from 8.3 to 6.3 at 0°C is calculated. The time period with blank is taken as T_0 and with enzyme is taken as T.

Expt. 128   *To estimate the presence of the enzyme carbonic anhydrase*

MATERIALS REQUIRED

Laboratory glassware

Laboratory equipment

Spectrophotometer

Enzyme

Dilute the enzyme to a concentration of 0.1 mg/ml in ice-cold distilled water. Store it in ice bath. Just before use, dilute the suspension to a concentration of 0.01 mg/ ml of ice-cold distilled water.

REAGENTS

a. 0.02M Tris HCl buffer

Adjust the pH to 8.0, store it at >4°C before and during use.

b. Carbon dioxide saturated water

Carefully bubble CO_2 through 200 ml ice-cold water for about 30 minutes. Store it at >4°C in an ice bath. Perform this procedure prior to the assay.

PROCEDURE

1. Take 6 ml of ice-cold 0.02M tris HCl buffer in a 50-ml beaker.

2. Make certain that the buffer solution is maintained at a temperature of 4°C and a pH of 8.3.

3. Using a syringe, carefully withdraw 4 ml of ice-cold CO_2-saturated water.

4. Dispense the same into the beaker using the stopcock.

5. Record the exact time taken by the CO_2-saturated solution to lower the pH from 8.3 to 6.3. Let it be T_1.

6. Similarly transfer 6 ml of ice-cold 0.02M tris HCl buffer into a 50-ml beaker. Adjust the temperature and pH as described above.

7. To the beaker, add 0.1 ml of diluted enzyme solution. Always prepare fresh before use.

8. Without delay, add 4 ml of ice-cold carbon dioxide-saturated water.

9. Note down the time required for the pH to drop from 8.3 to 6.3. Let it be T_2.

CALCULATION

$$\text{Amount of carbonic anhydrase (unit/mg)} = \frac{2 \times (T_0 - T)}{T \times E}$$

where,

T_0 – Time taken for the blank to change pH

T = Time taken for the sample to change pH

6.10 CELLULASE

Cellulase is a collective term for a group of enzymes which act on the substrate cellulose. They are abundant in fungi and bacteria. The reaction involves two steps.

1. *Prehydrolysis* Here the hydration of anhydroglucose takes place. The enzyme involved is Enzyme I. Usually they act on highly ordered cellulose substrates like cotton fibres. However they exhibit weak activity on soluble derivatives of cellulose.

2. *Hydrolysis* Breakdown of polymers occurs in the presence of the enzyme, (Enzyme x). The enzyme with exo- and endo-B-1,4-glucanases target on soluble derivatives of cellulose.

6.10.1 Principle

Action of cellulase on cellulose to liberate glucose is measured at 340 nm. One unit of activity releases 0.01 mg glucose/hour from cellulose substrate.

Expt. 129 *To estimate the presence of the enzyme cellulase*

MATERIALS REQUIRED

Laboratory glassware

Laboratory equipment

Spectrophotometer

Enzyme

Dissolve the enzyme to a concentration of 0.1–1.0 mg/ml of distilled water.

REAGENTS

a.　0.05M acetic acid (pH 5.0)

b.　Microcrystalline cellulose

c.　Glucose

PROCEDURE

1.　Take 0.2 g of microcrystalline cellulose in a clean dry test tube.

2.　To this add 4 ml of 0.05M acetic acid and 1 ml of diluted enzyme solution.

3.　To ensure complete mixing, place it in a shaker for 2 hours at 37°C.

4　Following this, transfer the test tubes to an ice bath and let the precipitate to settle.

5.　Separate the clear solution by centrifugation and decant the precipitate. To the cuvette add 3 ml of glucose reagent, maintain its temperature at 25°C and then measure absorbance in the spectrophotometer at 340 nm.

6.　Now add 0.1 ml of supernatant into the cuvette and measure the change in absorbance at 340 nm for 3–5 minutes.

7.　Record the stable value. Let this be S.

8.　Similarly, run the blank using 0.2 g microcrystalline cellulose, 4 ml of 0.05M acetic acid and 1 ml distilled water. Note the value obtained for change in absorbance as B.

CALCULATION

$$\text{Amount of cellulase (unit/mg)} = \frac{(S-B) \times 3.1 \times 180 \times 5}{6.22 \times 10^3 \times 0.1 \times 0.001 \times E}$$

where,

S = sample rate

B = blank rate

E = concentration of enzyme in the sample

6.11 LACTATE DEHYDROGENASE

Lactate dehydrogenase (LDH) is a ubiquitous metabolic enzyme catalysing the following reaction.

$$\text{Pyruvate} + NADH^+ \rightarrow \text{Lactate} + NAD + H^+$$

Five tetrameric isoenzymes of mammalian LDH consist of 2 subunits 'H' and 'M' with different compositions. These enzymes differ in catalytic, physical and immunological properties.

Isoenzyme 1 → H_4

Isoenzyme 2 → M_4

Isoenzyme 3 → H_3M

Isoenzyme 4 → H_2M_2

Isoenzyme 5 → HM_3

Heart muscle in which the 'H' subunit predominates catalyses the aerobic oxidation of pyruvate whereas skeletal and liver muscles in which the 'M' subunit predominates favour the anaerobic metabolism and pyruvate reduction. It gains clinical significance as its enhanced level in serum is an index of certain pathological conditions like leukemia, generalized cancer and acute cancer. It is abundant in heart, liver, kidney, skeletal muscles and erythrocytes. The concentration of LDH in heart muscle is reported to show remarkable increase soon after the onset of myocardial infarction and gets established after one week. Hence measurement of this enzyme level during the treatment period helps in the prognosis of the disease. The isoenzymes can be isolated when the serum is subjected to electrophoresis.

6.11.1 Principle

The reaction rate is determined by the decrease in absorbance at 340 nm, which in turn is dependent on the oxidation of NADH. The mechanism behind the oxidation of one micromole of NADH/min (at 25°C and pH 7.3) causes a drop by one in the unit of absorbance.

Expt. 130   *To estimate the presence of the enzyme lactate dehydrogenase*

Materials Required

Laboratory glassware

Laboratory equipment

   Spectrophotometer

Enzyme

   Dilute the enzyme to a concentration of 1.0–2.5 units/ml of distilled water.

Reagents

a.   0.2M Tris HCl (pH 7.3)

b.   6.6 mM NADH in about 0.2M tris buffer

c.   30 mM sodium pyruvate in about 0.2 tris buffer (pH 7.3)

Procedure

1.   To the cuvette add 2.8 ml of 0.2M tris HCl and 0.1 ml each of 6.6 mM NADH and 30 mM sodium pyruvate.

2.   Shake well and record the blank rate by measuring the absorbance value in spectrophotometer at 340 nm.

3.   Record the absorbance value.

4.   See to it that the initial value is 1.4 ± 0.1.

5. In the case of fluctuations, repeat the procedure with freshly prepared pure reagents.

6. To this add 10 µl of diluted enzyme solution containing 10 ml of LDH and record the sample rate in a clean dried test tube.

7. From the linear portion of the curve, calculate the OD value/min.

CALCULATION

Amount of LDH present in the sample (µg/mg) $= \dfrac{\text{OD value/min}}{6.2 \times E}$

where,

E = enzyme concentration in the sample

6.12 LIPASE

Lipase favours the hydrolysis of aggregate lipids such as ester molecules, monomolecular forms or micelles to yield long chain fatty acid and glycerol. It is substrate-specific and can act on glycerol at the α and β position. No isomerase is required and the enzyme activity is directly dependent on the concentration of substrate molecule. Origin of this enzyme is pancreas and therefore in acute pancreatitis, it is secreted along with amylase. It can be used to diagnose pancreatitis and pancreatic duct obstruction.

6.12.1 Principle

Under suitable temperature and pH, hydrolysis of olive oil emulsion releases fatty acid and one unit of the titrant releases one micromole of fatty acid/minute.

Expt. 131   *To estimate the presence of the enzyme lipase*

MATERIALS REQUIRED

Laboratory glassware

Laboratory equipment

Spectrophotometer

Enzyme

Dissolve the enzyme to a concentration of 1 mg/ml of 0.005M calcium chloride.

REAGENTS

a. 3.0 M sodium chloride

b. 0.0075M calcium chloride

c. 0.5% albumin (always prepare fresh)

d. Gum arabic–olive oil emulsion

Dissolve 16.5 g of gum arabic in 130 ml of distilled water. Make up to 165 ml using distilled water. To this add 20 ml of reagent grade olive oil and 15 g of crushed ice. Mix well with a homogenizer at low speed and then filter the emulsion through glass wool.

e. 0.005M calcium chloride

PROCEDURE

1. To the 50-ml beaker add the following reagents—5 ml each of gum arabic, oil emulsion and reagent grade distilled water; 2 ml each of sodium chloride and albumin and 1 ml of calcium chloride.

2. Maintain the temperature of reaction mixture at 25°C.

3. Mix well, set the pH of the mixture to 8.

4. Titrate it against NaOH and record the volume of titrant needed to maintain the already set pH (8), and the pH should remain stable for about 6 minutes.

5. Draw a standard graph by plotting the pH of the solution against the volume of titrant added per minute.

6. From the linear portion of the curve, calculate the blank rate. Let this value be B.

7. Similarly process the sample by adding enzyme solution instead of distilled water.

8. However before starting, the pH of the mixture should be adjusted to 8.0.

9. Titrate it against NaOH and record the volume of titrant needed to maintain the pH fixed for 3–4 minutes of the curve.

CALCULATION

Amount of lipase present in the sample (μg/mg) $= \dfrac{(S - B) \times N \times 1000}{E}$

where,

S = sample rate

B = blank rate

W = normality of titrant

E = amount of enzyme in the sample

6.13 GLUCOSE OXIDASE

Glucose oxidase is a flavo enzyme highly specific to B-D-glucose, D-mannose and D-galactose but with low activity.

$$B\text{-D-glucose} + FAD \rightarrow FADH_2 + S\text{-D-glucanolactone}$$

$$FADH_2 + O_2 \rightarrow FAD + H_2O_2$$

6.13.1 Application

1. It is an antibiotic.

2. It is used in analytical laboratories for glucose estimation.

Expt. 132   *To estimate the presence of the enzyme glucose oxidase*

MATERIALS REQUIRED

Laboratory glassware

Laboratory equipment

Spectrophotometer

Enzyme

Dissolve the enzyme to a concentration of 1 mg/ml in distilled water. For estimation, dilute further to 0.03–0.09 units/ml of distilled water.

REAGENTS

a.   0.1M potassium phosphate buffer (pH 6.0)

b.   Dianisidine (1%)

Dissolve 1 g of dianisidine in 100 ml of distilled water. Exercise great care while handling dianisidine, as it is a potent carcinogen in solid form.

c.   Peroxidase

Dissolve 0.2 g of peroxidase in 1 ml of distilled water.

d.   Glucose (18%)

Dissolve 18 g of glucose in 100 ml of distilled water. Store it overnight.

e.   Dianisidine buffer mixture

Dissolve 0.1 ml of 1% dianisidine in 12 ml of 0.1M potassium phosphate buffer. Adjust the pH to 6.0 and bubble oxygen through the solution.

PROCEDURE

1.   To the cuvette, add 0.1 ml of peroxidase, 2.5 ml dianisidine (1%) and 0.3 ml of glucose (18%).

2.   Shake well and record the blank rate by measuring the absorbance value in the spectrophotometer at 460 nm.

3.   Record the absorbance value.

4.   To this, add 10 µl of diluted enzyme solution and measure the absorbance of the sample for 5 minutes. Record the sample rate.

5.   Draw a standard curve and from the linear portion of the curve, calculate the OD value/min.

CALCULATION

Amount of glucose oxidase present in the sample (µg/mg)

$$= \frac{\text{OD value/min}}{11.3 \times E}$$

where,

E = enzyme concentration in the sample

6.14 CATALASE

Catalase is a haemoprotein and in combination with H_2O_2 forms intermediate compound and the reaction is as follows:

$$2H_2O_2 \rightarrow 2H_2O + O_2$$

Here the H_2O_2 acts both as substrate and hydrogen acceptor. It differs from peroxidase in that the latter requires a separate acceptor. All animal cells and aerobes are sources of catalase. It gains commercial significance in the food industry where H_2O_2 is used as germicide in pasteurizing milk prior to cheese-making.

6.14.1 Principle

Disappearance of H_2O_2 is measured at 240 nm. Under suitable conditions of temperature (28°C) and pH (7.0), one unit decomposes one micromole of H_2O_2 per minute.

Expt. 133   *To estimate the presence of the enzyme catalase.*

MATERIALS REQUIRED

Laboratory glassware

Laboratory equipment

Spectrophotometer

Enzyme

Dilute the enzyme to a concentration of 5–20 units/ml of 0.05M phosphate buffer. Adjust the pH to 7.0.

REAGENTS

a.   0.05M potassium phosphate (pH 7.0)

b.   0.059M hydrogen peroxide in 0.05M potassium phosphate. Adjust the pH to 7.0.

PROCEDURE

1.   Carefully dispense 1 ml of 0.059M hydrogen peroxide into a 5-ml test tube. Make up to 2.9 ml using distilled water.

2.   Mix the contents well and read the absorbance value in the spectrophotometer at 240 nm.

3.   To this add 0.1 ml of diluted enzyme solution and measure the decrease in absorbance.

4.   Draw a standard curve and from its linear portion calculate the absorbance/min.

5.   Run a blank using 1 ml substrate solution and 2 ml distilled water.

CALCULATION

Amount of catalase present in the sample (μg/mg) $= \dfrac{\text{OD value/min}}{43.6 \times E}$

where,

E = enzyme concentration in the sample

6.15 UREASE

Urease acts on the substrate urea and the following reaction takes place.

$$(NH_2)_2 CO + 3H_2 \rightarrow CO_2 + 2NH_4OH$$

It is concentrated in several species of yeast and a number of higher plants. It is very much essential for quantitative estimation of urea.

6.15.1 Principle

Under suitable conditions of temperature (25°C) and pH (7.6), one unit results in the oxidation of one micromole of NADH per minute.

Expt. 134 *To estimate the presence of the enzyme urease*

MATERIALS REQUIRED

Laboratory glassware

Laboratory equipment

Spectrophotometer

Enzyme

Dissolve the enzyme to a concentration of one mg/ml of 0.1M phosphate buffer. Just before use, dilute to a concentration 0.1–0.3 units/ml of phosphate buffer. Adjust the pH to 7.6.

REAGENTS

a.   0.1M potassium phosphate buffer, (pH 7.6)

b.   1.8M urea in phosphate buffer

c.   0.023M adenosine 5-diphosphate (ADP) in phosphate buffer

d.   0.0072M NADH in phosphate buffer

e.   0.026M 2-ketoglutarate in phosphate buffer

f.   Glutamate dehydrogenase (GLDH)

Dilute 500 units of glutamate dehydrogenase in 1 ml of 50% glycerol or phosphate buffer. Store it in a cool place during use.

PROCEDURE

1.   Dispense 2.4 ml of phosphate buffer and 1 ml each of ADP, NADH, ketoglutarate, distilled water and GLDH into a clean dry test tube.

2.   Measure the absorbance at 340 nm in the spectrophotometer. Absorbance value may fluctuate due to the trace amount of ammonia in reagent.

3.   Soon after its stabilization, add 0.1 ml diluted enzyme solution and record the decrease in its absorbance value for 8–10 min.

4.   Draw a standard curve and from its linear portion determine the absorbance value/min.

CALCULATION

$$\text{Amount of urease present in the sample } (\mu g/mg) = \frac{\text{OD value/min}}{6.2 \times E}$$

where,

E = enzyme concentration in the sample

RNA POLYMERASE

RNA polymerase is a holoenzyme that plays an important role in protein synthesis. Its function involves site location, chain initiation, elongation and termination in the transcription process.

6.16 RNA POLYMERASE I

6.16.1 Principle

Under specific temperature (37°C) one unit incorporates 1 n mole of AMP into an acid-precipitable product over 10 minutes incubation period.

Expt. 135   *To estimate the presence of the enzyme RNA polymerase I*

MATERIALS REQUIRED

Laboratory glassware

Laboratory equipment

Spectrophotometer

REAGENTS

a.   Assay cocktail

To prepare a reaction mixture, add the following reagent

0.04M tris HCl

0.01M $MgCl_2$

0.1 mM EDTA

0.1 mM DTT

0.15M KCl

0.5 mg/ml Bovine serum albumin

0.15 mM UTP, CTP, GTP

0.15 mM (3H) ATP

0.15 mg/ml calf thymus DNA

b.   Solution A

Prepare solution A by adding the following reagents.

4 parts saturated $Na_4P_2O_4$

4 parts saturated NaH_2PO_4

1 part 100% TCA

1 part 10 mM thymidine

c. Scintillation Fluid

Mix 640 ml of toluene, 320 ml of Triton x-100 and 40 ml of permafluor and store it for use.

PROCEDURE

1. Pipette 0.25 ml assay cocktail and 2 ml diluted enzyme solution into a silicon test tube. Swirl and let it stand for 10 minutes at 37°C.

2. After 10 minutes pipette 2 ml of solution into the test tube, swirl and place it in an ice-cold water bath for 15 minutes.

3. By using vacuum pump, filter the acid-soluble products in 0.45 micron filter discs.

4. Wet the discs with solution A prior to use.

5. Proceed with the filtration process by washing the tubes with 30 ml of 10% TCA for five times.

6. Now the filter disc is washed with 10 ml of 10% TCA and 5 ml of chloroform–methanol solution.

7. Allow the filters to dry.

8. Transfer the filter discs in glass liquid scintillation vials. To the vial add 1 ml of methyl celloslave and 10 ml of scintillation liquid.

9. Swirl the solution and determine the disintegration rate/minute (DPM_S).

10. Similarly process the blank and determine the disintegration rate/minute (DPM_B).

CALCULATION

Amount of RNA polymerase I present in sample (units/mg)

$$= \frac{DPM_S - DPM_B}{DPM \text{ nmol/s} \times E}$$

where,

DPM_S = Disintegration rate/min. of the substrate

DPM_B = Disintegration rate/min. of the blank

6.17 RNA POLYMERASE II

DNA-dependent RNA polymerase II is nucleoplasmic whose activity is favoured by the presence of manganese and ammonium sulphate. In contrast α-amanitin greatly inhibits its activity. It is this sensitivity that distinguishes RNA polymerase II from other nuclear RNA polymerases.

6.17.1 Principle

Under suitable conditions, 10 picomoles of uridine 5′ monophosphate is incorporated into an acid-precipitable product. This occurs using heat-denatured DNA and the reaction is completed in 15 minutes.

Expt. 136   *To estimate the presence of the enzyme RNA polymerase II*

MATERIALS REQUIRED

Laboratory glassware

Laboratory equipment

Spectrophotometer

REAGENTS

a.   Assay cocktail

To prepare a reaction mixture of 0.25 ml, add the following reagents

2.5 μ moles of Tris HCl

3.5 μ moles of manganese chloride

5 μ moles of magnesium chloride

31.3 μ moles of ammonium sulphate

100 n moles adenosine – 5′-triphosphate

100 n moles guanosine – 5′-triphosphate

100 n moles cytosine – 5′-triphosphate

100 p moles of tritiated uridine –5′-triphosphate

50 μg heat-denatured DNA

125 μg bovine serum albumin

After the addition of above contents adjust the pH of the solution to 7.9.

b.   Heat-denatured DNA

Dissolve 1 mg/ml of calf thymus DNA in 10 mM EDTA. Mix well and adjust the pH to 7.0. Heat the solution in a water bath at 100°C for about 15 minutes. Then transfer the solution in an ice water bath until it is cooled.

c.   Solution A

Prepare solution A by adding the following reagents.

4 parts of saturated NaH_2PO_4

4 parts of saturated NaH_2PO_4

1 part of 100% TCA

1 part of 10 mM thymidine

d.   Scintillation fluid

Add 640 ml of toluene, 320 ml of Triton X-100 and 40 ml of permafluor. Mix well and store it for use.

PROCEDURE

1.   Pipette 0.25 ml assay cocktail and 2 ml diluted enzyme solution into a silicon test tube. Swirl and let it stand for 10 minutes at 25°C.

2.   After 15 minutes, pipette 2 ml of solution into the test tube, swirl and place it in an ice-cold water bath for 5 minutes.

3. By using vacuum pump, filter the acid-soluble products in 0.45 micron filter discs.

4. Wet the discs with solution A prior to use.

5. Proceed with the filtration process by washing the tubes with 3 ml of 10% TCA five times.

6. Now the filter disc is washed with 10 ml of 10% TCA and 5 ml of chloroform–methanol solution.

7. Allow the filters to dry.

8. Transfer the filter discs in glass liquid scintillation vials. To the vial add 1 ml of methyl celloslave and 10 ml of scintillation liquid.

9. Swirl the solution and determine the disintegration rate/minute (DPM_S).

10. Similarly process the blank and determine the disintegration rate/minute (DPM_B).

CALCULATION

Amount of RNA polymerase II present in the sample (units/mg)

$$= \frac{DPM_S - DPM_B}{DPM \text{ nmol/s} \times E}$$

where,

DPM_S = Disintegration rate/min. of the substrate

DPM_B = Disintegration rate/min. of the blank

6.18 DNA POLYMERASE (Eukaryotic α)

It is a high molecular weight protein (100,000 to 250, 000) which plays an important role in growing cells. Other sources are nuclear and cytoplasmic fractions of calf thymus and other tissues. It acts as a catalyst in phosphodiester bond formation between complementary nucleotides. Thus it amplifies the template primer chain. It is active at a pH range between 7.3 and 7.65. For DNA synthesis, the free 3-hydroxyl end of the primer strand is annealed to the DNA template strand. Since the deoxynucleotide triphosphates forms base pairs with the template strand, addition of base pairs occurs towards 5′-3′ direction with the subsequent release of pyrophosphate. RNA–DNA hybrids and an RNA duplex may also serve as template primer.

6.18.1 Principle

Under a specific temperature of 37°C, acid-insoluble product is stuffed with one nanomole of dAMP and the activity is favoured by one unit activity.

Expt. 137   *To estimate the presence of eukaryotic DNA polymerase enzyme*

MATERIALS REQUIRED

Laboratory glassware

Laboratory equipment

Liquid scintillation counter

REAGENTS

a. Assay cocktail

Prepare a reaction mixture by adding the following reagents

 0.083M Tris pH 7.8

 20% glycerol

 670 mg/ml bovine serum albumin

 1.667 mM dithiothreitol

 10.0 mM magnesium acetate

 267 mg/ml polydeoxy thymidilic acid

 53 mg/ml oligoribo adenylic acid

 167 mM tritiated deoxyadenosine 5′-triphosphate

b. Solution A

To prepare solution A, mix the following

4 parts of saturated $Na_4P_2O_4$

4 parts of saturated NaH_2PO_4

1 part of 10 mM thymidine

c. Scintillation fluid

Add 640 ml of toluene, 320 ml of tritan X-100 and 40 ml of permafluor.

PROCEDURE

1. Add 15 μl assay cocktail, 8 μl potassium chloride and 2 μl diluted enzyme solution in a silicon test tube placed in an ice-cold water bath.

2. Then allow the reaction mixture to pass through a filter disc (0.45 m) and by a vacuum suction pump separate the acid-insoluble product from the reaction mixture. Soak the filter disc in solution prior to use.

3. Rinse the test tube with 3 ml TCA several times and filter the same.

4. Similarly wash the filter discs with 40 ml each of TCA and chloroform–methanol.

5. Dry the filters and transfer into glass liquid scintillation vials.

6. Add 1 ml of methyl cellulose so as to dissolve the filters.

7. Leave it to stand for 30 minutes.

8. Following this add 10 ml of scintillation fluid, shake well and determine the disintegration per minute using liquid scintillation counter.

CALCULATION

Amount of eukaryotic DNA polymerase present in sample (units/mg)

$$= \frac{DPM_S - DPM_B}{DPM \text{ nmol/s} \times E}$$

where,

DPM_S = Disintegration rate/min. of the substrate

DPM_B = Disintegration rate/min. of the blank

E = Protein concentration in enzyme

6.19 DNA POLYMERASE *(E. coli)*

In *E. coli,* DNA polymerase I is the predominant polymerizing enzyme with single disulphide bond and one sulphydryl group. It is associated with 3′- 5′ exonuclease activity. However the enzyme activity is influenced by the concentration of monovalent cations such as K^+, Rb^+, Cs^+ and NH_4^+. It degrades both single-and double-stranded DNA in the 3′ -5′ direction, the end product being 5′-mononucleotides.

6.19.1 Principle

Under specific temperature (37°C) and incubation period (30 minutes), acid-insoluble product is stuffed with 10 nanomoles of the total nucleotide.

Expt. 138 *To estimate the presence of DNA polymerase in E. coli.*

MATERIALS REQUIRED

Laboratory glassware

Laboratory equipment

Spectrophotometer

REAGENTS

a. Assay cocktail

Prepare 0.3 ml of reaction mixture by adding the following.

20 μ moles of potassium phosphate buffer (pH 7.4)

2 μ moles of magnesium chloride

0.3 μmoles of 2-mercaptoethanol

65 μg/ml activated calf thymus DNA

10 n moles deoxyadenosine 5′-triphosphate

10 n moles deoxyguanosine 5′-triphosphate

10 n moles deoxycytosine 5′-triphosphate

10 n moles of tritiated deoxythymidine 5′-triphosphate

b. Solution A

To prepare solution A, mix the following reagents

4 parts of saturated $Na_4P_2O_4$

4 parts saturated NaH_2PO_4

1 part of 100% TCA

1 part of 10 nm thymidine

    c.   Scintillation fluid

       Mix 640 ml of toluene, 320 ml of trition X-100 and 40 ml of permafluor.

PROCEDURE

1. Add 2 ml of diluted enzyme solution and 0.3 ml of assay cocktail in a silicon test tube (10 × 75 mm) placed in an ice-cold water bath.

2. Shake well and keep it in a water bath at 37°C for 30 minutes.

3. Following the incubation period, pipette 1 ml of solution A into the test tube, swirl and place it in ice-cold water bath (>4°C) for 15 minutes.

4. Then allow the reaction mixture to pass through a filter disc (0.45 m) and by a vacuum suction pump separate the acid-insoluble product from the reaction mixture. Soak the filter disc in solution prior to use.

5. Rinse the test tube with 3 ml TCA several times and filter the same.

6. Similarly wash the filter discs with 40 ml each of TCA and chloroform–methanol.

7. Dry the filters and transfer them into glass liquid scintillation vials.

8. Add 1 ml of methyl cellulose so as to dissolve the filter.

9. Leave it to stand for 30 minutes.

10. Following this, add 10 ml of scintillation fluid, shake well and determine the disintegration per minute using liquid scintillation counter.

CALCULATION

$$\text{Amount of DNA polymerase (units/mg)} = \frac{DPM_S - DPM_B}{DPM \text{ nmol/s} \times E}$$

where,

DPM_S = Disintegration rate/min. of the substrate

DPM_B = Disintegration rate/min. of the blank

E = Protein concentration in enzyme

6.20 DNA POLYMERASE (γ)

It is present in the nucleus and cytoplasm and is a high molecular weight compound (110,000–230,000). It favours the synthesis of DNA hybrids from ribohomopolymer template. It acts effectively at a pH range between 7.0 and 7.6. It is stable in the presence of glycerol, KCl, ETA and sulphydryl reagent. However substances like N-ethylmaleimide, p-hydroxy mercuric benzoate, sodium pyrophosphate, o-phenanthroline and pyridoxal phosphate inhibit its activity.

6.20.1 Principle

In this procedure one unit of activity is incorporated into one nanomole of dTMP to form an acid-insoluble product.

Expt. 139　*To estimate the presence of γ DNA polymerase*

MATERIALS REQUIRED

Laboratory glassware

Laboratory equipment

　Spectrophotometer

Enzyme

　100 μ/ml using 0.083M tris with pH 7.8 as enzyme diluent.

REAGENTS

a.　Assay cocktail

　Prepare reaction mixture by adding the following.

　　0.083M tris, pH 7.8

　　20% glycerol

　　670 μg/ml BSA

　　1.667 mM dithiothreitol

　　0.833 mM manganese chloride

　　0.167 mM ethylene dinitrillo (tetraacetic) acid

　　disodium salt (EDTA)

　　266 μg/ml polyriboadenylic acid (poly (rA))

　　53 μg/ml oligodeoxythymidilic acid

　　167 μM tritiated deoxythymidine 5′-triphosphate [(H^3)dTTP]

b.　Solution A

　　4 parts saturated $Na_4P_2O_7$

　　4 parts saturated NaH_2PO_7

　　1 part 100% TCA

　　1 part 10 mM thymidine

　Mix well.

c.　Scintillation fluid

　Prepare a solution containing 640 ml of toluene, 320 ml triton X-100, 40 ml permafluor 25 X.

PROCEDURE

1.　Pipette 15 μl of assay cocktail and 8 μl of 80 mM KCl and 2 μl sample and siphon into a silicon test tube.

2.　Incubate the test tube at 37°C in a water bath for about one hour.

3.　Following the incubation period, remove the tube and add solution A.

4.　Shake well again, incubate in an ice-cold water bath for about 15 minutes.

5.　Then allow the reaction mixture to pass through a filter disc (0.45 m) and by a vacuum suction pump separate the acid-insoluble product from the reaction mixture. Soak the filter disc in solution prior to use.

6.   Rinse the test tube with 3 ml TCA several times and filter the same.

7.   Similarly wash the filter discs with 40 ml each of TCA and chloroform–methanol. Dry the filters and transfer them into glass liquid scintillation vials.

8.   Add 1 ml of methyl cellulose so as to dissolve the filters.

9.   After an hour, add 10 ml of scintillation fluid to the tube and then mix.

CALCULATION

Amount of DNA polymerase (γ) in the sample (units/mg)

$$= \frac{DPM_S - DPM_B}{DPM \text{ nmol/s} \times E}$$

where,

DPM_S = Disintegration rate/min. of the substrate

DPM_B = Disintegration rate/min. of the blank

E = protein concentration in enzyme

IMMUNOTOXICOLOGICAL METHODS

Immunotoxicology is a subdiscipline of toxicology that focuses on the immune system as a target organ which is subjected to the deleterious effects of xenobiotics (environmental pollutants). The sensitivity of the immune system to environmental pollutants is due to the general properties of the chemical as to the complex nature of immune system. Assessment of effects of a pollutant on the immune system is therefore necessary to evaluate the toxic property of that pollutant. The assays used for such studies are called immunotoxicological techniques and they are as follows.

6.21 HAEMAGGLUTINATION ASSAYS

Haemagglutination assay is used to determine the anti-BSA or ARBC antibodies in the serum of blood by twofold serial dilutions. The antigen and antibodies agglutinate and form a mat-like structure. The highest dilution of the serum samples which shows detectable macroscopic agglutination in microtitre plate, is recorded and expressed as log2 titre of the serum. The microtitre plate wells 1, 2, 3, 4, 5 etc., which show highest dilution of agglutination are expressed as $\frac{1}{2}$, $\frac{1}{4}$, $\frac{1}{8}$, $\frac{1}{16}$ and so on respectively. This gives the production of antibodies in the blood serum against the particular antigen.

Expt. 140   *To conduct the haemagglutination assay*

ANTIGEN PREPARATION

Soluble BSA (S-BSA)

> Prepare S-BSA by overlapping the BSA (bovine serum albumin) on isotonic saline (0.15N) and allowing it to dissolve without agitation.

Sheep red blood corpuscles

> Collect the sheep blood in a glassware containing 1% EDTA from a nearby slaughterhouse without contamination. Then prepare SRBC

by washing sheep blood in isotonic saline (0.15N) and re-suspending to the required concentration (volume/volume) shortly before use.

IMMUNIZATION WITH ANTIGEN

1. Immunize the animals with optimum dose of antigen (0.2 ml/animal) by a single injection using one ml tuberculin syringe with a 24-gauge needle.

2. To avoid circadian rhythmic variations on the immune response, the antigen administration and serial bleeding of animals should be performed between 2 and 4 p.m.

ANTISERUM COLLECTION

1. Collect blood from immunized animals and keep in vials at room temperature for 15 minutes.

2. Following this period, store it overnight at 4°C. Separate the serum by centrifuging at 3000 rpm for 15– 20 minutes and keep at 57°C in a water bath for 30 minutes.

3. Store at –20°C until use.

COUPLING OF BSA WITH SRBC

1. Wash sheep erythrocytes thrice with isotonic saline (0.15N).

2. To one volume of packed SRBC, add equal volume of BSA at a concentration of 5 mg/ml isotonic saline and chromic chloride solution at a concentration of 1 mg/ml isotonic saline.

3. Agitate the mixture by hand shaking for 10 minutes and keep it at 37°C in a water bath for 15 minutes with intermittent agitation.

4. Abruptly stop the reaction by dilution with isotonic saline.

5. Wash the BSA-coupled cells thrice with isotonic saline and then suspend it to 1% isotonic saline. This is used as antigen in passive haemagglutination assay.

PASSIVE HAEMAGGLUTINATION ASSAY

1. Add 50 ml of isotonic saline to all the wells of microtitre plates. It has two controls—positive control and negative control

2. Add 50 ml of antiserum to the first and second wells.

3. From the second well, make twofold serial dilution of the antisera with saline in microtitre plates.

4. Add 25 ml of 1% BSA-coupled SRBC in isotonic saline to all wells of the microtitre plate.

5. Shake for effective mixing and incubate for an hour at 37°C and for another hour at 4°C.

6. Record the highest dilution of the serum samples that shows detectable macroscopic agglutination.

7. Express it as log2 antibody titre of the serum.

DIRECT HAEMAGGLUTINATION ASSAY

1. This assay is used to determine anti-SRBC antibodies in the serum.
2. Similar to that of passive haemagglutination assay, make twofold serial dilution of the antisera with isotonic saline (0.15N) in microtitre plate.
3. Add to each well, 2.0 ml of 1% SRBC.
4. Record the highest dilution that shows detectable agglutination.
5. Express the result in log2 antibody titre of the serum.

6.22 MACROPHAGE MIGRATION INHIBITION TEST

6.22.1 Principle

Antigen-stimulated sensitized lymphocyte releases a number of soluble substances (lymphokines) that affect other cell populations. The migration inhibitory factor (MIF) inhibits the migration of normal animal peritoneal macrophages. Related to the MIF is the leukocyte inhibitory factor, which inhibits the migration of human puffy coat leukocytes (mononuclear and polymorphonuclear leukocytes) from a capillary tube into a tissue culture medium. These factors are found in the cell-free supernatant of sensitized lymphocyte cultures exposed to the specific antigen. The method presented has two phases.

1. Sensitized lymphocytes are cultured with the specific antigen and are thus stimulated to produce the MIF, which is found in the supernatant.
2. The MIF is assayed on leukocytes in capillary tubes embedded in positive supernatant tissue culture medium. The inhibition of migration is read at 18–24 hours. Under the conditions of the test, normal leukocytes migrate from the capillary tube into tissue culture medium for a visible distance. This migration area is measured after 24 hours (control) and is compared with the inhibited migration area. The migration inhibition is the result of MIF produced by sensitized and challenged t-lymphocytes and is an indicator of the acquisition of cellular immunity to an antigen. The migration inhibition is expressed as percent inhibition or as migration index.

Expt. 141   *To perform the macrophage migration inhibition test*

MATERIALS REQUIRED

Laboratory glassware

    Microcapillary tubes (75 mm length and 1.1–1.2 mm diameter)

Planchette (clear plastic chamber with a volume of 0.5 ml with circular sterile coverslip)

Clay

Silicone grease

REAGENTS

a. Ammonium chloride (0.83%)
b. Lactated Ringer's solution
c. Eagle's mini essential medium (MEM) in Eagle's basic salt solution with L-glutamine, vitamins, penicillin, streptomycin and mycostain.

PROCEDURE

1. Draw 30 ml of venous blood into a heparinized plastic syringe.

2. Allow to stand in an upright position until the red cells and plasma separate. Aspirate eukaryocyte-rich plasma, centrifuge at 150 g (gravity) for 10 minutes and aspirate plasma.

3. Wash cell pellets in 0.83% ammonium chloride and four times in lactated Ringer's solution.

4. Suspend leukocytes in MEM to obtain a final suspension of 10,000,000, cells/ml. Determine viability of cells with tryptan blue stain.

5. Draw final suspension of cells into microhaematocrit tube to 20 ml mark and seal with clay.

6. Centrifuge at 600 g for 5 minutes.

7. Edge capillary with file and break at the level of the culture medium and cell interphase.

8. Place the cell portion in a planchette, holding it in place by a drop of silicone grease.

9. Using injection port add tissue culture medium (MBM) to planchette, place a coverslip and seal it with silicon grease.

10. Incubate at 37°C for 24 hours under 5% concentration.

11. Measure the project areas of migration on the paper with a planimeter.

12. Repeat test by injecting MIF-containing MEM into the chamber and measure migration for each test group and take the average results.

13. Relate average migration area of MIF-containing culture to average migration area of control cultures without antigen and express as migration index.

CALCULATION

$$\text{Migration index} = \frac{\text{average area of migration with antigen}}{\text{average area of migration without antigen}}$$

6.23 ROSETTE TEST

Rosette formation is a phenomenon of immunocyte adherence. It was first noted by Reiss et al (1950). He reported that the bacteria is specifically bonded with the lymph node cell surface obtained from rabbits immunized with *Salmonella* or *Brucella* antigens. Among the techniques that have been developed to detect cells with this antigen-binding capacity, the rosette assay employing sheep erythrocytes is one of the most commonly used method. This is due to the fact that sheep erythrocytes may serve directly as the antigen or they may be used as carriers for other antigens coated onto their surface.

Rosette forming cells (RFC) are principally members of the lymphocyte series and binding is due to antibodies or antibody-like receptors present on the cell surface. These receptors appear to be synthesized by the cells, rather than acquired positively. RFC are specific for a particular antigen and are necessary for an immune response to that antigen. Following antigenic stimulation, depletion of a cell population of RFC reactive with sheep erythrocyte antigen induces the specific immune responsiveness of that population so that the number of specifically reacting RFC enhances markedly. Both T and B cells have been shown to be rosette-forming. In addition, macrophages can form rosette with sheep red cells, possibly via cytophilic anti-erythrocyte antibody. Special type of rosette (E rosette), identifies human T lymphocytes, which spontaneously bind sheep erythrocytes at 4°C. However, such spontaneous rosettes are not formed by human lymphocytes with B cell markers. E rosette has been used extensively to identify and quantify peripheral human thymus-derived cells and thus has aided in the definition of immunodeficiency states and other conditions affecting the T cell system.

Expt. 142 *To perform the Rosette test*

MATERIALS REQUIRED

Laboratory glassware

Test tubes (10 × 75 mm)

Measuring pipettes (5 ml, 1 ml and 0.2 ml)

Pasteur pipettes with bulbs

Laboratory equipment

i. Refrigerated centrifuge, with rotor accommodating 10 × 75 test tubes

ii. Improved Neubaur–Levy–Hausser chamber

The improved Neubauer–Levy–Hausser chamber is divided into nine 1 mm 2 areas. There are sixteen subdivisions in each of the four corner areas and twenty-five subdivisions in the central area. The large squares of the central area, which are separated by double lines, are each divided into sixteen small squares. The space between the counting grid and the cover glass, that is the depth of the fluid when the chamber is filled is 0.1 mm.

iii. Microscope, with 16X and 40X objectives and 10X oculars

iv. Mouse (18–20 g) immunized intraperitoneally 4 days previously with 0.2 ml of 20% SRBC suspension in saline

v. Sheep red blood cells (SRBC), photometrically standardized with 1% suspension PBS

REAGENTS

a. Acetic acid (1%)

b. PBS

c. Phosphate-buffered saline (PBS), 0.001M, pH 7.2 to 7.3

d. Tryptan blue dye solution, containing 4 parts 0.2% tryptan blue in distilled water plus 1 part 4.25% NaCl in distilled water, freshly mixed.

PROCEDURE

1. Carry out all the steps at ice-cold temperature with pre-cooled glassware and reagents.

2. Use siliconized glassware throughout the experiment.

PREPARATION OF SPLEEN CELL SUSPENSION

1. Prepare a single cell spleen suspension from a mouse previously immunized with SRBC. Wash thrice and suspend the cells in PBS.

2. After resuspending the cells in PBS to a volume of 2 ml, determine the concentration of nucleated cells by haemocytometer count.

STANDARDIZATION OF THE SPLEEN CELL SUSPENSION PREPARED FOR THE ROSETTE ASSAY

1. Label two 10 × 75 mm test tubes A and B.

2. Pipette 2.4 ml and 4.9 ml of 1% acetic acid into tubes A and B respectively.

3. Pipette 0.1 ml cell suspension into each tube. Mix gently but thoroughly. Now tube A is a 1 : 25 dilution of the spleen cell suspension and tube B is a 1 : 50 dilution.

4. Load a haemocytometer chamber with tube A suspension.

5. Use a Pasteur pipette for the transfer.

6. Allow the cells to settle for 2 minutes.

7. Prepare the haemocytometer by placing the cover glass on the lateral supports so that an equal amount of the glass projects beyond the edges of the chambers.

COUNTING OF SPLEEN SUSPENSION CELLS

1. Using suction, fill a Thoma-type blood-diluting pipette for spleen cell suspension to the level of the 11 marks.

2. If blood is drawn slightly past the 11 marks, adjust the meniscus by moving the tip of the pipette across a nonporous surface.

3. Dilute with acetic acid, effecting a 1 : 10 dilution.

4. Remove the rubber tubing from the pipette, seal both ends with thumb and forefinger and shake the action.

5. Allow the cells to settle for about 3 minutes.

6. Place the haemocytometer on the microscope stage.

7. Using 10X oculars, bring the 16 objectives as close to the cover glass as possible and move it upward until the grid is in focus.

8. Then switch to the 40X objective and count the cells in the sixteen large squares in each of the four corner areas. A minimum of 50 to 60 cells should be present in each of the four 1–mm^2 areas. If the number of cells is too low, prepare a 1 : 10 dilution. Generally, a 1 : 25 or 1 : 50

dilution of a 2-ml cell suspension from one mouse spleen will provide the number of cells needed for the haemocytometer count.

9. Use the formula given for blood leukocyte counts to calculate the concentration of nucleated cells per millilitre of the undiluted suspension. One spleen should yield 0.7 to 1.0×10^8 cells.

10. Dilute the suspension to 2×10^7 cells/ml in PBS. It is important to know the viability of a cell suspension before assessing its performance in a test such as the rosette assay.

11. Live and dead cells can be distinguished on the basis of their ability to exclude tryptan blue dye.

12. The membrane of a living cell is impermeable to this dye and the cell appears colourless in the presence of tryptan blue. In a 10×75-mm test tube, place 0.3 ml of diluted cell suspension (dilution as determined in step 2.)

13. Add 0.1 ml tryptan blue dye solution and mix gently.

14. Wait for 5 minutes to allow sufficient time for dead cells to take up dye. Count live and dead nucleated cells to a minimum combined total of 200 to 300. Calculate viability as follows:

$$\text{Percent viable cells} = \frac{\text{Number of live cells counted} \times 100}{\text{Total number of cells counted}}$$

15. In calculating the cell concentration per cubic centimetre, the total number of cells counted in four 1 mm^2 areas is adjusted to $1 \text{ mm}^2 \times 1/4$, multiplied by the dilution ($\times 10$) and then by 10^4.

$$\text{Spleen suspension cells (cells/cm}^3) = \text{total cells counted} \times \text{dilution} \times 10^4 \text{ number of 1 mm}^2 \text{ areas counted}$$

Rosette Assay Procedure

1. The rosette assay is performed in duplicate.

2. Label two 10×75 mm test tubes as A and B into each tube, pipette 0.2 ml of the spleen cell suspension adjusted to 2×10^7 cells/ml.

3.   Then add 0.2 ml 1% SRBC suspension. Mix gently.

4. Centrifuge for 12 minutes at 500 g. After centrifugation, place the tubes in an ice bucket for 5 minutes before resuspending the pellet with a Pasteur pipette.

5. Fill both chambers of the haemocytometer with samples from tube A and scan the entire area of both grids for RFC. A nucleated cell that has bound 5 or more SRBC is scored as an RFC.

6. Also count the RFC in tube B. Calculate the RFC/mm.

7. Using the general formula given for blood leukocyte counts, average the RFC/mm for tubes A and B. Since the rosette mixture is a 1:2 dilution of the original spleen cell suspension, RFC/mm equals RFC per 10^7 nucleated cells.

8.   Express the results of a rosette assay as the number of RFC per 10^6 viable nucleated cells. It is calculated according to the formula.

$$\text{RFC per ml in rosette mixture} = \frac{\text{RFC} / 10^6 \text{ viable nucleated cells}}{\text{Viability} \times 10}$$

6.24 PRECIPITATION OF IMMUNOGLOBULINS

Immunoglobulins can be isolated from other serum proteins by repeated precipitation with various concentrations of ammonium sulphate starting from 100 % saturation solution. In general, three precipitations are enough to separate immunoglobulins and at 40% ammonium sulphate concentration, pure form of immunoglobulin-G can be obtained.

Expt. 143   *To isolate immunoglobulins by ammonium sulphate precipitation*

MATERIALS REQUIRED

Laboratory glassware

Beakers

Serological tubes

Graduated centrifuge tubes

Micropipettes

Laboratory equipment

pH meter

Magnetic stirrer

Centrifuge

Refrigerator

Dialysis membrane 600 (Hi media)

REAGENTS

a.   Serum sample

b.   Ammonium sulphate (100% saturated, pH 7.2)

c.   40% ammonium sulphate solution

d.   Sodium hydroxide

e.   PBS

PROCEDURE

1.   To 1 ml serum add 1 ml of PBS buffer.

2.   To this 2 ml, add 1.6 ml of (100% saturated) ammonium sulphate solution drop by drop. The ammonium sulphate concentration in this 3.64 ml mixture is 45%.

3.   Keep this mixture in an icebox (at 4°C) for 30 minutes and stir intermittently.

4.   At the end of 30 minutes, a cloudy white precipitate is seen.

5.   Centrifuge the precipitate at 2500 rpm for 15 min. and discard the supernatant (first precipitation).

6.   Wash the pellet with 1.64 ml of ammonium sulphate solution and recentrifuge it at 2500 rpm for 15 min. and discard the supernatant.

7.   Redissolve the pellet in 1 ml of PBS buffer and centrifuge at 2500 rpm for 10 min. Mix the supernatant with 100% saturated ammonium sulphate so as to achieve a 40% ammonium sulphate solution.

8.   Centrifuge the precipitate formed at 2500 rpm for 10 min and discard the supernatant.

9.   Wash the precipitate with equal volume (i.e. the volume of serum used for precipitation) of 40% ammonium sulphate solution (second precipitation).

10. After centrifuging discard the supernatant. Mix the precipitate with minimum volume of PBS (<1 ml).

11. Dialyse the sample with PBS with five changes at 4°C.

6.25 DIALYSIS

Dialysis is a method used to concentrate the enriched immunoglobulins after ammonium sulphate fractionation or precipitation.

Expt. 144   *To get a concentrate of immunoglobulins by dialysis*

MATERIALS REQUIRED

Laboratory glassware

Beaker

Glass rod

Dialysis bag

Twine

REAGENTS

EDTA

Sodium carbonate

Double-distilled water

PROCEDURE

1.   Dissolve 5 mM EDTA of sodium bicarbonate in 100 ml distilled water. Heat the solution to boil. Keep the dialysis bag (dialysis membrane 60, Himedia, India) above the boiling solution for about 5 minutes (This will activate the dialysis membrane).

2.   Remove the dialysis bag with a forceps, wash it with distilled water.

3.   Tie one end of the bag with a twine (before, use keep the twine in boiling water for few minutes) Then load the sample into the bag through its open end. After loading tie the open end with a twine.

4. Carefully knit both the twines in a glass rod at equal distances. Keep the glass rod over a beaker containing PBS. The rod should be adjusted in such a way that the bag is suspended enough to be completely immersed in PBS. See to it that the bag does not touch the bottom of the beaker.

5. After dialysis is completed, collect the immunoglobulin and store it under freezing condition for future use.

CONFIRMATION OF IMMUNOGLOBULIN

Prepare a 1:20 dilution of sample and determine the peak absorbance using UV–VIS spectrophotometer. The immunoglobulins show maximal absorbance at 280 nm.

6.26 ANTI-ENA HAEMAGGLUTINATION TECHNIQUE

6.26.1 Principle

Anti-ENA-based (extractable nuclear antigens) haemagluttination technique is used to detect antibodies produced against two types of saline-soluble nuclear antigens—Sm and RNP (ribonucleoprotein). Antibody to Sm is present exclusively in systematic lupus erythematosus (SLE), perhaps indicating that this is the marker for the disease. Antibody to RNP occurs in a number of rheumatic diseases such as SLE. Most often high titre RNP antibodies are associated with "mixed connective tissue" which has a relatively low prevalence in renal disease and good response to treatment with corticosteroids.

Sm is saline-soluble and easily extracted from the nucleus and is a non-histone nuclear protein that is devoid of nucleic acid. Its antigenicity is not destroyed by RNase and this is the basis for distinguishing between antibodies to RNP and Sm in the haemagglutination test. Tanned sheep erythrocytes are coated with an extract of rabbit thymus containing RNP (RNase-resistant) antigens. A portion of the treated cells is incubated with RNase to selectively remove the RNP antigens.

A passive hemagglutination test is performed in a microtitre system with the use of parallel dilutions of the patient's serum against both the untreated and treated cells. A positive haemagglutination test with the untreated cells and a negative one with the treated cells indicate antibody to RNP.

Expt. 145   *To perform the anti-ENA haemagglutination technique*

MATERIALS REQUIRED

Laboratory glassware

Plate sealers

Go-on-go blotters

Pipette droppers (0.0025 ml calibration)

Microdiluters

Disposable V-shaped microtitre plates

Centrifuge tubes

Laboratory   equipment

Water bath

Centrifuge

REAGENTS

a.   Phosphate buffered saline (PBS) (pH 7.2 ± 0.10)

Add 30.6 g NaCl (0.1M), 2.9 g Na_2HPO_4 and 0.84 g KH_2PO_4 (0.01 m PQ) to demineralized or distilled water bringing total volume to 4 litres. Stir on magnetic stirrer to ensure adequate mixing. Check pH of each batch of PBS prepared. Store at 40°C.

b.   Mcllvaline buffered saline (pH 7.8)

Add 1.139 of 0.02M Na_2HPO_4 and 3.5 g of 0.15M NaCl to demineralized or distilled water making the volume to 4 litres. Adjust pH to 7.8 with 0.1M citric acid (1.92 g in distilled water).

c.   Tannic acid, (0.0012% w/v)

Prepare fresh just before use. Weigh 0. 012 g (12 mg) of tannic acid and add to 1 d1 PBS. Warm to 37°C and then dilute to 0.0012% tannic acid with warm (37°C) PBS (90 ml PBS + 10 ml 0.012% tannic acid)

d.   Foetal calf serum (FCS)

Inactivate foetal calf serum at 56°C for 30 min and absorb with sheep red cells (FCS–ABS), 1% (v/v). Add 1 ml of FCS–ABS to 99 ml PBS.

e.   Normal human serum

Inactivate normal human serum at 56°C for 30 min and absorb with sheep red cells in the same manner as FCS–ABS. Add 0.4 ml NHS–ABS to 19.6 ml PBS.

f.   RNAse solution

1 mg of RNase in 1 ml of 1% FCS will treat cells for one set of plates.

g.   Rabbit thymus acetone powder

- To each 60 mg of rabbit thymus powder add 1 ml of PBS. Extract for 4–12 hr at 40°C on a magnetic stirrer using a very low speed to avoid denaturation (a rheostat aids in decreasing the speed of the stirrer).

- Centrifuge at 10,000 ×g for 2 min (9300 rpm)

- Determine the protein concentration of the supernatant fluid.

- Adjust the concentration of antigen to 1 mg/ml.

- Store saline extract of rabbit thymus, which contains Sm and RNP antigens at −70°C and is stable for at least 10 weeks.

- Always use clear extract containing antigens by centrifugation before for coupling to sheep red cells.

PROCEDURE

1. Prepare a 1 : 10 (v/v) dilution of the exposed animal sera and control sera in PBS (0.05 ml of serum + 0.45 ml of PBS).

2. Inactivate samples at 56°C for 30 min.

3. Meanwhile wash four times with PBS so as to get sufficient sheep erythrocytes to give a packed volume of 0.5 ml for serum and 0.5 ml for tannic acid treatment.

4. Absorb each serum with an equal volume (0.5 ml) of packed cells for 30 min at room temperature.

5. Centrifuge samples and remove supernatant serum for use in testing.

6. Resuspend the remaining 0.5 ml packed cells in PBS to give a 2.5% (v/v) suspension (0.5 ml of packed sheep cells + 19.5 ml of PBS).

7. Mix gently equal volume of the 2.5% suspension PBS (20 ml) of sheep erythrocyte and 0.0012% tannic acid in PBS (20 ml).

8. Incubate for 10 min in a 37°C water bath with gentle swirling at 5 and 10 min.

9. Centrifuge and wash twice with PBS (this should be at least 20 times the volume of packed sheep erythrocytes) and resuspend to 2.5% suspension.

10. Remove 2 ml of tanned sheep erythrocytes into each of the three centrifuge tubes designated as control, test, and RNAse, centrifuge and discard supernatant.

11. Resuspend each packed cell volume with 10 ml McIlvaline saline buffer.

12. Add 2 ml of McIlvaline saline buffer to the tube containing the control cells.

13. Add 2 ml of antigen solution to the test and RNase cells for each 0.05 ml of packed sheep erythrocytes.

14. Prepare sufficient antigen in McIlvaline saline buffer to have 1 mg of rabbit thymus extract in 2 ml.

15. Mix gently and incubate the tubes at 37°C for 30 min. Swirl or shake gently at 10-min intervals to facilitate easy mixing.

16. Centrifuge and wash cells three times with 1% FCS in PBS.

17. Add 5 ml of 2% NHS (normal human serum) in PBS for each 0.05 ml of control cells and test cells.

18. These cells are ready to use. Add 4 ml of 2% NHS in PBS and 1 ml of ribonuclease solution containing 1 mg of RNase/ml for each 0.05 ml of packed sheep erythrocytes to RNase cells.

19. Incubate at 37°C in a water bath for 30 min with gentle mixing at 10-min. intervals.

20. These RNase-treated cells are now ready for use.

TEST PERFORMANCE

1. Label plates, allowing 24 wells for each antigen. Therefore each patient's sera will need two rows of 24 wells (one for RNP and one for Sm). Sera are titrated to 23 wells. The 1st well is labelled as antibody control. Three wells are labeled for cell controls—tanned cells, test cells, and RNase-treated cells.

2. With the pipette dropper, add 0.0025 ml of diluent (1% FCS in PBS) to each well.

3. Add with a diluter 0.0025 ml of patient's and/or control serum to the proper antibody well (serum control).

4. Mix well by swirling about 10 times.

5. Remove and blot.

6. Add 0.0025 ml of the patient's and/or control serum to the first well of the plate using the same diluter.

7. Repeat this procedure until all sample diluters are in the first wells.

8. Make serial dilutions of these samples simultaneously. Mix well by swirling about 10 times in each well.

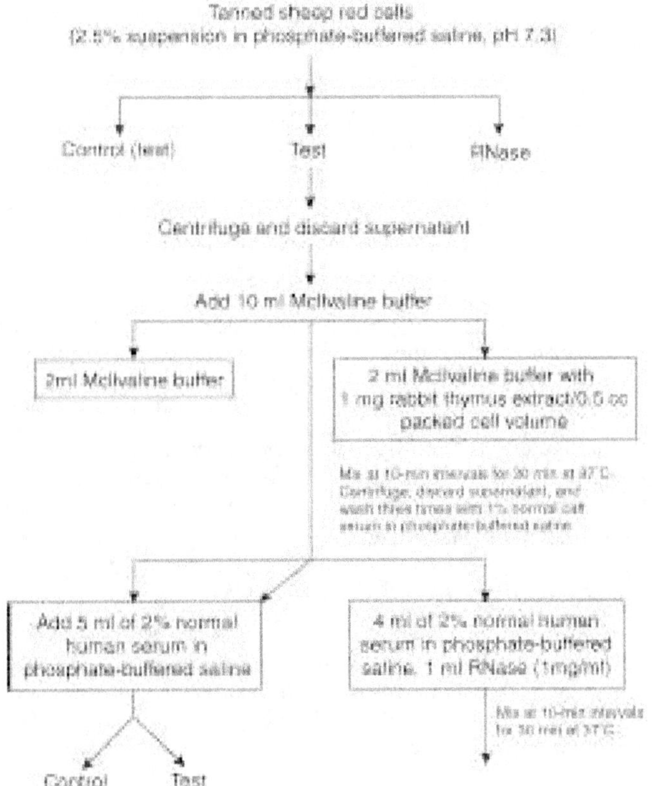

9. Blot the loops, and when the dilutions are finished, rinse in distilled water, flame to incandescence, and plunge into distilled water. Blot again and pre-wet in dilution.

10. Add 0.0025 ml of control cells with a pipette dropper to the row of antibody control (serum control) wells and to the cell control for tanned sheep cells.

11. Add 0.0025 ml of the test cells with a pipette dropper to the appropriate serum dilutions and to the cell control labelled for test cells.

12. Add 0.0025 ml of RNase-treated cells with a pipette dropper to appropriate serum dilutions and to cell control labelled for RNase test cells.

13. Tap plates gently to mix. Cover with plate sealers.

14. Allow cells to settle for at least 2 hours at room temperature and observe haemagglutination reaction.

INTERPRETATION

Results positive   a smooth mat of cells covering the bottom of the well(+).

Results negative   a clearly defined button at the bottom of the well (−).

Results intermediate   incomplete settling of cells with no clearly defined button (the reactions are noted (±) but not reported).

End point   Highest dilution giving a positive reaction.

If haemagglutination is present in wells with test cells RNase-treated cells, this indicates antibody to RNase-resistant antigens (Sm). There could also be RNase-sensitive antigens (RNP) in lower titers. If haemagglutination is present only with test cells this shows antibody to RNase-sensitive antigen (RNP) only. This is considered especially in high titers, to be highly suggestive of MCTD (mixed connective tissue disease). No haemagglutination with body antigens indicates a negative test.

BIOCHEMICAL METHODS

Knowledge of basic concepts of biochemistry is the fundamental need for the students to excel in various fields of life sciences such as microbiology, biotechnology, bioinformatics and health sciences. Therefore efforts have been put forth to present basic biochemical techniques in this chapter.

6.27 ESTIMATION OF PROTEINS

The word protein was coined by J. J. Berzelius in the year 1838 (Proteious-first rank). Proteins are made up of amino acids and the combination of amino acids differ between the proteins and therefore they can simply be defined as the polymers of different amino acids and the peptide bond links the amino acids, i.e., between a carboxylic group of one amino acid and an amino group of another amino acid.

$$NH_2 \!-\! \overset{\displaystyle R_1}{\underset{\displaystyle H}{C}} \!-\! CO\,\boxed{O}H \;+\; \boxed{H_2}N \!-\! \overset{\displaystyle R_2}{\underset{\displaystyle H}{C}} \!-\! COOH$$

H₂O

Amino acid 1   Amino acid 2

They form the essential constituents of protoplasm, cell membrane and nuclear material. They exist as simple and complex proteins. In complex forms, proteins combine with lipids and nucleic acids. Basic function of proteins, in general, is to serve as building blocks for cellular and organic structures. Besides they are involved in several biological functions, i.e., few of them act as enzyme to catalyse biooxidation reactions. In addition they can function as hormones, regulators of body metabolism.

Proteins can be estimated by several methods such as

1.   Biuret method
2.   Lowry method
3.   Bradford method
4.   Fluorescence method

6.28 BIURET METHOD FOR ESTIMATION OF PROTEINS

6.28.1 Principle

The test is specific for peptide bonds and hence evaluates the number of peptide bonds in the given sample. The copper atom (Cu^{2+}) in the copper sulphate solution complex with nitrogen atom of the peptide bond to give a purple colour at high temperature and this when measured at 535–560 nm gives accurate result.

Expt. 146   *Estimation of protein by Biuret Method*

MATERIALS REQUIRED

Laboratory glassware

Since even a fingerprint would add about 0.1 mg of protein, mishandling of glassware should be strictly avoided. Clean the test tubes. Then soak them in acid (conc. HCl 1 : 1 part of conc. HNO_3 acid) for 10 hours. Wash them clean and dry in a hot-air oven. Never touch the inside of the test tubes after cleaning.

REAGENTS

a.   Bovine serum albumin stock solution

Dissolve 10 mg of BSA in 10 ml double-distilled water. Shake well and after complete dissolution, measure the absorbance at 280 nm. To find out the exact concentration of BSA in mg/ml, divide it by 0.66.

b.   Biuret reagent

In 500 ml distilled water, add 1.5 g of copper sulphate ($CuSO_4 \cdot 5H_2O$) and 6.0 g of sodium potassium tartrate. Mix well and store it under room temperature. Presence of dissolved oxygen in the distilled water may convert the Cu^{2+} of copper sulphate to CO^+ which results in the formation of a reddish orange precipitate. To prevent this, heat the distilled water to boiling point and then cool it before use.

PROCEDURE

1. Siphon out different volumes of bovine serum albumin standard solutions in a series of test tubes from 50 µl to 500 µl interval.

2. Make up the volume of each test tube to 0.5 ml using double-distilled water.

3. Mix well and add 0.5 ml of the biuret reagent to all the test tubes.

4. Again mix well and boil the test tubes in a water bath at 80°C for five minutes.

5. Cool the test tubes immediately under running water.

6. After cooling, allow it to stand for another ten minutes.

7. Measure the absorbance at 535 nm.

8. Similarly process the sample and measure its absorbance value.

9. Draw a standard curve by plotting the absorbance value of standard solutions against their concentrations.

10. Evaluate the concentration of sample by using the standard curve.

6.29 LOWRY METHOD FOR ESTIMATION OF PROTEINS

6.29.1 Principle

Colour intensity of the sample depends on the aromatic amino acid content of the protein. Tyrosine and tryptophan residues of protein reduce the phosphomolybdate and phosphotungstate anions in the folin ciocolteau phenol reagent which when combined with copper of the copper sulphate gives a blue coloured complex (heteropolymolybdenum blue and tungsten blue) at 750 nm.

Expt. 147　　*Estimation of protein by Lowry method*

MATERIALS REQUIRED

Laboratory glassware

Follow the same procedure described in the biuret method.

REAGENTS

a. 2% Na_2CO_3 in 0.1N NaOH.

Take 2 g of sodium hydroxide in 400 ml of double distilled water. Mix well and then add 10 g of anhydrous sodium carbonate. Shake well and make up to 500 ml using distilled water.

b. 2% $CuSO_4 \cdot 5H_2O$

Dissolve 1 g of copper sulphate in 50 ml of distilled water.

c. 2% sodium potassium tartrate

Dissolve 1 g of sodium potassium tartrate in 50 ml distilled water.

d.   Add 0.5 ml of reagent b with 0.5 ml of reagent c. To this mixture add 99 ml of reagent a. Follow the same order as described here and always prepare the solution fresh.

e.   Folin ciocolteau reagent

It is commercially available and prepared using double-distilled water. Store the reagent at 4°C.

PROCEDURE

1.   Pipette out different volumes of bovine serum albumin stock solution into a series of test tubes (10 μl, 20 μl, 30 μl, 40 μl, 50 μl, 60 μl, 70 μl, 80 μl, 90 μl, 100 μl).

2.   Make up the volume of each test tube to 0.2 ml.

3.   To each test tube, add 1 ml of reagent d, mix well and allow it to stand for ten minutes.

4.   Again add 0.3 ml of reagent e, mix well and leave it undisturbed for 60 minutes.

5.   Measure the absorbance at 660 nm.

6.   Run a blank using distilled water.

Note   Shake the test tubes well after the addition of folin ciocolteau reagent to get better results.

6.30. BRADFORD METHOD FOR ESTIMATION OF PROTEINS

6.30.1 Principle

Coomassie brilliant blue when combined with positively charged and hydrophobic group of proteins gets converted from pale orange and to a blue coloured complex.

Expt. 148   *Estimation of proteins by Bradford method.*

MATERIALS REQUIRED

Laboratory glassware

Spectrophotometer

REAGENTS

Bradford reagent

Dissolve 50 mg of coomassie brilliant blue in 50 ml of ethanol/methanol. For complete mixing, place it in a shaker for 60 minutes. To this add 50 ml of 88% orthophosphoric acid, mix well and make up to 250 ml using double-distilled water. Filter the content using Whatman No. 1 filter paper. From the filtered solution, pipette out 50 ml in a beaker, make up to 100 ml and measure the absorbance at 550 nm.

PROCEDURE

1.   Pipette out different volumes of bovine serum albumin stock solution into a series of test tubes (10 μl, 20 μl, 30 μl, 40 μl, 50 μl 60 μl 70 μl, 80 μl, 90 μl and 100 μl).

2.   Make up the content of each test tube to 100 µl.

3.   To each test tube add 1.5 ml of bradford reagent and shake well.

4.   Similarly process the blank using distilled water.

5.   Allow it to stand for five minutes.

6.   Measure the absorbance at 595 nm.

6.31 FLUORESCENCE METHOD FOR ESTIMATION OF PROTEINS

6.31.1 Principle

It is a very sensitive method to measure the protein concentration from 0.1 µg and therefore small peptides with molecular weight less than 10,000 can be detected by this method. Fluorescamine readily reacts with primary amines to form a fluorescent product which in turn reacts with the n-terminal amino group available in the free form. During this reaction, amino groups of the side chains may get ionized and to prevent this, the pH of the solution is maintained between 8.5 and 10 using borate buffer.

Expt. 149   *Estimation of proteins by fluorescence method*

MATERIALS REQUIRED

Laboratory glassware

Spectrophotometer

REAGENTS

a.   Borate buffer (pH 8.5)

Dissolve 1.5 g of boric acid and 2.4 g of sodium tetraborate in 100 ml distilled water.

b.   Fluorescamine solution (0.03%)

Dissolve 7.5 mg of fluorescamine in 25 ml of acetone.

PROCEDURE

1.   Pipette out different volumes of bovine serum albumin stock solution into a series of test tubes. (0.1 µl, 0.2 µl, 0.4 µl, 0.6 mg, 0.8 µl, 1 µl, 1.2 µl, 1.4 µl 1.6 µl, 1.8 µl and 2.0 µl).

2.   Make up the content of each test tube to 100 µl with distilled water.

3.   To each test tube, add 0.9 ml of borate buffer and mix well.

4.   After rapid mixing, add 0.1 ml of flourescamine solution and mix well.

5.   Measure the fluorescence at 465 nm.

6.   Similarly process the blank using distilled water.

Note   For the estimation of proteins, the sample should not possess any compounds that may interfere either with the reagents or with the end products. Therefore based on the type of interfering compound the sample should be processed before use. For example compounds like carbohydrates, pigments, lipids and polyphenols are known to interfere with the estimation of proteins and the extraction procedure of these compounds are discussed below.

Carbohydrates and pigments

    a.   Add equal volume of ice-cold 10% TCA to the sample.

    b.   Leave it to stand for ten minutes.

    c.   Centrifuge it and discard the supernatant.

    d.   Draw off the pellet.

    e.   Dissolve it on 0.1N NaOH.

Lipids

    a.   Extract the sample with ethanol and ether mixture (2 : 1) and wait for the layer formation.

    b.   Discard the upper organic phase.

    c.   Subject the sample to extraction with ether once again.

    d.   Discard the ether phase.

    e.   To get rid of the solvent, air-dry the sample.

    f.   Again extract the sample with equal volume of ice-cold TCA.

    g.   Protein is then precipitated out which can then be used for extraction.

Polyphenols from plant extracts

    a.   To the extraction buffer, add 1% polyvinyl pyrrolidone.

    b.   Add the mixture to the sample and carry out the extraction process.

    c.   After centrifugation, draw off the pellets which is a combination of polyphenols and polyvinyl pyrrolidone.

For low concentration samples (<100 µg/ml)

    a.   Prepare 0.1% of sodium deoxycholate in distilled water.

    b.   Pipette out 0.15 ml of this solution and add to 1 ml of the sample.

    c.   Leave it to stand for 10 minutes.

    d.   Then add 0.2 ml of 50% ice-cold TCA and allow the content to remain on ice for another 10 minutes

    e.   Centrifuge the content and discard the supernatant.

    f.   Carefully invert the ependorff tube over a tissue paper so that the pellets are transferred onto the paper.

    g.   Wipe off the droplets with tissue paper.

    h.   Collect the pellets and dissolve it in 20 µl of 10 mM NaOH before use.

6.32 ESTIMATION OF AMINO ACIDS

6.32.1 Principle

Estimation of amino acids using ninhydrin is a widely accepted method. It can react with 2 amino acids, primary amines, imino acids and ammonia and give a purple coloured complex which can be measured at 550 nm. With proline and hydroxyproline it gives a yellow coloured complex and it can be measured at 440 nm.

Expt. 150 *Estimation of amino acids*

MATERIALS REQUIRED

Laboratory glassware

Laboratory equipment

Spectrophotometer

REAGENTS

a. 0.5M acetate buffer (pH 5.5)

b. Ninhydrin reagent

Dissolve 2.0 g ninhydrin in 30 ml of anthrone and mix well. To this add, 20 ml of 0.5M acetate buffer. Always prepare fresh.

c. Amino acid stock solution

Dissolve 0.01 g of the required amino acid in 100 ml distilled water.

PROCEDURE

1. Pipette out several volumes of amino acid stock solution into a series of test tubes (10 μl, 20 μl, 30 μl, 40 μl and 50 μl).

2. Make up the volume of each test tube to 4 ml using distilled water.

3. To this add, 1.0 ml of ninhydrin reagent.

4. Mix well and boil the test tubes in the water bath for ten minutes.

5. Cool the test tubes by placing them in a trough with tap water and then add 1 ml of 50% ethanol.

6. Mix well and leave it to stand for 5 minutes.

7. Measure the colour at 550 nm. This should be done immediately as the colour does not persist for a long time.

6.33 ESTIMATION OF CARBOHYDRATES

6.33.1 Principle

In the alkaline pH, reducing sugar is converted to enediol which in the presence of oxidizing agent 3, 5-dinitrosalicylic acid is oxidized to sugar acid.

$$\text{Reducing Sugar} \xrightarrow{\text{alkaline pH}} \text{Enediol}$$

$$\text{Enediol} \xrightarrow{\text{3, 5-dinitrosal icylic acid}} \text{Sugar acid}$$

Expt. 151 *To estimate the amount of carbohydrates*

MATERIALS REQUIRED

Laboratory glassware

Laboratory equipment

Spectrophotometer

a.   Dinitrosalicylic acid reagent

Dinitrosalicylic acid 1 g

Phenol   0.2 g

Sodium sulphite   0.05 g

NaOH (1N)   5 ml

Dissolve the chemicals mentioned above in 80 ml of distilled water. Add the chemicals one after the other and shake well after each addition.

b.   Standard glucose solution

Dissolve 0.1 g of glucose in 100 ml of double-distilled water.

PROCEDURE

1.   Prepare different volumes of standard glucose solution in a series of test tubes ranging from 0.2 ml to 3 ml at 0.2 ml interval.

2.   Make up the content of each test tube to 3 ml using distilled water.

3.   Add 3 ml of dinitrosalicylic acid to each test tube and immediately cover the test tubes with aluminum foil.

4.   Place the test tubes in a beaker containing boiling water for 5 minutes.

5.   After 5 minutes, cool it under tap water while doing this. Take care not to allow water into the test tubes.

6.   Allow it to stand for five minutes.

7.   Measure the colour at 575 nm.

6.34 NELSON SOMOYGYI'S METHOD FOR ESTIMATION OF CARBOHYDRATES

6.34.1 Principle

In the presence of copper sulphate, reducing sugar is oxidized and the reduced cuprous oxide (red-orange precipitate) in combination with arsenomolybdate reagent gives green coloured complex at 520 nm and the colour persists for about 18 hours.

Copper sulphate + reduced sugar → cuprous oxide + oxidized sugar

Cuprous oxide + arsenomolybdate reagent → Green coloured complex

Expt. 152   *Estimation of carbohydrates by Nelson Somoygyi's method*

MATERIALS REQUIRED

Laboratory glassware

Laboratory equipment

Spectrophotometer

REAGENTS

a.   Copper reagent A

Solution 1 Dissolve 12 g of anhydrous sodium carbonate and 6 g of sodium potassium tartrate in 150 ml distilled water.

Solution 2 Dissolve 2 g of copper sulphate in 20 ml of distilled water.

Solution 3 With continuous stirring, carefully add (b) to (a). Then add 8 g of $NaHCO_3$ and mix well.

Solution 4 Dissolve 90 g of sodium sulphate in 250 ml distilled water and boil. This will get rid of the dissolved oxygen in the solution which would otherwise reduce cupric ions. Cool the solution.

Add reagent (d) to reagent (c) and make up to 500 ml using distilled water. Store it for a week time. Then filter it through Whatman filter paper and remove the precipitate. Use the clear solution for the experiment.

b. Arsenomolybdate colour reagent

Dissolve 12.5 g of ammonium molybdate in 225 ml distilled water. To this, add 21 ml of concentrated sulphuric acid, stir well. After the solution becomes clear, add sodium arsenate solution (1.5 g/12.5 g distilled water) and make up to 250 ml. Incubate it for 48 hours.

c. Standard sugar solution

Dissolve 0.1 g glucose in 100 ml distilled water.

PROCEDURE

1. Prepare different volumes of standard sugar solution in a series of test tubes from 0.2 ml to 1 ml at 0.2 ml interval.

2. Make up the content of each test tube to 2 ml using distilled water.

3. Then add 2 ml of copper reagent to each test tube.

4. Place the test tubes in a beaker containing boiling water for ten minutes.

5. Cool it under tap water.

6. Then add 2 ml of arsenomolybdate colour reagent, mix well and measure the absorbance at 520 nm.

6.35 ANTHRONE PHENOL METHOD FOR ESTIMATION OF CARBOHYDRATES

6.35.1 Principle

This method is used to evaluate total sugars in the sample (reducing + non-reducing sugars). In the presence of concentrated sulphuric acid, sugar gets dehydrated to form either furfural (pentoses) or 5 hydroxymethyl furfuryl (hexoses). This when allowed to react with anthrone/phenol gives rise to a coloured compound and the absorbance can be measured at 625 nm.

$$\text{Pentose + conc. } H_2SO_4 \rightarrow \text{Furfural}$$

$$\text{Hexose + conc. } H_2SO_4 \rightarrow \text{5-hydroxymethyl furfural}$$

Expt. 153 *Estimation of carbohydrate by anthrone phenol method*

MATERIALS REQUIRED

Laboratory glassware

Laboratory equipment

Spectrophotometer

REAGENTS

a. Anthrone reagent (0.2%)

Dissolve 0.2 g of anthrone in 5 ml of ethanol. Add slowly 75% of sulphuric acid till the mark reaches 100 ml.

b. Phenol reagent

Dissolve 5 g of phenol in 100 ml of distilled water.

c. Standard sugar solution (0.1%)

Dissolve 0.1 g of sugar in 100 ml of distilled water.

PROCEDURE FOR ANTHRONE REAGENT

1. Prepare different volumes of standard sugar solution in a series of test tubes from 0.1 to 1 ml at 0.1 ml interval.

2. Make up the content of each test tube to 1 ml using distilled water.

3. To each test tube add 5 ml of cold anthrone reagent and mix well.

4. Do not leave the tubes open after the addition of anthrone reagent.

5. Close it with aluminium foil and place it in a boiling water bath for about ten minutes.

6. Then cool the test tubes under tap water and measure the absorbance at 625 nm.

7. Process the blank similar to the method described above.

PROCEDURE FOR PHENOL REAGENT

1. Prepare different volumes of standard sugar solution in a series of test tubes from 0.1 ml to 1 ml at 0.1-ml interval.

2. Make up content of each test tube to 1 ml using distilled water.

3. To each test tube, and 5 ml of concentrated sulphuric acid and 1 ml of phenol reagent.

4. Allow the tubes to remain in a water bath at 37°C for about 20 minutes.

5. Measure their absorbance at 490 nm.

ADVANCED METHODS IN BIOTECHNOLOGY

Advanced techniques such as isolation and estimation of genetic materials, DNA, RNA, transformation, gel electrophoresis, southern transfer, PCR technique, ELISA and gel diffusion techniques are discussed below.

6.36 ESTIMATION OF DNA—PHENYLAMINE METHOD

6.36.1 Principle

This method is specific for the deoxyribose components of purine nucleotides, i.e., adenine and guanine. Under acid pH, deoxyribose moieties form hydroxylevulinic acid which when combined with diphenylamine produces blue coloured complex at 595 nm.

$$\text{Deoxyribose component of DNA} \xrightarrow{\text{acid pH}} \text{Hydroxylevulinic acid}$$

$$\text{Hydroxylevulinic acid} + \text{diphenylamine} \rightarrow \text{Blue coloured complex}$$

Expt. 154   *Estimation of DNA by phenylamine method*

MATERIALS REQUIRED

Laboratory glassware

Laboratory equipment

Spectrophotometer

REAGENTS

a.   Diphenylamine reagent

Dissolve 0.5 g of diphenylamine in 48.7 ml glacial acetic acid. Add slowly 2.5 ml of concentrated sulphuric acid. Stir well and store for future use.

b.   DNA stock solution

Dissolve 5 mg of calf thymus DNA in 25 ml of 5 mM NaOH. Mix well and store for future use.

PROCEDURE

1.   Prepare different volumes of DNA stock solution in a series of test tubes from 0.05 ml to 1 ml at 0.01 interval.

2.   Make the content of each test tube to 1 ml using distilled water.

3.   To each test tube, add 5 ml of diphenylamine reagent, mix well.

4.   Close the test tubes with aluminium foil and keep it firm by rubber bands.

5.   Place the test tubes in a beaker containing boiling water for ten minutes.

6.   Measure the absorbance at 595 nm.

Note   For the reaction to occur, dissolution of diphenyl amine reagent in sample is essential. Since it is insoluble in water, rinse the test tubes with ethanol before use.

6.37 ESTIMATION OF RNA—ORCINOL METHOD

6.37.1 Principle

Under acidic conditions, ribose moieties of ribonucleic acids form furfural which reacts with orcinol to give rise to a brilliant green coloured complex.

Expt. 155   *Estimation of RNA by Orcinol method*

MATERIALS REQUIRED

Laboratory glassware

Laboratory equipment

Spectrophotometer

REAGENTS

a. Orcinol solution (1.0%)

   Dissolve 1g of orcinol in 5 ml ethanol taken in a 100-ml volumetric flask. Make up the content of the flask to 100 ml using distilled water.

b. Conc. hydrochloric acid

c. Ferric chloride solution (10%)

   Dissolve 10 g ferric chloride in 100 ml distilled water.

d. Orcinol reagent

   Transfer 10 ml of 10% ferric chloride solution to 390 ml of conc. hydrochloric acid. Mix this solution to 100 ml of orcinol solution. Continuously stir the mixture while adding and store it for future use. Always prepare fresh.

e. RNA stock solution

   Dissolve 5 mg of RNA stock solution in 50 ml of distilled water. Store it at $-20°C$.

PROCEDURE

1. Prepare a series of working standard solutions with their concentration ranging from 20 mg/ml to 100 mg/ml. Label them appropriately.

2. Take 2 ml of the sample in a separate test tube.

3. Make the volume of all the test tubes to 3 ml using distilled water.

4. To the sample and standards, add 3 ml of orcinol reagent.

5. Similarly prepare a blank using distilled water (3 ml distilled water + 3 ml orcinol).

6. Heat them in a water bath for 10 minutes and the time period starts after the water reaches the boiling point.

7. Observe for the development of green colour and after cooling, measure the absorbance at 665 nm.

8. Draw a standard graph by plotting the concentration of the standard solution against the absorbance value.

9. From the standard curve, evaluate the amount of RNA in unknown sample.

6.38 ISOLATION OF GENOMIC DNA FROM EUKARYOTES

6.38.1 Principle

Mechanical shearing process breaks open the cells and the sodium dodecyl sulphate favours the discharge of cell contents. Phenol chloroform solution ensure the settling of proteins and lipids in the organic phase and DNA in the aqueous phase. The DNA-containing aqueous phase is separated and centrifuged. It is collected in the form of pellets and suspended in TE buffer.

Expt. 156   *To isolate genomic DNA from eukaryotes*

MATERIALS REQUIRED

Laboratory glassware

Centrifuge

REAGENTS

a.   TEA Buffer

50 mM Tris	0.604 g
100 mM EDTA	3.722 g
200 mM NaCl	1.16 g
Distilled water	100 ml
pH 9	

b.   0.3M sodium acetate

c.   20% sodium dodecyl sulphate

Dissolve 20 g of SDS in 100 ml distilled water.

d.   Phenol chloroform solution (1 : 1)

e.   TE buffer.

PROCEDURE

1.   Weigh accurately 0.1 g of tissue, grind well in a homogenizer using 0.5 ml TEA buffer.

2.   Carefully transfer the homogenate in a microcentrifuge tube.

3.   To this, add 0.5 ml of TEA buffer and 30 µl of sodium dodecyl sulphate.

4.   Keep the tubes inverted for about 5 minutes.

5.   After this, hold it upright and add phenol and chloroform solution in the ratio 400 ml : 400 µl.

6.   Again invert the tubes for about 5 minutes.

7.   Centrifuge the tubes for fifteen minutes at 1500 rpm. During centrifugation, maintain the temperature at 4°C.

8.   Transfer the aqueous phase in a separate centrifuge tube and add 0.3M sodium acetate and 0.8 ml ethanol.

9.   Mix well and centrifuge again as described in point 7.

10. Wash the pellets formed with 70% ethanol several times. See to it that the pellet is allowed to dry between each subsequent washing.

11. Finally transfer the pellet in a tube containing TE buffer and store it at −85°C for future use.

6.39 ISOLATION OF CELLULAR DNA

6.39.1 Principle

Cell lysis followed by protein separation isolates the DNA.

Expt. 157   *To isolate cellular DNA*

MATERIALS REQUIRED

Laboratory glassware

Centrifuge

Mortar and pestle

REAGENTS

a.   Sodium chloride solution (0.9%)

Dissolve 0.9 g of sodium citrate in 50 ml distilled water.

b.   Sodium citrate solution (0.5%)

Dissolve 0.25 g of sodium citrate in 50 ml distilled water.

PROCEDURE

1.   Weigh accurately 0.2 g of the tissue sample. Suspend it in a 5 ml saline citrate solution and then homogenize.

2.   After homogenization, make up the volume to 10 ml using the same saline citrate solution.

3.   Now transfer the entire solution into a microcentrifuge tube and spin it at 3000 rpm for about 8 minutes. Remove off the supernatant.

4.   To the pellets add 5 ml of saline citrate solution and again homogenize.

5.   Make up the content to 10 ml using saline citrate solution and repeat the procedure mentioned in point 3.

6.   Transfer the pellets to 10 ml of sodium chloride and spin it at a fast rate (10,000 rpm). Maintain the temperature at 4°C and the duration of centrifugation is about 10 minutes.

7.   Collect the supernatant in a separate test tube having 3 ml absolute alcohol.

8.   Shake the tube well so that the white fibrous DNA is precipitated.

9.   Carefully wind a clean, sterile bent, glass rod over the white fibrous DNA so that it is spooled out.

6.40 ISOLATION OF GENOMIC DNA OF BACTERIA

6.40.1 Principle

Cellular lysis followed by the release of cellular contents are the two basic steps involved in this procedure CTAB extraction process favours the separation of polysaccharides and residual proteins. Then inoculation of cellular contents with phenol chloroform/isoamyl alcohol completely removes the contaminating molecules thus isolating the DNA.

Expt. 158   *To isolate genomic DNA from bacteria*

MATERIALS REQUIRED

Laboratory glassware

REAGENTS

a.   TE Buffer

Tris Cl 10 mM (pH 7.5)

EDTA – 1 mM

b.   Sodium dodecyl sulphate 10%

Dissolve 10 g of sodium dodecyl sulphate in 100 ml distilled water.

c.   Proteinase K (20 mg/ml of distilled water)

d.   5 mM NaCl

e.   CTAB/NaCl solution

Prepare 10% of CTAB solution and mix with 0.7M sodium chloride.

f.   Chloroform/isoform alcohol/phenol

Prepare this solution in the ratio 1 : 24 : 25

g.   Chloroform/Isoamyl alcohol

Prepare chloroform isoamyl alcohol in the ratio 24 : 1.

h.   Isopropanol

i.   Ethanol 70%

PROCEDURE

1.   Prepare a minimum quantity of a liquid broth (10 ml) and sterilize it.

2.   By an inoculation loop, transfer the bacterial strain of interest into the broth and incubate it overnight.

3.   Observe for the bacterial growth and pipette out 1.5 ml from this culture.

4.   Transfer the culture into the microcentrifuge tube.

5.   Centrifuge it for 3 minutes (500 rpm) and collect the supernatant to discard.

6.   To the pellet, add the following reagent

a.   567 µl of TE buffer

b.   30 µl of 10% sodium dodecyl sulphate

c.   3 µl of proteinase K

7. Shake well and keep it for incubation. Incubation period should not be continued beyond one hour.

8. To the incubated pellet, add 100 µl of sodium chloride and 80 µl of CTAB-NaCl solution. Shake the tube well between every addition of reagent.

9. Continue the incubation for another 10 minutes at 65°C.

10. Centrifuge the tube for 4-5 minutes after adding equal proportion of chloroform isoamyl alcohol mixture. This will form a clear white interface.

11. Then using the phenol : chloroform : isoamyl alcohol mixture, carefully extract the aqueous phase. To ensure the extraction process, centrifugation should be done.

12. Collect the supernatant in a separate tube. To precipitate the suspended nucleic acids, add 0.6 ml of isopropanol.

13. Again collect the precipitated white DNA in a tube.

14. Moisten the pellet with 70% ethanol and centrifuge the suspension at 10,000 rpm for 5 minutes.

15. Discard the supernatant and allow the pellets to air-dry. After few minutes, transfer the pellets to a tube containing 100 µl TE buffer and store it for further use.

6.41 QUANTIFICATION OF HUMAN DNA

6.41.1 Principle

Based on the amount of DNA in the sample, different methods are adopted to quantify it. For example, DNA if present in low quantities can be estimated by ethidium bromide binding whereas for DNA in considerable amounts, spectrophotometric determination is the best option and subsequently its purity can be assessed by UV light absorption method. Since the rate of absorption of incident radiation depends on the area or unit volume of solute in the cuvette, denatured DNA exhibits noticeable hyperchromicity than native DNA.

Expt. 159   *To quantify human DNA*

MATERIALS REQUIRED

Laboratory glassware

REAGENTS

a. TE Buffer

    Tris   1.08 g

    EDTA   0.044 g

Dissolve in 100 ml distilled water

b. Standard DNA solution

Mix 10 mg of DNA in 10 ml of TE buffer

   c.   Working standard DNA solution

Take 1 ml of standard DNA solution and make it up to 5 ml using TE buffer.

PROCEDURE

1. Withdraw 1 ml of the working standard solution and carefully siphon into the quartz cuvette.

2. Maintain a blank using TE buffer solution.

3. Measure the absorbance value for blank and standard between 200 and 300 nm.

4. Transfer the blank and standards in separate ependorff tubes and heat them in a water bath. After the temperature of the water bath reaches the boiling point, keep the tubes for ten minutes.

5. Again replace the solutions in their respective cuvettes and measure the absorbance value between 200 and 300 nm. Note down the value.

6. Draw a standard graph by plotting the wavelength against the absorbance.

6.42 ISOLATION OF PLASMID DNA (ALKALINE LYSIS METHOD)

6.42.1 Principle

Even on the exposure to alkaline solution, plasmid DNA does not separate and it is this characteristic property that makes its isolation possible by alkaline lysis method. The procedure involves the following steps.

a. Cell lysis

b. Denaturation of chromosomal DNA and proteins

c. Discharge of plasmid DNA

Isolation of plasmid DNA is made easier as the denatured chromosomal DNA, proteins and unwanted contaminants of the cell are trapped in anionic detergent (SDS) coated complexes which are then precipitated by ion exchange between sodium and potassium. This method can be successfully adopted for all strains of *E. coli* whose sample size range from 1 ml to 500 ml. Separated native DNA is subjected to electrophoresis or restriction endonuclease digestion process for purification. To be used as templates in DNA sequencing reactions, they have to be further purified with polyethylene glycol (PEG).

Expt. 160   *To isolate plasmid DNA by alkaline lysis method*

MATERIALS REQUIRED

Laboratory glassware

REAGENTS

a. Alkaline lysis solution (1)

      Glucose 20%   22.5 ml

      1.0 M Tris (pH 8.0)   12.5 ml

0.2 M EDTA (pH 8.0)   25.0 ml

Dissolve the contents in 500 ml double-distilled water, mix well and store it at 4°C.

b.   Alkaline lysis solution (2)

Sodium dodecyl sulphate   25.0 ml

5 MnOH 20.0 ml

To the 500 ml double-distilled water add the contents and mix well. Always prepare fresh.

c.   Alkaline lysis solution (3)

Add 300 ml of 5 M potassium acetate to 200 ml of glacial acetic acid.

d.   Ethanol

e.   Phenol chloroform solution

Prepare phenol chloroform solution in the ratio 1:1.

f.   TE Buffer solution

10 mM Tris HCl

1 mM EDTA

To 1 ml of the solution add 20 µg of RNase. Adjust the pH to 8.0

g.   Lactose broth

h.   Ampicillin antibiotic

Prepare 50 mg of antibiotic in 1 ml of distilled water, filter and store at –20°C.

i.   CSCI–TE buffer (10 mg/10 ml)

PROCEDURE

1.   Sterilize 10 ml of lactose broth, add required amount of antibiotic and inoculate the desired *E. coli* strain into it.

2.   Allow the culture to grow at 30°C shaking overnight.

3.   Observe for the growth of the culture and if the results are found to be positive, transfer 10 ml medium to the 2-litre flask containing 1 litre of lactose broth+antibiotic. Allow it to grow overnight. Adjust the shaker speed form 200 to 250 rpm.

4.   Following the incubation period, centrifuge the medium (8000 rpm) for about 10 minutes. Collect the pellets in separate bottles. Continue to centrifuge until the pellets collected is equal to one litre. Discard the supernatant.

5.   Suspend the pellets in alkaline solution (1) in such a way that 1 l of pellets are released into 40 ml of solution. Perform the suspension in a progressive manner, i.e., the pellets in small quantities are suspended in 10 ml of the alkaline solution (1); half of the entire cell volume is added to the solution (1). Now add 25 ml of solution. Finally the remaining pellet culture is introduced followed by the addition of 5 ml of solution (1). To this add 200 mg of lysosome and mix well.

6. To the content of the bottle, add 80 ml of alkaline solution (2) shake well and incubate in ice for about 10 minutes.

7. Add 40 ml of alkaline solution (3). Ensure complete mixing by shaking the bottle upside down several times.

8. Observe for the appearance of white coloured floc formation. Soon after this, incubate the bottle in ice for another 10 minutes.

9. Following the incubation period, add carefully 10 ml of double-distilled water. Shake well and centrifuge for 20 minutes (8000 rpm).

10. Allow the supernatant to pass through a cheese cloth, collect it in a volumetric flask and note down its volume.

11. Then transfer the flask content into a 500-ml bottle. To this add 2-propanol and the volume of propanol should be equal to the 0.6th portion of the total volume of flask content.

12. Shake the bottle well and incubate it in ice for about 5 minutes. Then centrifuge it for about 10 minutes (8000 rpm).

13. Discard the supernatant. Partially dry the pellet either by vacuum or by passage of air.

14. Collect the pellet and suspend it in little amount of TE buffer. Using Pasteur pipette, add Tris base (2M) in drops. While adding, stir the buffer and monitor the pH. Stop the addition of Tris base, when the pH reaches 7.5. Make up the total volume to 4 ml.

15. To this, add 4.2 g of caesium chloride, mix well and add 0.2 ml of 10 mg/ml ethidium bromide. Centrifuge it in vTi 65.2 rotor tubes (55,000 rpm) at 25°C overnight.

16. Following the incubation period, illuminate the gradient using UV light (366 nm). DNA can be seen as bands. In the case of two band formations the second one from above is the plasmid DNA. Carefully remove the plasmid DNA by needle.

17. Transfer it into an ultracentrifuge tube containing caesium chloride-Tris EDTA buffer to 3/4th of its volume. Add 50 ml of Ethidium bromide solution (10 mg/ml) and shake gently.

18. Rotate the content at 55,000 rpm overnight. Observe for the band separation and remove the lower band and transfer it into a ependorff tube.

19. Ethidium bromide solution can be removed by extraction with TE saturated butanol and dialyse the isolated plasmid DNA against TE buffer at 4°C.

6.43 TRANSFORMATION

Transformation process can simply be defined as the transfer of exogenous DNA/rDNA/ plasmid vectors into the living bacterial cells or higher organisms. This is based on the fact that bacterial cells when exposed to an ice-cold solution of calcium chloride and a mild heat-shock develop a chance to get transfected with bacteriophage lambda DNA.

Under these conditions they exhibit a competence state and the uptake of provided DNA is the possible consequence. Certain chemical substances are said to increase the transformation and they are dimethyl sulphoxide and hexamine cobalt chloride. In general, transformation process can be demonstrated by three methods.

Method I   Based on calcium chloride (figure 5.1)

Method II   Based on polyethylene glycol (figure 5.2)

Method III   Based on electroporation

Method III is not in practice for two major reasons—highly expensive electroporation apparatus and the storage temperature requirement of 700°C for the process.

Expt. 161   *To demonstrate the process of transformation in E. coli*

MATERIALS REQUIRED

Laboratory glassware

REAGENTS

    a.   *E. coli* strain

    b.   X-gal (40 mg/ml)

    c.   Calcium chloride solution

        Dissolve 60 mM of calcium chloride solution in 51% glycerol. To this add 10 mM PIPES, mix well. Adjust the pH to 7.0.

    d.   20 mM glucose

    e.   Ampicillin (50 mg 1 ml)

    f.   IPTG (10 mM stock)

    g.   PEG Buffer solution

    h.   Blue col plasmid

PROCEDURE

Method I

1. Grow the *E. coli* strain in 50 ml lactose broth at 37°C overnight.

2. Transfer the grown culture into another test tube containing 25 ml LB broth.

3. Again grow it at 37°C for 4 hours and when the absorbance of the grown culture is measured at 600 nm it should be between 0.3 and 0.4.

4. Take 5 ml aliquots in five separate test tubes. Centrifuge the culture content of each test tube at 4500 rpm for about 10 minutes at 4°C. Discard the supernatant.

5. Transfer the pellets to 0.5 ml ice-cold polyethylene glycol buffer. While performing these steps, see to it that the test tube is maintained at ice-cold temperature.

6. Meanwhile, sterilize 5, 10-ml test tubes and label them as 1, 2, 3, 4 and 5. Transfer the culture (10-ml aliquots) to each of the test tube.

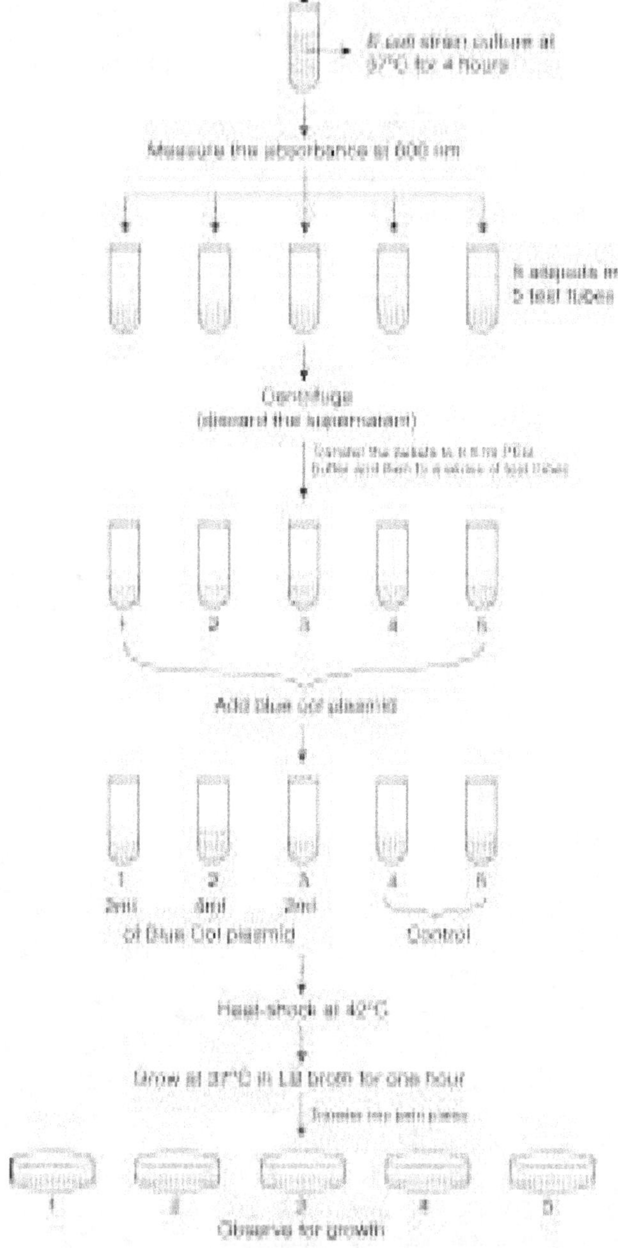

Figure 6.1   Steps involved in the experiment to demonstrate transformation process in *E. coli* (Method I)

7. Now add the blue col plasmid to 3 of the test tubes.

    Test tube 1, 3   2 ml

    Test tube 2   4 ml

    Test tubes 4,5   control

8. After addition, maintain these test tubes at 4°C for about one hour.

9. To each test tube add, 0.9 ml of LB broth and 30 mM of glucose, shake well and incubate it for 2 hours.

10. Prepare 3 sets of LB agar medium with varied composition. Let the volume of each set be 50 ml.

    Set I

    LB agar medium 50 ml

    Ampicillin   200 ml

    IPTG   30 ml

    X gal   70 ml

    Set II

    LB agar medium 50 ml

    Ampicillin   200 ml

    SET III

    LB agar medium 50 ml

11. Take 5 sterilized petri plates and label it as 1, 2, 3, 4 and 5. To the petri plates 1, and 2 pour set I LB agar medium, to the petri plates 3 and 4 pour SET II LB agar medium and to the petri plate 5, pour set III LB agar medium.

12. Transfer the culture from each test tube into the appropriate petri plates.

13. Streak the poured culture and incubate it at 37°C for 24 hours.

Method II

1. Grow the *E. coli* strain in 50 ml LB broth at 37°C overnight.

2. Transfer the grown culture into another test tube containing 25 ml LB broth.

3. Again grow it at 37°C for 4 hours and when the absorbance of the grown culture is measured at 600 nm it should be between 3.0 and 0.4.

4. Take 5 ml aliquots in five separate test tubes. Centrifuge the culture content of each test tube at 4500 rpm for about 7 minutes at 4°C. Discard the supernatant.

5. Transfer the pellet in 5 ml ice-cold calcium chloride solution and rotate it at 3600 rpm for about 5 minutes. Discard the supernatant.

6. Transfer the pellet in 10 ml ice-cold calcium chloride solution and from this stock, take 200 µl aliquots in ice-cold microtubes. Store the tubes at −20°C.

7. Meanwhile, sterilize 5, 10-ml test tubes and label them as 1, 2, 3, 4 and 5 and add 100 ml of competent cells to each tube.

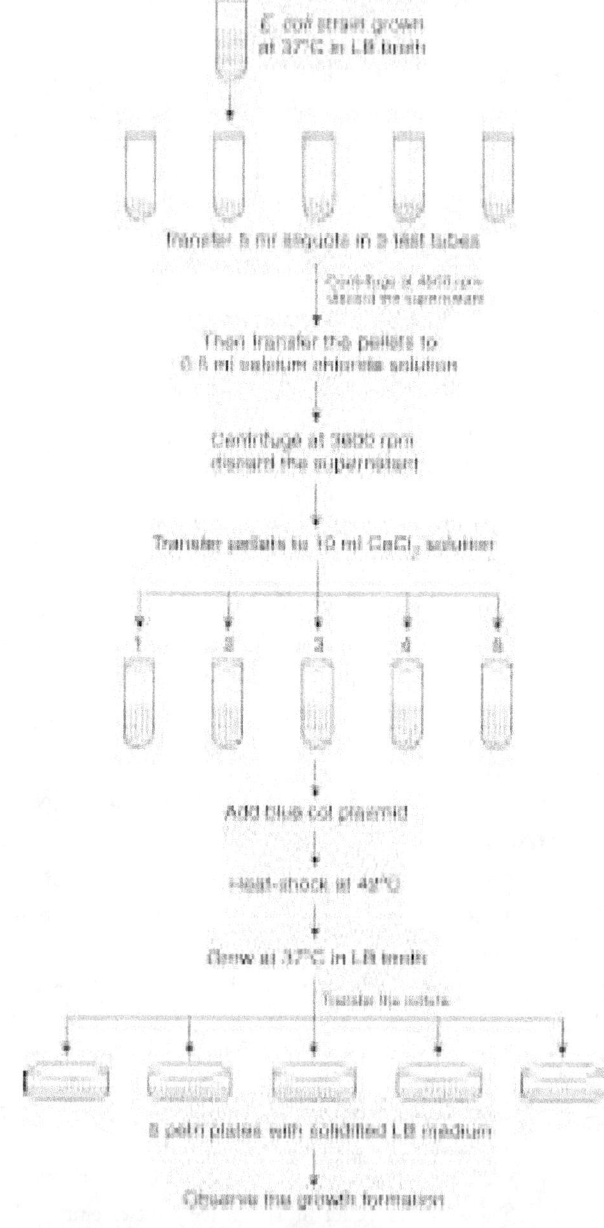

Figure 6.2 Steps involved in the experiment to demonstrate transformation process in *E. coli* (Method II)

8. Now add the blue col plasmid to 3 of the 5 test tubes.

    Test tubes 1, 3   2 ml

    Test tube 2   4 ml

    Test tube 4,5   control

    Mix well and incubate them in ice for 10 minutes.

9. Expose the cells to heat-shock by keeping them in a hot water bath (42°C) for 2 minutes.

10. To facilitate the growth of transformed cells, add 0.9 ml of LB broth to each tube.

11. Incubate all the test tubes at 37°C for about one hour.

12. Prepare 3 sets of LB agar medium with varied composition. Let the volume of each set be 50 ml.

    Set I

        LB agar medium   50 ml
        Ampicillin   200 ml
        IPTG   30 ml
        X gal   70 ml

    Set II

        LB agar medium   50 ml
        Ampicillin   200 ml

    SET III

        LB agar medium   50 ml

13. Take five sterilized petri plates and label them as 1, 2, 3, 4 and 5. To the petri plates 1 and 2 pour set I LB agar medium, to the petri plates 3 and 4 pour set II LB agar medium and to the petri plate 5, pour set III LB agar medium and to the petri plate 5, pour set III LB agar medium.

14. Transfer the culture from each test tube into appropriate petri plates.

15. Streak the poured culture and incubate it at 37°C for 24 hours.

6.44 AGAROSE GEL ELECTROPHORESIS

Agarose gel electrophoresis is a technique adopted to

a. Separate the molecules in DNA population.
b. Determine the purity and degradation and size of DNA fragments.
c. Detect the contamination of RNA.
d. Separate the protein molecules in an effective manner.

Here the agarose is melted and then cooled in an aqueous solution so that gel is formed by hydrogen bonding. The movement of DNA fragments along the gel under the influence of an electric field helps it to group at the anodic end. This is because DNA fragments are negatively charged molecules. Since the migration of DNA fragment depends on its size, a standard graph is drawn by plotting the log of mobility against the molecular weight of the DNA fragment in concern. From the calibration curve, molecular weight of unknown DNA fragments can be determined.

Expt. 162 *To separate the DNA molecules by the method of agarose gel electrophoresis*

MATERIALS REQUIRED

Laboratory glassware

Laboratory equipments

Heat block at 65°C

Gel electrophoresis apparatus (horizontal)

Whatman filter (22A)

Microcentrifuge

Microwave oven

UV transilluminator

Water bath

REAGENTS

a. Agarose

b. Bovine serum albumin (10 mg/ml)

c. Dithiothreitol 20X – 20 mM

d. Ethidium bromide (10 mg/ml)

e. λ Hind III size markers (0.5 µg/ml)

f. pRY 121 plasmid DNA (maxi prep)

g. pRY 121 plasmid DNA (mini prep)

h. Tracking dye solution (6X)

Sucrose	40 g
Bromophenol blue	250 mg
Distilled water	60 ml

Add the above contents and shake well for about 15 minutes. This stock solution is added to 1/6 of the final volume of DNA. Store at 4°C until it is used.

9. Restriction enzyme reaction buffer

10. Restriction enzymes

Hind III

Pst I

Bam H I

Eco RI

11. 200 mM EDDA (pH 8.0)

12. Tupperware containers

13. TEA buffer

Tris base	121 g
Sodium acetate	10.2 g

EDTA	18.6 g
Distilled water	750 ml

Add the above contents, mix well and make up to one litre using glacial acetic acid. Adjust the pH to 7.8.

PROCEDURE

1. Quantify the DNA to be digested and calculate the amount of restriction enzyme (1 µl) required to digest the quantified DNA.

2. Dilute the volume of the digestion reaction with X buffer so that its total volume after dilution is equal to 5% glycerol content where the restriction enzyme is present. Since the enzyme is supplied with 50% of glycerol content, 1:10 dilution is performed. Selection of reaction buffer depends on the ability of enzyme to be active at its salt concentration. Usually dilution of enzyme is not performed and it is added in excess (10 µml).

3. Add appropriate amounts of reaction buffer and dithiothreitol to the reaction mixture in such way that the total amount of both these reagents is exactly 1X.

4. To accomplish digestion process, certain enzymes require stabilizing agents like BSA (Bovine Serum Albumin) and therefore depending on the nature of the enzymes, add 2 µl of Bovine serum to the reaction mixture.

5. Make up the total volume of the reaction mixture to 2 µl.

6. Based on the concentration of enzymes, determine the amount of plasmid DNA to be added and the different types of plasmid DNA added to their respective restrictive enzymes are as follows.

   PRY 121 (mini prep) – Bam HI

   PRY 121 (maxi prep) – Hind III, Eco RI, Pst I, Bam HI

7. Carefully transfer the reaction mixture into a 50 µl centrifuge tube. Use separate tips for each addition. Avoid contamination by sterilizing them before use.

8. Give a gentle shake to the centrifuge tube and then centrifuge it at maximum revolution per minute. Do not extend the centrifugation beyond three minutes.

9. Incubate the tubes in a water bath at 37°C for about 1 hour.

10. Meanwhile dissolve 0.8 g of agarose in 100 ml 1X TEA buffer solution. Ensure complete dissolution of gel and then melt it in a microwave oven.

11. Transfer the molten agar in the trough and without any time interval insert the comb. Do not entertain any air bubbles. If formed, remove it gently by means of Pasteur pipette. See to it that the solidified gel is about 75 mm thick. With great care pull out the comb.

12. Now fill the buffer reservoirs of the trough with 1X TEA buffer. Continue until the solution spreads over the entire gel.

13. Condense the pellets by centrifugation at maximum rpm.

14. Pipette the sample into a 50 µl centrifuge tube. Dilute it with 15 ml distilled water. Similarly add 1 ml of restriction enzyme to 19 µl of distilled water.

15. To the sample tube, add 4 ml of 6X tracking dye solution. Since the dye is viscous, mixing is favoured by gently pipetting the mix upward and downward. Then centrifuge the tubes for about 3 seconds. This will completely remove the air bubbles in the mix..

16. Disruption of DNA aggregates is done by heating the sample (65°C) in a heat block for 2 minutes. After this, fill the wells of the agarose gel with samples. Follow an order while sample loading i.e., from left or from right so as to note down the position of wells loaded with different samples. Immediately record this in a notebook. Sampling should be done immediately.

17. Provide power supply to the apparatus so that the gel runs from positive side to the negative side (80 volts). Continue to run the gel for about 1 hour.

18. After one hour, draw off the buffer solution and cut the unwanted edges of the gel using razor blade. Carefully transfer the gel in tupperware container and stain it by 5 µl of ethidium bromide.

19. Wash out the stain of the gel by adding 100 ml of distilled water. Shake the gel gently for 5 minutes and this action completely removes the stain from the gel.

20. Do not discard the stain in the sink. As it is highly toxic, collect it in a waste container and as a safety measure, use gloves.

21. Exhibit the gel under a UV transilluminator and observe for band formation.

22. Focus the band by placing a red filter over the camera lens (Polaroid MP 4 camera).

23. Measure the size of the fragments by comparing it with the λ Hind III size markers.

6.45 SOUTHERN TRANSFER

Southern transfer technique is used to identify the genetic sequence of a specific DNA in a sample mixture, and the technique is schematically represented in figure 6.3.

Expt. 163   *To identify the genetic sequence of a specific DNA using southern transfer technique*

MATERIALS REQUIRED

Laboratory glassware

Laboratory equipments

Gel electrophoresis apparatus

Microwave oven

UV transilluminator

Water bath

Heat sealer

Stratalinker crosslinking apparatus

Vacuum oven 70°C

UV light source

Accessories

Gloves

Nylon transfer membranes (0.2 μm)

Paper towels exactly to the size of gel

Paraffin

Plastic boiling bags

Plastic wrap

Razor blades

Filter paper (3 MM)

Sample

Reagents

a. Ethidium bromide (10 mg/ml)

b. TEA buffer 25X

c. Tracking dye 6X

d. Ethanol

e. Neutralization buffer 1X

Tris	0.5M
NaCl	1.5M
pH	7.6

f. SSC stock 20X

| NaCl | 3M |
| Sodium citrate | 0.3M |

pH 7.0

Procedure

1. Dissolve 1 g of agarose gel in 100 ml of TEA buffer solution. Melt it in a bunsen burner and bring it to the working temperature by cooling them in a water bath at 45°C.

2. Pour the molten agar in the gel mould till the level reaches 1 cm thicknes. If the temperature of the agar exceeds 45°C, it shrinks during solidification. Similarly if temperature is less them 45°C it solidifies before pouring. Therefore better gel formation requires the optimum temperature.

3. Soon after gel pouring, insert the comb and do not entertain air bubbles. Allow the gel to solidify.

4. Fill the buffer reservoirs with TEA buffer solution and see to it that the level of buffer solution is above the agarose gel. Now carefully remove the comb.

5. Add 4 ml of 6X tracking dye to the DNA samples and heat it at 65°C for about 2 minutes. Following this, fill the wells of the gel with samples. While sample loading, follow an order, i.e., from left to right or from right to left so that the position of wells loaded with different samples can be noted down. Register this in a laboratory note book. To avoid diffusing of sample, loading should be done immediately.

6. Provide power supply to the apparatus (25 volts). Run the gel for 16 hours. At the end of 16 hours draw off the buffer solution.

7. Keep in identity for orientation by cutting the small edge of the gel in the upper right corner.

8. Carefully transfer the gel in a glass baking dish and stain it with 25 μl ethidium bromide (10 mg/ml).

9. Using distilled water, destain the gel. Place the gel in distilled water for about 15 minutes. Slight agitation of the gel completely removes the stain.

10. Do not discard the stain in the sink. Follow the procedure as given in note.

11. Exhibit the gel under a UV transilluminator and observe for band formation.

12. Extend the exposure period to about 30 seconds. This favours nick formation in the DNA and helps to transfer it to nylon membranes.

13. Add 400 ml of 1X denaturation buffer to the gel. Agitate the gel for about half an hour. Draw off the buffer solution. Again flood the gel with distilled water, discard it.

14. Expose the gel with 400 ml of 1X neutralization buffer for about 20–30 minutes.

15. Wet the nylon transfer membranes in 10X SSC. The size of the wet membrane should be exactly to that of gel.

16. Take Whatman 3 MM filter paper and prepare

    a. two pieces 1/8 inch large than the membrane

    b. a small wick

17. Similarly prepare four paraffin strips (1 inch x 6 inch) and a stack of paper towels (6 inch). The size of the paper tower should be equal to that of gel.

18. Arrange these accessories as described below

    • Transfer a piece of plexiglass in a glass baking dish.

- The height of the plexiglass is 2 inch. It is about 1 inch long and wider than the agar sufficient enough to hold it.

- Fill the dish with SSC buffer solution and set so that the level of the buffer solution is below the top layer.

- Wet the filter paper wicks with SSC buffer and place them on either side of the plat form . They are held in such a way that one end of the wick is on the top layer and the remaining portion is allowed to overhang so that the other edge is completely immersed in buffer solution.

- Place the gel on the platform and keep the nylon membrane over the gel. After placing the membrane, do not alter it as it may interfere with DNA transfer.

- Above the membrane, place two filter papers without air bubbles. In case, the arranged layers touch the buffer solution, the buffer tend to short circuit and to avoid this, place the paraffin strips on the filter wick.

- After arranging 3–4 inches of paper towels over the filter paper, keep a weight (1½ pound worth) so that the arranged layers are held compact.

19. Cover the whole setup with a plastic and leave it undisturbed overnight.

20. Following the incubation period, carefully remove the paper towels and paraffin strips. Do not disturb the filter wick-gel-nylon membrane filter paper sandwich. Hold it tight and invert. Now pull out the wicks and mark

    - DNA fragments orientation

    - Well position with lead pencils

21. Remove the gel. Place it under UV transilluminator and observe for ethidium bromide stained DNA fragments. Absence of fragments confirms complete transfer. Discard the gel cautiously as it contains EBr.

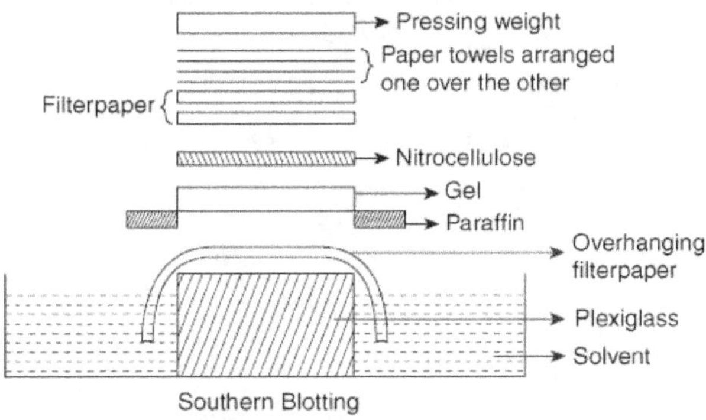

Figure 6.3 Southern blotting

22. Excess gel portions are removed with ethanol-sterilized razor blade.

23. Fix the DNA in nylon membranes by exposing it in short wave UV light for about 20 seconds. Flooding the membrane with 2 × 55C will remove the excess gel particles. Then place them between two pieces of Whatmann filter paper and allow it to dry.

24. To emit uniform quantity of UV light use Stratolink cross linking apparatus. Now the fixed membranes can be stored in booking plastic leads. To prevent contamination the bags are heat-sealed.

6.46 PCR TECHNIQUE

Polymerase chain reaction is a method by which a specific DNA fragment is amplified into several copies. The reaction can therefore produce the photocopy of the original in billions. The principle of this technique is similar to that of DNA replication process that occurs naturally. Amplification is done in three steps (Figure 6.4).

1. Denaturation of desired DNA fragment
2. Annealing of the primer to the target sequence
3. Extension of DNA sequence complementary to the target sequence

The process is repeated several times and the number of repetitions depends on the number of copies needed. This technique gains more importance in the field of clinical medicine and research. Even a trace level of specimen is sufficient enough to produce a test sample whose genetic sequence can be analysed for the presence of any microorganisms (bacteria/virus). This technique is highly applicable to detect the infectious organisms causing AIDS and hepatitis and to evaluate chromosomal aberrations in human genes. In future its applicability may be extended to a wide variety of diseases. The characteristic property of this technique is that gene sequences even with minute variations can be differentiated.

Expt. 164 *To perform the PCR technique for DNA amplification*

MATERIALS REQUIRED

Thermal cycler

Primer for targeted sequence

Taq DNA polymerase

PROCEDURE

Denaturation of desired DNA fragment

1. Denaturation is a process in which the heat required (>90°C) for the separation of double-stranded DNA is provided. As a result, two single strands of DNA are formed.

2. Since the hydrogen bonds interconnecting the nitrogen bases of double-stranded DNA are susceptible to breakage at high temperature, they get disconnected so as to separate the strands.

3. However the deoxyriboses and phosphates that are held together by the covalent bonds remain intact.

Annealing of the primer to the target sequence

1. It should be noted that the technique implies to replicate only the target sequence specific to that organism and therefore the first step in the annealing process is to mark the ends of the target sequence. This is approximately 100–600 base pairs long.

2. Identification of target sequence is done by a short synthetic single-stranded DNA sequence called primer. They are about 20–30 bases long.

3. Two primers, one for the target sequence of each complementary single strand is introduced. Based on the length and base sequence of primers, temperature is altered between 40°C and 65°C so that the primers added are annealed to the target sequence.

4. Now the temperature is gradually increased to 72°C and the enzyme Taq DNA polymerase favour the DNA replication process. This enzyme isolated from the organism *Thermus aquaticus* is enzymatically active even at high temperatures. It proceeds the synthetic process by keeping the region marked by the primer as initiation point.

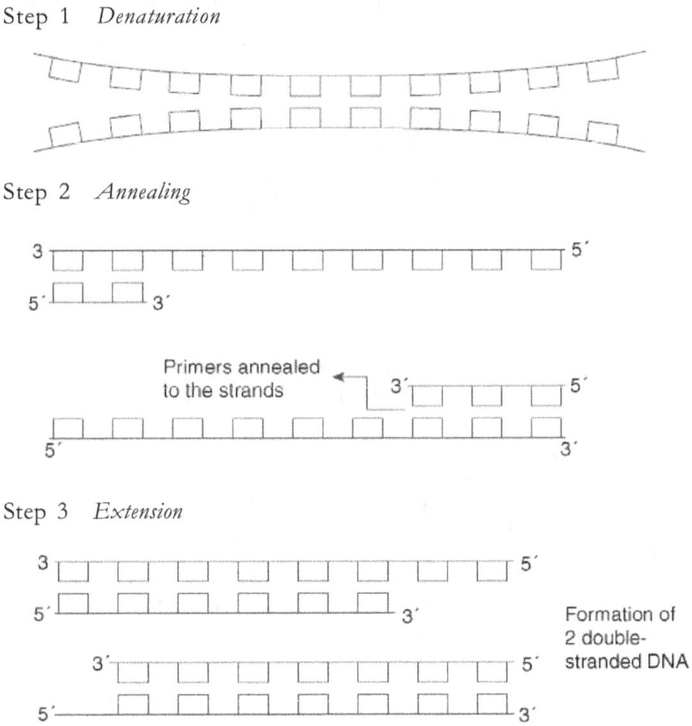

Step 1   *Denaturation*

Step 2   *Annealing*

Step 3   *Extension*

Figure 6.4   Diagrammatic representation of PCR technique

5. After the completion of targeted sequence, complementary strands bind together to form double-stranded DNA. Thus a copy of original DNA is formed.

6. Since the synthetic process is directed from 5′ end to 3′ end, free nucleotides are incorporated to the 3′ end. However polymerase enzyme cannot identify the end point. This is because a primer with 5′ and 3′ at its end continues to add fresh nucleotides.

7. However when the process is repeated for several cycles definite strands serve as template and replication of DNA by this template has a limited length by possessing 5′ at both ends. These templates are called amplicon.

8. Repetition of cycles leads to the production of large number of DNA strands that are similar to the target sequence, and can be collected and used.

6.47 ISOLATION OF IMMUNOGLOBULINS

6.47.1 Serum Sample Preparation from Whole Blood

1. Draw the blood to the required volume through veins.
2. Transfer it into a glass container.
3. Allow it to coagulate and wait for clot formation.
4. Transfer the content into a centrifuge tube and spin at 1000 rpm for about 15 minutes.
5. As a result of centrifugation, the clotted cells precipitate and the straw-coloured supernatant can be collected in a separate test tube for the isolation of immunoglobulins.

Expt. 165   *To isolate immunoglobulins*

METHOD I

1. Supernatant collected from the above process is the sample for immunoglobulin isolation.

2. Transfer the supernatant into a conical flask and using distilled water, make up the volume to 100 times that of sample volume.

3. Centrifuge the sample at 15,000 rpm for about 60 minutes.

4. Draw off the supernatant and to the pellets add 50 ml of borate buffer.

5. Carefully load the sample in a sephacryl IS-200 HR column. Prior to the sample loading, see to it that the column is neutralized with the sample buffer.

6. Allow the sample to flow into the column and the separated immunoglobulins can be detected by peaks in UV spectroscopy (280 nm).

7. Collect the separated immunoglobulins and store at 4°C.

METHOD II

MATERIALS REQUIRED

Laboratory glassware

UV spectrophotometer

REAGENTS

Ammonia solution

Dilute ammonia solution

Serum

Sodium chloride solution 0.14M

Phosphate buffered saline (PBS)

PROCEDURE

1. To one litre distilled water, add 1 kg of ammonium sulphide. Place the beaker in a hot water bath at 50°C and stir well. Let it stand for 10 hours at room temperature. Neutralize the solution to pH 7.2 using dilute ammonia solution.

2. Prepare serum saline solution in the ratio 2 : 1 and to this add an equal volume of saturated ammonium sulphite solution so that the concentration of the final solution is 45% saturated w/v.

3. Ensure complete dissolution by the continuous stirring and after 30 minutes centrifuge the solution at 40°C for about 15 minutes.

4. Decant the supernatant, collect the precipitate and wash it with 45% saturated ammonium sulphate solution.

5. Again centrifuge the washed precipitate and decant the supernatant. To the precipitate, add PBS whose volume is equal to that of original serum value.

6. Now centrifuge so that the insoluble materials can be easily separated out.

7. Add 40% of saturated ammonium sulphate solution to the centrifuged sample, centrifuge again and collect the precipitate.

8. Suspend the precipitate in a small amount of PBS.

9. Centrifuge it and discard the precipitate. Dilute the solution using PBS (1 : 20) and measure its absorbance at 280 nm using UV spectrophotometer.

METHOD III

MATERIALS REQUIRED

Laboratory glassware

Centrifuge

Conductivity meter

pH meter

REAGENTS

Saturated ammonium sulphate

Polyethylene glycol 24%

Phosphate buffered saline (PBS)

Phosphoric acid 1M

PROCEDURE

1. Take 100 ml of plasma in a beaker. To this add 42 ml of saturated ammonium sulphate and stir well for about 30 minutes.

2. Centrifuge the beaker content for 20 minutes.

3. Collect the supernatant and to this add 50 ml of saturated ammonium sulphate, stir well for about 30 minutes.

4. Again centrifuge the solution for about 30 minutes. Now remove the supernatant.

5. Carefully transfer the pellet in 100 ml of 50% saturated ammonium sulphate. Centrifuge this again for 20 min.

6. Again decant the supernatant. Suspend the pellet in distilled water and stir well.

7. Adjust the pH of the solution to 6.5–7.0 using 1M phosphoric acid and conductivity to 80 mU-1/cm^3.

8. Then add 24% of polyethylene glycol to the protein solution in the ratio (1 : 3).

9. After gentle stirring, centrifuge and decant the supernatant.

10. Collect the pellets and suspend in 10 ml of phosphate buffer solution, stir well and ensure complete mixing.

11. Remove the residual PEG and SAS by dialysing.

6.48 GEL DIFFUSION TECHNIQUES

It is a simple inexpensive technique used to detect soluble antigens. Passive diffusion of antibody and soluble antigen (to be detected) in a solid agar gel medium results in the formation of insoluble complex where they meet and react with each other. This complex is called precipitin. Presence of identical antigens in a mixture leads to the union of two precipitin lines to form a single one. Factors that influence the results of this technique are

1. pH
2. Temperature
3. Concentration of antigen and antibody

The third factor plays an important role because in some cases, precipitating band formation demands excess of antibodies whereas antigen in appropriate amount is very much required for the rapid band formation. However the main drawback of this technique is that it is expensive and in some cases it does not work well with unpurified antigens, e.g. fish.

Expt. 166 *To perform gel diffusion technique*

MATERIALS REQUIRED

Laboratory glassware

Gel puncher

REAGENTS

a. Antigen

b. Immune serum

c. Phosphate buffer saline (7.4)

$NaCl$	8.5 g
Na_2HPO_4	1.07 g
$NaH_2PO_4 \cdot 2H_2O$	0.39 g
Distilled water	1 litre

d. Coomasie brilliant blue

Prepare distilled water–glacial acetic acid–methanol mixture in the ratio 5 : 1 : 5. To make the stain solution, add 0.5 g of coomasie brilliant blue to 100 ml of the above solution.

e. Barbitone buffer solution (0.05M)

Dissolve 9.21 g of 5, 5-diethyl barbituric acid in 3 l distilled water. To this add 51.54 g sodium barbitone and 5 g sodium azide. Stir well and ensure complete mixing after addition of every reagent.

PROTOCOL

1. Dissolve 1 g of agarose in 100 ml of barbitone buffer solution. Pour this on glass slides. Allow it to solidify. Using gel puncher and template, set the well in the gel medium.

2. Load the wells with antigen and antibody. While loading see to it that the wells are not disturbed.

3. Cover the gel in a humid box overnight.

4. Flood the gel with saline, leave it undisturbed for several hours. Then dry the gel using hair dryer.

5. After drying, transfer the gel into a beaker containing coomasie brilliant blue stain.

6. Destain the gel until the precipitin bands appear distinct against the colourless background.

6.49 ENZYME-LINKED IMMUNOSORBENT ASSAY (ELISA)

Enzymes, when chemically conjugated with either antigen or antibody, help in the easy detection of immunocomplex which in turn confirms the presence or absence of specific antigen or antibody. The enzyme-linked immunoabsorbent assay termed as ELISA is a simple and sensitive test and is found to be highly applicable in clinical diagnoses. Steps involved in the technique are as follows:

6.49.1 Binding of Antigen and Antibody to the Titre Plates

Majority of antigens tend to bind themselves to the plastic surfaces and hence when they are loaded in the wells of titre plates, they exhibit a strong binding to it. Similarly antibodies in spite of their antigen-binding activity attack the solid surface. Now the excess and unwanted reagents can be removed by detergent buffer solution.

6.49.2 Formation of Immune Complex

Incubation of the bound antigen and antibody for a specific period of time favour the immunocomplex formation.

6.49.3 Enzyme Conjugation

Enzymes like HRP (horse radish peroxidase) and alkaline phosphatase are generally used. They can be allowed to conjugate either with antigen or antibody. These enzymes do not in anyway interfere with the biological property of antigen/antibody.

6.49.4 Colour Development

As soon as the antigen binds with the antibody to form immune complex, the enzyme is let free. Introduction of substrate with chromogen induces the enzyme to act on the substrate. This reaction ultimately liberates the coloured compound to give colour. The colour signal can be determined by optical density measurement.

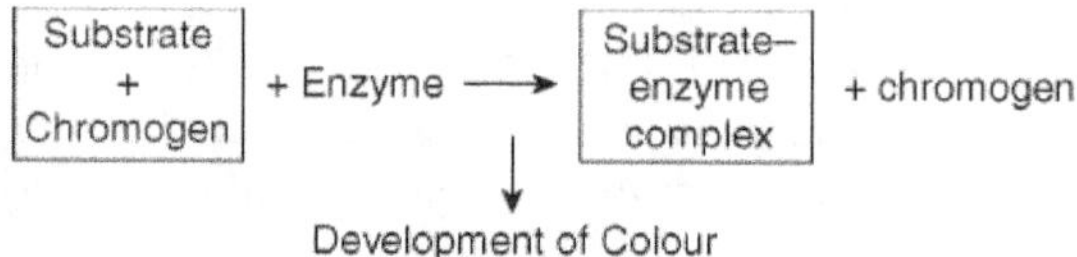

Step 1   Binding of antigen to the titre plate
Step 2   Binding of antibody to the titre plate.
Step 3   Conjugation of enzyme with antigen/antibody
Step 4   Immunocomplex formation
Step 5   Liberation of enzyme
Step 6   Reaction of enzyme with substrate

Expt. 167   *To perform ELISA test*

MATERIALS REQUIRED

Laboratory glassware

Elisa titre plates

REAGENTS

a.   Coating buffer

Dissolve 1.5 g of sodium carbonate and 2.9 g of sodium bicarbonate in one litre of distilled water. Adjust the pH of the solution to 9.6.

b.   Phosphate buffer saline solution

NaCl   0.15M

Phosphate buffer 50 mm

pH 7.4

    c.   PBS Tween 20

        To the prepared phosphate buffer saline solution add 0.5 ml/l of Tween 20.

    d.   Blocking solution

        To the phosphate buffer saline solution add 25% casein.

    e.   Substrate buffer

    f.   Substrate, antigen, antibody

    g.   Enzyme conjugated antiglobulin

PROCEDURE

1. To the well of Elisa plates, add 100 µl of antigen. Wrap the titre plate and incubate overnight.

2. Remove the coating solution by agitating the titre plate over a disposable tray.

3. Invert the plate over a paper towel so as to remove excess solution.

4. Repeat the process several times.

5. Load each with 100 µl of blocking solution. Let it stand for one hour.

6. Repeat step 2. Wash the plates thrice with PBS tween solution and repeat step 3.

7. Now dilute the test antiserum and fill each well with 100 µl of the serum. Again wrap the plate and incubate for about one hour.

8. Follow step 2 and 3 once again.

9. As per the instruction of the supplier, dilute the enzyme and load 100 µl of the same to each well. Incubate it for about one hour.

10. Repeat steps 2 and 3.

11. Finally add 100 µl of the substrate to each well and let it to stand for 1 hour. Observe for the colour formation.

EXERCISES

1. Amino acids in protein are linked by _______________ bond.

2. In agarose gel electrophoresis, DNA fragments migrate on the basis of their _______________.

3. Gel diffusion techniques are used to detect _______________.

4. Elisa test is highly applicable in the field of _______________.

5. In the PCR technique, free nucleotides are incorporated at the _______________ end.

6. Haemagluttination assay is to determine _______________ or _______________ in the blood serum.

7. Direct haemagluttination assay is used to determine _______________.

8. _____________ enzyme favours the hydrolysis of internal α 1-4-glucon links in polysaccharides.

9. Presence of acid phosphatase enzyme in the blood is an index of _____________.

10. DNA polymerase γ is present in _____________ and _____________.

11. Peroxidase is a _____________.

12. Lipase is used to diagnose _____________.

13. Glucose oxidase is specific to _____________.

14. RNA polymerase II(B) is favoured by the presence of _____________ and _____________ salts.

15. Cellulose enzyme is specific for _____________ substrate.

16. Absorbance value in LDH estimation depends on _____________.

17. Standard method to separate, identify and purify DNA fragments is _____________.

18. Mention a few biological functions of proteins.

19. Which method is specific for the deoxyribose component of purine nucleotide?

20. What is the principle of RNA estimation using orcinol method?

21. Which method is suitable to estimate minimum amount of DNA?

22. Define transformation process.

23. Give an account of the application of agarose gel electrophoresis.

24. What is the principle of PCR technique?

25. What are the influencing factors in gel diffusion technique?

26. What is the primary step in annealing?

27. What is denaturation?

28. What is the objective behind the addition of ice calcium chloride solution and a mild shock to bacterial cells in transformation process?

29. What is acquired immunity?

30. What is the origin of stem cells?

31. What is the function of primary lymphoid organs?

32. How does the immune system respond to xenobiotics?

33. How will you prepare antigen for haemagglutination assay?

34. What is the principle behind macrophage migration inhibition test?

35. How will you calculate migration index?

36. Which serves as an antigen in Rosette test?

37. How will you calculate percent viable cells in Rosette test?

38. What is the application of dialysis method?

39. Expand anti-ENA, Sm and RNP.

40. How will you isolate immunoglobulins from serum proteins?

41. In haemagluttination assay, how will you immunize the animal with antigen?

42. What is the main source of acid phosphatase enzyme?

43. Which enzymatic group is involved in digestion and intermediary metabolism?

44. What is the functional role of alkaline phosphatase in human?

45. Write down the principle behind the evaluation of lipase enzyme?

46. What are the sources of DNA polymerase (α)?
47. List out the significant properties of enzyme.
48. Give an example of transferase enzyme?
49. What are coenzymes?
50. What is the role of activators/kinase in enzymatic reaction?
51. Correlate enzymatic action and temperature?
52. Explain the clinical significance of LDH.
53. What are the applications of peroxidase enzyme?
54. What is the role of carbonic anhydrase in animals?
55. What are isoenzymes?
56. Which subunit of LDH predominates in heart muscle?
57. What is the commercial application of catalase?
58. Where is polymerase (γ) enzyme present?
59. Give examples for enzymes involved in transferase process.
60. What are reactive sites?
61. Interaction between enzyme and substrate is temporary. Explain.
62. Give few example of coenzymes.
63. What happens when enzymes with sulphydryl groups are exposed to oxygen/air?
64. List out the various forms of carbonic anhydrase in human erythrocytes.
65. How will you calculate cellulose activity in the given sample?
66. How will you measure H_2O_2 in the given sample?
67. What is the principle of amylase test?
68. How will you calculate RNA polymerase I in the given sample?
69. What makes the peroxidase enzyme suitable for the preparation of enzyme-conjugated antibodies?
70. In the isolation of genomic DNA, which step makes the cell break open?
71. What is amplicon?
72. What are the applications of PCR technique?
73. Mention the steps in PCR technique.

REFERENCES

Abramson, S., R. G. Phillips and R. A. Phillips, 1977. "The identification in adult bone marrow of pluripotent and restricted stem cells of the myeloid and lymphoid systems," *J. Exp. Med.*, 145. 1567.

Alan Wild (ed.), 1993. *Soils and the Environment: An Introduction,* Cambridge University Press, Cambridge, p. 276.

Arceivala, S. J., 1981. *Wastewater Treatment and Disposal,* Marcel Dekker Inc., New York.

Ardern, E. and W. T. Lockett, 1914. "The Oxidation of Sewage without the Aid of Filter," *J. Soc. Chemical Industries,* 33, pp. 1122–24.

Bick, P. H., 1985. "The immune system: Organisation and function," In: *Immunotoxicology and immunopharmocology,* Dean, J. H., Luster, M. I., Munson, A. E. and Anos, H. (eds.) Raven Press, New York.

Bolton, R. L. and L. Killen, 1971. *Sewage Treatments: Basic Principles and Trends,* Butterworth and Co. Ltd, pp. 9–37.

Chakraborty A. M., 1982. *Biodegradation and detoxification of environmental pollutants,* C. R. C. Press, Florida, U. S. A.

Chen, S. J., C. T. Li and W. K. Sheikh, 1985. "Performance of the Anaerobic Fluidised Bed System. II. Biomass Holdup and Characteristics," *J. Chem. Tech. Biotechnol,* 358, pp. 183–90.

Chopra, A. K. and N. J. Patrick, 1994. "Effect of Domestic Sewage on self Purification of Ganga Water of Rishikesh. 1, Physiochemical Parameters," *Ad. Bios* B (1), pp. 75–82.

Clair N. Sawer, Perry L. McCarty and Gene F. Parkin, (Eds.), 1994. *Chemistry for Environmental Engineering,* McGraw Hill International Editions, New York, p. 6.

Clark, J. A., C. A. Burger and L. E. Sabatinos, 1982. "Characterisation of Indicator Bacteria in Municipal Raw Water, Drinking Water and Main Water Samples," *Canadian J. Microbiology,* 28(9), pp. 673–678.

Cooke, G. D., R. T. Heath, R. H. Kennedy, and M. R. McComas, 1978. "The Effect of Sewage Diversion and Sulphate Application on two Eutrophic Lakes," *Ecol. Res. Ser,* EPA, 600/3-78-003, Cinn, OH, p. 101.

Dean, J. H., M. J. Murray and E. C. Ward, 1986. "Toxic respnse of the immune system," In: *Toxicology: The basic science of poisons,* Klaassen, C. D., Amudhan, M. O. and Doull, J. (eds). Macmillan, New York, 245.

Dold, P. L., G. A. Ekama and G. V. R. Marais, 1980. "A general model for the Activated Sludge Process," *Proc. Wat. Tech,* 12(6), pp. 47–77.

Droste, R. L. and W. A. Sanchez, 1983. "Comparison of Wastewater and Process Parameters on Energy Consumption in Aerobic and Alternative Anaerobic Treatement," Proc. International Assoc. on Water Pollution, Research Conference on Energy Savings. In. *Water Pollution Control,* Paris.51

Ecken Felder, W. W. T., 1963. "Trickling Filtration Design and Performance Transactions," *American Society of Civil Engineers*, (128), Part III, pp. 371–384.

Fang, H. H. P., H. K. Chui, Y. Y. Li and T. Chen, 1994. "Performance and granule characteristics of UASB Process treating waste water with hydrolysed proteins," *Wat. Sci. Tech.* 30(8) 55–63.

Flick, E. W., 1991. *Water Treatment Chemicals, An Industrial Guide*, Noyes Publications, Park Ridge, NJ.

Fox, J. C., P. R. Fitzgerald and Chue-Hing, 1981. *Sewage microorganisms. A Color Atlas*, Lewis Publishers, Chelsea, MI.

Fox, P., M. T. Suidan and J. T. Bandy, 1990. "A Comparison of Media Types in Acetate- Fed Expanded Anaerobic Bed Reactors", *Water Research*, 24(7), pp. 827–35.

Gregory, R. and T. F. Zabel, 1990. *Sedimentation and Floatation in Water Quality and Treatment*. 4th ed., (F. E. Pontius Ed.,) McGraw Hill, Toronto, pp. 367–453.

Gujer, W. and M. Boller, 1978. "Basis for the Design of Alternative Chemical Biological Waste Water Treatment Process," *Proc. Water Tech.*, 10(5/6), pp. 741–758.

Hardenbergh, W., (Ed.), 1950. *Sewerage and Sewage Treatement*, International Textbook Company, Pennsylvania, p. 454.

Hickey, R. F. and R. W. Owens, 1981. "Methane Generation from High Strength Industrial Wastes with the Anaerobic Fluidised Bed," Biotechnol. Bioeng. Symp, No. 11, pp. 399–411.

Huang, J. Y., M. D. Cheng and J. T. Mueller, 1985. "Oxygen Uptake Rates for Determining Microbial Activity and Application," *Nat. Res*, (19), pp. 373–81.

Iza, J., 1991. "Fluidised Bed Reactors for Anaerobic Waste Water Treatment," *Water Science and Technology*, 24(8), pp. 109–32.

Jeffrey, M. Becker, Guy A. Caldwell and Eve Annzachgo, (Eds.)., 1996. *Biotechnology: A Laboratory Course*, Academic Press, Inc., California, p. 252.

Kavitha, K. and A. G. Murugesan, 2004. "Performance evaluation of paper mill effluent in a granular bed UASBR," *Indian J. Environ and Eco Plan*. 8 (3).

King, D. L. and R. C. Bull, 1964. "A Quantitative Biological Measure of Stream Pollution," *J. Wat. Pollut. Control Fed.*, 36, pp. 650–53.

Lawrence, A. W. and P. L. McCarty, 1970. "Unified Basis for Biological Treatment Design and Operators," *J. San. Engg. Division, ASCE*, Vol. 96, No. SA3, pp. 757–78.

Lehr, J. H., T. E. Gass, W. A. Pettyjohn and J. Damaree, 1980. *Domestic Water Treatment*, McGraw Hill Book Co.

Lettinga, G., A. F. M. Vanentsen, S. M. Hobma, W. Dezeeu and A. Klapwijk, 1980. "Use of Upflow Sludge Blanket (USAB) Reactor Concept for Biological Waste Water Treatment," *Biotechnol. Bioeng.*, 22, pp. 699–674.

Mara, D., 1976. *Sewage Treatment in Hot Climates*, John Wiley and Sons, New York.

Milton Owen, Ahmed Riaz, 1992. "Biodegradation of hazardous wastes: The bioslurry reaction process," In: *Chemical Weekly*, pp. 155–158.

Murray, M. G. and W. F. Thompson, 1980. *Nucleic Acid Research,* 8: 4321–25.

Murugesan, A. G. and M. A. Haniffa, 1992. "Histopathological and histochemical changes in the oocytes of the air breathing fish, *Heteropneustes fossilis* (block) exposed to textile mill effluent," *Bull. Environ. Contom. Toxicol.,* 48: 929–936.

Murugesan, A. G., C. Rajakumari, S. Sakthivelmurugan and N. Sukumaran, 2004. "Behavioral Changes of freshwater fishes exposed to pulp and paper mill effluent at different temperatures," *Journal of Theoretical and Experimental Biology.* 1: 2533.

Murugesan, A. G. and C. Rajakumari, 2000. "Perspectives of Advanced Waste Water Treatment Technologies in the Treatment of Pulp and Paper Mill Effluent," *Proc. Workshop. Impact. Ind. Eff. Water, Qual.* 2003.

Murugesan, A. G. and M. A. Haniffa, 1985. "Effects of textile mill effluent on haematological changes of the obligatory air breathing fish, *Anabas testudineus* (block)," *Proc. Symp. Assess. Environ. Pollut.,* 121–128.

Murugesan, A. G., C. Rajakumari and N. Sukumaran, 2000. "Perspectives of acvanced biological treatment technologies in the treatment of pulp and paper mill effluents," *Proc. Workshop on Industrial effluent on Water quality,* 30–36.

Murugesan, A. G., M. I. Zahir Hussain and N. Sukumaran, 2000. "Toxic impact of mercury effluent on immune and haemato system of common carp *Cyprinus carpio,*" *Proc. VIIIth National Conference,* Tamilnadu Sci. Cong.

Murugesan, A. G., R. Ramsankar, K. Karthikeyan and N. Sukumaran, 2000. "Performance and evaluation of upflow anaerobic sludge blanket reactor (UASBR) for treating distillery spent wash," *Proc. IPC, National Sem. Indust Poll. Control,* 177–185.

Nahle, C., 1991. "The Contact Process for the Anaerobic Treatment...Waster Water Technology Design and Experiences," *Wat. Sci. Tech.,* Vol. 24, No. 8, pp. 179–191.

Rajakumari C., A. G. Murugesan and N. Sukumaran, 2004. "Avoidance behaviour of fresh water prawn, *Caridna levis,* to pulp and paper and distillery effluents," *Poll Res.,* 23n(1)n: 135–140.

Rajakumari, C. and A. G. Murugesan, 2000. "High rate anaerobic reactors for the treatment of industrial effluents," *Proc. Sem. Workshop on recent Trend in Environmental Biotechnology:* 37–50.

Rajakumari, C. and A. G. Murugesan, 2000. "Highrate Anaerobic Reactors for the Treatment of Industrial Effluents," *Proc. Sem. Workshop on Recent Trends in Environmental Biotechnology,* pp. 26–36.

Rajamani, S., R. Suthamarajan, E. Ravindranath, A. Mulder, J. W. Vancroenes Jijin and T. S. A. Langerwert, 1995. "Treatment of Tannery Waste and Biogas Generation using UASB reactor," *Journal IAEM,* Vol. 22, pp. 240–3.

Ramachandra, N. D., 1998. *Resource Journal of Science Education,* Vol. (3): 71-73.

Reiss, E., E. Mertens and W. E. Ehrich, 1950. "Agglutination of bacteria by lymphoid cells in vitro," *Proc. Soc. Exp. Biol. Med.,* 74. 732.

Rich, L. G., 1977. "Rational Design of Aerbic Digestion System," *Water and Sewerage Works,* Vol. 124.4, pp. 94–95.

Rickert, D. A. and J. V. Hunter, 1971. "General Nature of Soluble and Particulate Organics in Sewage and Secondary Effluent," *Wat. Res*, 5, pp. 421–36.

Rittman, B. E. and V. C. Snoeyink, 1984. "Achieving Biologically Stable Water," *J. American Water Works Association*, 76 (10), pp. 106–14.

Robinson, P. K., 1988. "Biotechnological Involvement and advances in waste water treatment," In: *Resources and Applications of Biotechnology – The new wave*, Greenshields Rod (ed.) The Macmillan Press Ltd.

Sambrook, J., E. F. Fritsch, and T. Maniatis, 1989. *Molecular Cloning: A Laboratory Manual*, 2nd ed., Cold Spring Harbor Laboratory, Cold Spring Harbor, New York.

Scanders, V. M. and A. E. Scanders, 1985. "Role of alpha adrenoceptor activation in modulating the murine primary antibody response in vitro," *J. Pharmacol Exp. Ther.*, 231. 527.

Schmitz, R. J., 1996. *Introduction of Water Pollution Biology*, Gulf Publishing Co., Texas.

Shah, D. B. and A. Coulman, 1978. "Kinetics of Nitrification and Denitrification," *Biotech and Bioenergy*, 22, pp. 42–72.

Suja P. Devipriya, A. G. Murugesan and N. Sukumaran, 1999. "Effect of pulp and paper mill effluent in the physicochemical properties of soil in an agro-ecosystem," In: *Sustainable Environment*, N. Sukumaran (ed.), 284–289.

Surampalli, R. Y. and E. R. Baumann, 1989. "Supplemental Aeration Enhanced Nitrification in a Secondary RBC Plant," *J. Water Pollution Control Federation*, 61(92), pp. 200–07.

Sutton, P. M., A. Li, 1983. "Single Phase and Two Phase Anaerobic Stabilisation in Fluidised Bed Reactors," *Water Science Technology*, (15), pp. 333–44.

Tehobanoglous, G. and R. Elisaaen, 1970. "Filtration of Treated Sewage Effluent," *J. Sanitary Engineering Division*, ASCE, 96, SA2, pp. 243–65.

Vesiland, P. A., 1979. "Treatment and Disposal of Waste Water Sludges," *Ann Arbor Science*, Arbor, MI.

Worthington, (Ed.), 1997. *Enzymes and Related Biochemicals*, Worthington Biochemical Corporation, New Jersey, p. 208.

Yao, K. M., 1970. "Theoretical Study of High Rate Sedimentation," *J. Water Pollution Control Federation*, 42(2), 218–28.

Young J. C. and B. C. Yang, 1989. "Design Considerations for full Scale Anaerobic Filters," *J. Water Pollution Control Federation*, 61, pp. 9–10, 1576–87.

INDEX